Classroom Companion: Economics

The Classroom Companion series in Economics features fundamental textbooks aimed at introducing students to the core concepts, empirical methods, theories and tools of the subject. The books offer a firm foundation for students preparing to move towards advanced learning. Each book follows a clear didactic structure and presents easy adoption opportunities for lecturers.

Giulio Bottazzi

Advanced Calculus for Economics and Finance

Theory and Methods

 Springer

Giulio Bottazzi ⓘ
Scuola Superiore Sant'Anna
Pisa, Italy

ISSN 2662-2882   ISSN 2662-2890 (electronic)
Classroom Companion: Economics
ISBN 978-3-031-30315-9   ISBN 978-3-031-30316-6 (eBook)
https://doi.org/10.1007/978-3-031-30316-6

This Springer imprint is published by the registered company Springer Nature Switzerland AG
The registered company address is: Gewerbestrasse 11, 6330 Cham, Switzerland

To Cilli and Gigi

Preface

This document started as a collection of notes prepared for the courses of advanced calculus for undergraduate and graduate students of economics held at the Scuola Sant'Anna. It has been a work in progress for more than ten years. There are multiple purposes of this work. First, the treatment of some topics, most notably topological and metric spaces, but also measure theory, is developed in a rather axiomatic way, starting from general definitions and then analysing what these definitions imply under different conditions. The axiomatic approach is an excellent tool for training students in logical deduction, logical consistency, and formalisation of ideas. Attention is paid here not only to the final results, but also to how those results are derived and what type of logical relation connects them. For this purpose, I decided to clearly separate the analysis of topological, normed, and metric spaces, which are often merged in introductory and middle-level calculus textbooks. Second, while presenting material which is already part of basic introductory courses on calculus, such as numerical sequences, series, and differential calculus, I follow a theorem-proving approach, stressing the boundary of application of the different theorems, often attempting to provide thought-provoking counterexamples. I also include notable results that are often used in applied mathematics books, but that I have found to be constantly missing in basic- and mid-level mathematical courses. For many topics, this book provides far more results than the usual book on calculus, even if not the kind of coverage one could find in a specialised publication. The idea is to teach the reader the appropriate language and notions about each topic so that they can understand where and how to look for more specific discussions should the need arise. The main criterion that drove the selection of topics is usability in applications, in particular, applications in economics and social sciences. This is why I have avoided almost entirely any geometric consideration, which is important in a course for hard scientists and engineers, and I have insisted on mathematical results mostly useful for optimisation theory and statistics. The two final appendices, on the initial value problem for systems of differential equations and on the Brower fixed-point theorem, exemplify two possible domains of application of the material covered in this book.

I have several people to thank. First, the Allievi of Scuola Sant'Anna, who, at different levels, have followed my courses. They asked the questions that this document was designed to answer and gave me invaluable feedback on the text itself.

Pietro Battiston, Francesco Cordoni, Pietro Dindo, Carolyn Phelan, and Davide Pirino provided useful comments at different stages and helped me decide what to include and what to leave out. This is always a very difficult choice when designing a manual. Marta Talevi and Matteo Quagliotto helped me with the typewriting of part of the material.

Pisa, Italy   Giulio Bottazzi
February 2023

Contents

1 Preliminaries 1
  1.1 Sets, Equivalence Relations, and Functions 1
  1.2 Order Relations, Supremum, and Infimum 4
  1.3 Countable Sets 6
  1.4 Operations and Fields 7
  1.5 Real Numbers 9
  1.6 Convexity and Concavity 18
  Exercises 21

2 Topology 25
  2.1 Definition and Basic Properties 25
  2.2 Base of a Topology 29
      2.2.1 Countability 31
      2.2.2 Euclidean Topology 32
  2.3 Cover and Compactness 34
      2.3.1 Hausdorff Spaces 35
      2.3.2 Compactness in Euclidean Topology 38
      2.3.3 The Extended Real Number System 40
  2.4 Connectedness 40
  2.5 Limit of Functions and Continuity 42
      2.5.1 Continuity in Euclidean Topology 47
  Exercises 50

3 Metric Spaces 55
  3.1 Definition and Basic Properties 55
  3.2 Metric and Topology 57
  3.3 Uniform Continuity 59
  3.4 Lipschitz Continuity 60
  Exercises 61

4 Normed Spaces 65
  4.1 Definition and Basic Properties 65
  4.2 Example of Normed Spaces 67
      4.2.1 Euclidean Norm in $\mathbb{R}^n$ 67

	4.2.2	p-Norm in $\mathbb{R}^n$	69
	4.2.3	Operator Norm	71
4.3	Finite-Dimensional Normed Spaces		73
	4.3.1	Equivalence of Norms in $\mathbb{R}^n$	76
	Exercises		78

5 Sequences and Series 81
5.1	Sequences in Topological Spaces	81
	5.1.1	Subsequences	82
	5.1.2	Sequences and Functions	83
	5.1.3	Uniqueness of Limit	84
5.2	Sequences in Metric Spaces	85
	5.2.1	Cauchy Sequences and Complete Spaces	85
5.3	Sequences in $\mathbb{R}$	91
	5.3.1	Upper and Lower Limits	99
	5.3.2	Infinity and Infinitesimals	101
5.4	Sequences in Normed Spaces	105
5.5	Series in $\mathbb{R}$	108
	5.5.1	Series with Decreasing Terms	110
	5.5.2	Tests Based on the Asymptotic Behaviour of Terms	112
5.6	Sequences and Series of Functions	116
	5.6.1	Uniform Convergence	117
	Exercises		119

6 Differential Calculus of Functions of One Variable 125
6.1	Limit of Real Functions	125
6.2	Continuity of Real Functions	127
6.3	Differential Analysis	130
	6.3.1	Higher-Order Derivatives	136
	6.3.2	Derivatives and Function Behaviour	136
	6.3.3	Derivatives and Limits	140
6.4	Taylor Polynomial and Power Series Expansion	143
	Exercises		151

7 Differential Calculus of Functions of Several Variables 155
7.1	Limits and Continuity in $\mathbb{R}^n$	155
7.2	Differential Analysis in $\mathbb{R}^n$	157
7.3	Mean Value Theorems	170
7.4	Higher-Order Derivatives and Taylor Polynomial	171
	7.4.1	Local Maxima and Minima	175
7.5	Inverse Function Theorem	178
7.6	Implicit Function Theorem	181
	7.6.1	Real Functions of Two Variables	182
	7.6.2	Real Functions of Several Variables	184

	7.6.3	Vector Functions of Several Variables	186
	7.6.4	Dependent and Independent Functions	188
7.7	Constrained Optimisation		189
	7.7.1	One Dimensional Problems	190
	7.7.2	Two Dimensional Problems	193
	7.7.3	Theorems of the Alternatives	195
	7.7.4	First-Order Conditions	197
	7.7.5	Second-Order Conditions	202
	7.7.6	Envelope Theorem	208
	7.7.7	Minimisation Problems	211
Exercises			211

8 Integral Calculus 215
- 8.1 Definite Integrals 215
  - 8.1.1 Properties of the Definite Integral 219
  - 8.1.2 Riemann Integrable Functions 221
  - 8.1.3 Improper Integrals 223
  - 8.1.4 Integral of Vector-Valued Functions 225
- 8.2 The Fundamental Theorem of Calculus 226
- 8.3 Riemann–Stieltjes Integral 232
  - 8.3.1 Stieltjes Integrable Functions 234
  - 8.3.2 Properties of the Stieltjes Integral 239
- Exercises 241

9 Measure Theory 245
- 9.1 Algebras, Measurable Spaces, and Measures 245
  - 9.1.1 Complete Measure Space 249
  - 9.1.2 Borel σ-Algebra 250
  - 9.1.3 Lebesgue Measure 251
- 9.2 Measurable Functions 256
  - 9.2.1 Measurable Real-Valued Functions 259
- 9.3 Lebesgue Integral 261
  - 9.3.1 Lebesgue and Riemann Integral on $\mathbb{R}$ 274
- 9.4 Product Measure Space 277
  - 9.4.1 Product σ-Algebra 278
  - 9.4.2 Product Measure 279
  - 9.4.3 Multiple Integrals 283
- 9.5 Probability Measure 287
  - 9.5.1 Multiple Random Variables 292
  - 9.5.2 Banach Space of Square Summable Random Variables 294
- Exercises 296

Appendix A: Cauchy Initial Value Problem 301

Appendix B: Brouwer Fixed Point Theorem 305

Index 309

Preliminaries

1

1.1 Sets, Equivalence Relations, and Functions

A *set* is a collection of objects that are *elements* or *points* of the set. A subset of a set is a collection of elements which all belong to the original set. If B is a subset of A, I will denote it by writing $B \subseteq A$. In general, I will use capital letters, such as A, to denote sets and small capital letters, such as a, to denote elements. The intersection of set A and set B, denoted $A \cap B$, is the set composed of the elements common to the two original sets. Their union, denoted $A \cup B$, is the set composed of all elements that belong to at least one of them. Two sets are *disjoint* if they do not have elements in common. In this case, their intersection is the empty set, denoted with $\emptyset$. If $B \subseteq A$, but $B \neq A$, I will write $B \subset A$.

I will write $a \in A$ if the element a belongs to the set A. The expression $\exists a \in A$ means that "there exists an element a belonging to A ...", while the expression $\forall a \in A$ means "for any elements a of A ...". So, for example, if $\mathbb{N}$ is the set of integer numbers and we want to express the fact that for any integer, there is another integer that is exactly the first integer plus one, we can write $\forall n \in \mathbb{N}$, $\exists m \in \mathbb{N}$, such that $m = n + 1$.

The collection of all subsets of A is the *power set* of A and is indicated by 2^A. The name and notation 2^A come from the fact that if A contains a finite number of elements n, then the number of elements of 2^A is 2^n. This can be easily proved by induction.

Theorem 1.1 (Number of subsets) *If a finite set has n elements, then it has 2^n subsets.*

Proof This serves as an example of a proof by induction. If the set A is empty, then $2^A = \{\emptyset\}$. If the set A contains one element, $A = \{a\}$, then $2^A = \{A, \emptyset\}$. Thus, the statement is true for $n = 1$. Suppose the statement is true for a set with $n - 1$ elements and add one element to obtain a set with n elements. The power set of the new set contains all the 2^{n-1} subsets of the previous set, plus all the 2^{n-1} subsets obtained

© The Author(s), under exclusive license to Springer Nature Switzerland AG 2023
G. Bottazzi, *Advanced Calculus for Economics and Finance*, Classroom
Companion: Economics, https://doi.org/10.1007/978-3-031-30316-6_1

by adding the new element to one of the old subsets. Now $2^{n-1} + 2^{n-1} = 2^n$ and we have proved that the statement is valid also for sets with n elements. Thus, the statement is valid for any n.   □

Consider two sets A and B. Let $A \times B$ denote the *Cartesian product* of the two sets, that is, the set of ordered pairs (a, b) with $a \in A$ and $b \in B$. The very general notion of a relation is introduced in the following.

Definition 1.1 (*Relation*) A *relation* R between A and B is a collection of ordered pairs, $R \subseteq A \times B$. If $(a, b) \in R$, then a is said to be in relation to b.

The sets A and B can be different or they can be the same set. If $(a, b) \in R$, one can write $a R b$. The set of all elements of A in relation to some element of B, $\mathbb{D}_R = \{a \in A \mid \exists b \in B$ such that $(a, b) \in R\}$, is the *domain* of R. The set of all elements of B in relation to some element of A, $\mathbb{I}_R = \{b \in B \mid \exists a \in A$ such that $(a, b) \in R\}$, is the *range* or *image* of R. The *image* of a subset $A' \subseteq A$ are all elements of B in relation to at least one element in A'. The *inverse image* or *preimage* of a subset $B' \subseteq B$ are all elements of A in relation to at least one element of B'. For any relation $R \subseteq A \times B$, we can define an *inverse relation* $R^{-1} \subseteq B \times A$ with the stipulation that $(b, a) \in R^{-1}$ if and only if $(a, b) \in R$.

Although important, the concept of a relation is, in general, too generic for practical use. It becomes more meaningful by imposing some additional requirements. These generally take the form of the inclusion or exclusion of specific ordered couples in the set R. The most commonly used relation which we come across is the equivalence relation.

Definition 1.2 (*Equivalence Relation*) A relation $R \subseteq A \times A$ is an *equivalence relation* on the set A if it is

- *reflexive*, that is $\forall a \in A, (a, a) \in R$;
- *symmetric*, that is if $(a, b) \in R$, then $(b, a) \in R$;
- *transitive*, that is if $(a, b) \in R$ and $(b, c) \in R$, then $(a, c) \in R$.

An equivalence relation is generally denoted with $\sim$ (or other similar symbols such as $=$), so if R is an equivalence relation and $(a, b) \in R$ one writes $a \sim b$. The definition of an equivalence relation on a set A naturally groups its elements into classes.

Definition 1.3 (*Equivalence Class*) Consider an equivalence relation $R \subseteq A \times A$ defined over a set A. Consider a generic element $a \in A$. The set $[a] = \{x \in A \mid x \sim a\}$ is the *equivalence class* of a. The *quotient set* A/R is the set of all equivalence classes.

The equivalence class $[a]$ is the image of the element a in the equivalence relation. Equivalence classes partition the set A into disjoint subsets.

Theorem 1.2 *The different equivalence classes are disjoint, and their union is the entire A.*

Proof Consider two equivalence classes $[a]$ and $[b]$. If $\exists x \in [a] \cap [b]$, then $x \sim a$ and $x \sim b$, and, for the transitive property, $a \sim b$. That is, $[a] = [b]$. For the second part of the theorem, note that $A = \cup_{a \in A}[a]$.   □

The following example defines an equivalence relation on the set of natural numbers, including zero, $\mathbb{N}_0 = \{0, 1, 2, 3, \ldots\}$, and identifies its equivalence classes.

▶ **Example 1.1** Modulo p equivalence Consider a natural number $p > 1$. On the set $\mathbb{N}_0$, define the following relation R: two numbers n and m are related if they are equal or their difference, the greater minus the smaller, is a multiple of p. This relation is reflexive and symmetric by construction. In addition, if $m R n$ and $n R l$, then $m R l$ (you can prove it considering all possible orders of the three numbers). Thus R is an equivalence relation on $\mathbb{N}_0$. The equivalence classes are denoted with $[0], [1], \ldots, [p-1]$. The class $[k]$, with $k = 0, \ldots, p-1$, contains the numbers $\{k, k+p, k+2p, k+3p, \ldots\}$.

A commonly used type of relation is that of functions.

Definition 1.4 (*Function*) Consider the relation $R \subseteq A \times B$. If $\forall a \in \mathbb{D}_R, \exists! b \in B$ such that $(a, b) \in R$, then R is a *function* or *map*.

A function from A to B is a relation that assigns exactly one element of B to each element of its domain $\mathbb{D}_R \subseteq A$. Functions enjoy a special notation. A function from A to B is often denoted by $f : \mathbb{D}_R \subseteq A \to B$ and if $(a, b) \in f$ one writes $f(a) = b$.

Definition 1.5 Consider a function $f : A \to B$. The function f is *injective* or *into* if $\forall b \in \mathbb{I}_f, \exists! a \in \mathbb{D}_f$ such that $f(a) = b$. The function f is *surjective* or *onto* if $\mathbb{I}_f = B$. The function f is *one-to-one* or *bijective* if it is injective and surjective.

In other terms, a function is injective if any element of the image has a preimage made of a single element, while it is surjective if it has the largest possible image. With some abuse of notation, I will denote by $f(A')$ with $A' \subseteq A$, the set of images of points in A', $f(A') = \{b \in B \mid \exists a \in A', f(a) = b\}$, and by $f^{-1}(B')$ with $B' \subseteq B$, the set of points in A that are preimages of points in B', $f^{-1}(B') = \{a \in A \mid \exists b \in B', f(a) = b\}$. In general, the inverse of a function is not a function. However, if f is injective, then $\forall y \in \mathbb{I}_f$, the set $f^{-1}(y)$ contains a single element so that f^{-1} is a function.

1.2 Order Relations, Supremum, and Infimum

Another widely used type of relation is the order relation.

Definition 1.6 (*Order Relation*) A relation $R \subseteq A \times A$ is an *order relation* if it is

- *reflexive*, that is, $\forall a \in A$, $(a, a) \in R$;
- *antisymmetric*, that is, if $(a, b) \in R$ and $(b, a) \in R$, then $a = b$;
- *transitive*, that is, if $(a, b) \in R$ and $(b, c) \in R$, then $(a, c) \in R$.

The order relation is said to be *total* if $\forall a, b \in A$ is $(a, b) \in R$ or $(b, a) \in R$.

An order relation is denoted by $\leq$ or $\geq$ (note the use of the non-strict inequalities). A reflexive and transitive relation R is a *preorder relation*. It becomes an order relation by setting $a = b$ if $(a, b) \in R$ and $(b, a) \in R$. In this way, an equivalence relation is induced in the set by the relation R and R is by construction antisymmetric with respect to this equivalence relation. Any order relation also has a strict version.

Definition 1.7 (*Strict Order Relation*) Let $R \subseteq A \times A$ be an order relation. A *strict order relation* $R' \subset R$ is defined by requiring that if $(a, b) \in R'$ then $(a, b) \in R$ and $a \neq b$.

Order relations allow for the definition of "boundedness".[1] In what follows, the couple $(A, \leq)$ stands for an ordered set, that is, a set and an ordered relation defined over it.

Definition 1.8 (*Upper bound and supremum*) A subset $C \subseteq A$ is *bounded above* if there is an $a \in A$ such that $\forall x \in C$ is $x \leq a$. Then a is an *upper bound* of C. Let $U(C)$ be the set of all upper bounds of C. If $\exists a \in U(C)$ such that $a \leq a'$, $\forall a' \in U(C)$, then a is the *least upper bound* or *supremum* of C, denoted by sup C.

An analogous definition is in place for the lower bound.

Definition 1.9 (*Lower bound and infimum*) A subset $C \subseteq A$ is *bounded below* if there is a $a \in A$ such that $\forall x \in C$ is $a \leq x$. Then a is a *lower bound* of C. Let $L(C)$ be the set of all lower bounds of C. If there exists an element $a \in L(C)$ such that $a' \leq a$, $\forall a' \in L(C)$, then a is the *greatest lower bound* or *infimum* of C, denoted by inf C.

If $C = \emptyset$, then $L(C) = U(C) = A$. For any set C, $C \subseteq L(U(C))$ and $C \subseteq U(L(C))$. If $a = \sup C$, then from the definition of supremum it follows that $a \in L(U(C))$. But since $a \in U(C)$, a is greater equal than all elements of $L(U(C))$, so that

---

[1] A more general definition based on the notion of distance will be introduced in Chap. 3.

$a = \inf U(C)$. Thus, the supremum of a set is the infimum of the set of its upper bounds. Analogously, the infimum of a set is the supremum of the set of its lower bounds. Especially useful are those ordered sets in which one is always sure to find the supremum or infimum of any bounded subset.

Definition 1.10 (*Least upper bound property and greatest lower bound property*) An ordered set $(A, \leq)$ has the *least upper bound property* if every bounded above subset has a supremum and it has the *greatest lower bound property* if any bounded below subset has an infimum.

The following theorem clarifies that the two properties above are actually the same.

Theorem 1.3 *An ordered set $(A, \leq)$ has the least upper bound property if and only if it has the greatest lower bound property.*

Proof We prove the theorem in one direction only since the symmetric proof is basically identical. Let us assume that $(A, \leq)$ has the least upper bound property. Consider a bounded below subset C and let $L(C)$ be the set of all its lower bounds. Since $L(C)$ is bounded above, it has the supremum. Let us call it a. Now a is greater equal than any lower bound of C, then $a = \inf C$.   □

If the supremum of a set exists and belongs to the set, it is the *maximal element* or *maximum* of the set. If the infimum of a set exists and belongs to the set, it is its *minimal element* or *minimum*.

A function whose image belongs to an ordered set is *bounded above* or *bounded below* if its image possesses one of those properties. Generally, a function is said to be *bounded function* if it is bounded above and below. For such bounded functions, we can ask where the function reaches its highest or lowest values.

Definition 1.11 (*Maximum and minimum*) Consider a function $f : X \rightarrow A$ from a set X to an ordered set $(A, \leq)$. The element $x \in X$ is a *maximum* of the function f in X if $f(x) = \sup f(X)$ and it is a *minimum* of the function f in X if $f(x) = \inf f(X)$.

Note that for the maximum or minimum to exist, three conditions must be met: the image set $f(X)$ is bounded above or below, it has the supremum or infimum, and the supremum or infimum belongs to the set itself, that is, they are the image of some element of X.

Theorem 1.4 (Max–min inequality) *Consider a function $f : X \times Y \rightarrow Z$ that sends ordered pairs $(x, y) \in X \times Y$ to an ordered set $(Z, \leq)$ with the least upper bound property. Then for any $A_X \subseteq X$ and $A_y \subseteq Y$ such that $f(A_X \times A_Y)$ is bounded,*

$$\inf_{x \in A_X} \sup_{y \in A_Y} f(x, y) \geq \sup_{y \in A_Y} \inf_{x \in A_X} f(x, y).$$

Proof For any $x \in A_X$ define $u(x) = \sup_{y \in A_Y} f(x, y)$ and for any $y \in A_Y$ define $l(y) = \inf_{x \in A_X} f(x, y)$. These functions exist because the image of f is bounded and the order relation has the least upper bound and the greatest lower bound properties (see Theorem 1.3). Then $\forall (x, y) \in A_X \times A_Y$ it is $u(x) \geq f(x, y) \geq l(y)$. This implies $\inf_{x \in A_X} u(x) \geq \sup_{y \in A_Y} l(y)$, which proves the assertion. $\qquad\square$

As the name of the above inequality suggests, it can be applied to maximum and minimum, instead of supremum and infimum, if they exist.

1.3 Countable Sets

Depending on the number of their elements, sets can be finite or infinite. Among the latter, of particular relevance are the sets that contain a number of elements equal to the natural numbers.

Definition 1.12 The infinite set A is *countable* if there exists a one-to-one map f from A to $\mathbb{N}$ (the set of natural numbers).

The function f associates an index $n \in \mathbb{N}$ with any element of A. Notice that the one-to-one map of the previous definition actually induces a total order relation in the countable set: if $a_1, a_2 \in A$, $f(a_1) = n_1$, and $f(a_2) = n_2$, one might say that $a_1 \geq a_2$ if $n_1 \geq n_2$. This observation is related to a useful property of countable sets.

Theorem 1.5 *Every infinite subset of a countable set is countable.*

Proof Let A be countable, $f : A \to \mathbb{N}$ a one-to-one map, and $B \subseteq A$ infinite. Order the elements in B according to the order relation induced by f on A. Assign an index to the elements of B according to this ordering: b_k is the kth largest element in B. This is a one-to-one map from B to $\mathbb{N}$. $\qquad\square$

Notably, the Cartesian product of countable sets is countable.

Theorem 1.6 *If A and B are countable, then $A \times B$ is countable.*

Proof Consider two one-to-one maps $f_A : A \to \mathbb{N}$ and $f_B : B \to \mathbb{N}$. These maps exist because the two sets are countable. Order the elements of A and B according to these functions and arrange them along the row and column headings of a table, as shown:

$$
\begin{array}{c|cccc}
 & b_1 & b_2 & b_3 & \dots \\
\hline
a_1 & (a_1, b_1) \rightarrow (a_1, b_2) & (a_1, b_3) & \dots \\
 & \swarrow & \nearrow & \downarrow \\
a_2 & (a_2, b_1) & (a_2, b_2) & (a_2, b_3) & \dots \\
 & \downarrow & \nearrow & \swarrow \\
a_3 & (a_3, b_1) & (a_3, b_2) & (a_3, b_3) & \dots \\
 & \swarrow \\
\dots & \dots & \dots & \dots & \dots & \dots & \dots
\end{array}
$$

Now, count the elements of the Cartesian product according to the depicted arrows. This counting covers all $A \times B$ and is one-to-one to $\mathbb{N}$. □

We are now ready to prove a fundamental fact of mathematics.

Theorem 1.7 *The set of rational numbers $\mathbb{Q}$ is countable.*

Proof Let $p, q \in \mathbb{N}$ denote the rows and columns of a matrix like the one defined in the proof of Theorem 1.6 with the set of natural numbers $\mathbb{N}$ as row and column headings, reporting $(p, q) = p/q$ in each entry. Count the entries in that matrix as described there, but skip a position if the fraction p/q has already been met. In this way, a one-to-one relation $n(q)$ is defined between positive rationals $q \in \mathbb{Q}_{>0}$ and natural numbers. Consider any $q \in \mathbb{Q}$ and set $n'(q) = 2 * n(q) + 1$ if $q > 0$, $n'(q) = 2 * n(-q)$ if $q < 0$ and $n'(0) = 1$. n' defines a one-to-one relation between $\mathbb{Q}$ and $\mathbb{N}$. □

Not all infinite sets are countable, as the following example shows.

▶ **Example 1.2** An uncountable set To build an uncountable set, consider the set of all binary infinite sequences $S = \{(s_1, s_2, s_3, \dots, s_k, \dots)\}$ with $s_k \in \{0, 1\}$. Now, take any countable subset of S, $\bar{S} \subseteq S$. By hypothesis, the elements of $\bar{S}$ can be counted, $\bar{S} = (s^1, s^2, \dots, s^n,)$. Consider a new sequence $s' \in S$ such that $s'_j = 1 - s^j_j$, where s^i_j is the jth element of the ith sequence in $\bar{S}$. By construction, s' is different from any element of $\bar{S}$ but $s' \in S$. Then, it cannot be that $\bar{S} = S$. The elements in S are actually in a one-to-one relation with real numbers.

1.4 Operations and Fields

A field is a structure that is commonly encountered in linear algebra. For the sake of completeness, we provide its definition here. To this end, we need the concept of an operation on elements of the set.

Definition 1.13 (*Binary operation*) A *binary operation* O on a set A is a function from $A \times A$ to A.

An operation maps an ordered couple of elements of A, say (a, b), to a third element of A, generically denoted $O(a, b)$, which is the result of the operation. The following is a list of possible properties of an operation.

Definition 1.14 Let O and O' be two operations on A, and a, b, and c, generic elements of A. Then

- O is *associative* if $O(O(a, b), c) = O(a, O(b, c))$;
- O is *commutative* if $O(a, b) = O(b, a)$;
- O' is *distributive* with respect to O if $O'(O(a, b), c) = O(O'(a, c), O'(b, c))$;
- a is the *neutral element* of O if $\forall x \in A$, $O(x, a) = O(a, x) = x$;
- a is the *absorbing element* of O if $\forall x \in A$, $O(x, a) = O(a, x) = a$;
- assume the operation O has the neutral element, a. Then if for $x \in A$, $\exists y \in A$ such that $O(x, y) = O(y, x) = a$, the element y is said to be the *inverse x*.

A subset $B \subseteq A$ is *closed* with respect to operation O, if $\forall a, b \in B$, $O(a, b) \in B$. Using the notion of binary operations, a field can be formally defined.

Definition 1.15 (*Field*) A *field* $(A, +, *)$ is a set A together with two binary operations, an *addition* or *sum* operation $+$ and a *multiplication* or *product* operation $*$ such that

- $+$ is associative, commutative, has a neutral element, denoted by 0, and any element of a admits an inverse, often called the *opposite*;
- $*$ is associative, commutative, and has a neutral element, denoted by 1. All elements apart from 0 have an inverse, often called *reciprocal*;
- $*$ is distributive with respect to $+$.

From the very definition of a field, a few properties follow. First, for any a, the distributive property implies that

$$a = a * 1 = a * (0 + 1) = (a * 0) + (a * 1) = (a * 0) + a$$

so that $a * 0 = 0$. That is, the neutral element of the sum is the absorbing element of the product. Next, consider the neutral element of the product, 1, and denote its opposite (its inverse with respect to addition) by -1. Then, for any a,

$$0 = a * 0 = a * (1 + (-1)) = (a * 1) + (a * (-1)),$$

so that $a * (-1) = -a$. That is, the opposite of a is obtained by multiplying a by the opposite of 1. From Definition 1.15, it is clear that any field should contain at least two elements: 0 and 1. The following example shows that this is also enough.

▶ **Example 1.3** (*The smallest field*) Consider the set $A = \{0, 1\}$ with the following binary operations:

$$
\begin{array}{c|cc}
+ & 0 & 1 \\
\hline
0 & 0 & 1 \\
1 & 1 & 0
\end{array}
\qquad\qquad
\begin{array}{c|cc}
* & 0 & 1 \\
\hline
0 & 0 & 0 \\
1 & 0 & 1
\end{array}
$$

The operations are defined using a table: the entry relative to a specific row and column is the result of the operation between the element associated with that row and that column. The multiplication table is completely defined by the properties seen above. The addition is defined using the usual rules that we know from arithmetic with the only exception that now $1 + 1 = 0$. Note that if we put $1 + 1 = 1$, then the element 1 would not have an opposite, violating the requirement of Definition 1.15. You can directly check that these operations define a field on A.

A class of fields with a finite number of elements can be built by considering *integers modulo p*.

▶ **Example 1.4** (*The field of integers modulo p*) Consider the set of the first p natural numbers $Z_p = \{0, 1, 2, \ldots, p - 1\}$. Define the operations $+$ and $*$ by adding or multiplying the elements with the usual arithmetic rules as if they were integer numbers, but dividing the result by p and taking the remainder. You can easily check that these operations are both associative and distributive. The thornier question is the existence of the inverse.

First, notice that all elements have an opposite: indeed the opposite of $n < p$ is just $p - n$. Now, suppose that there are two positive integers m and n, different from 1 and p, such that $m * n = p$. This means, using the remainder rule, that $m * n = 0$. Now assume that both m and n have an inverse, then $m^{-1} * m * n * n^{-1} = m^{-1} * 0 * n^{-1} = 0$ but also $m^{-1} * m * n * n^{-1} = 1 * 1 = 1$ which implies $1 = 0$, contradicting Definition 1.15. In order to avoid this issue, we have to consider values of p without integer divisors. In fact, if p is a prime number, $(Z_p, +, *)$ is a field.

1.5 Real Numbers

The next definition introduces three alternative ways of thinking about the set of real numbers: as an ordered field, as an extension of rational numbers including ordered couples of "contiguous" sets, or as a collection of elements defined using the decimal representation. I will refer to the different definitions when necessary, but I will not prove that they are equivalent.

Definition 1.16 (*Real Numbers*) The set $\mathbb{R}$ of *real numbers* can be defined in three alternative ways:

1. A field $(\mathbb{R}, +, *)$ with an ordered relation $\leq$ compatible with its operations and with the least upper bound and the greatest lower bound properties (see Definition 1.10).
2. Consider all couples (A, B) of subsets of $\mathbb{Q}$ such that

   - if $x \in A$ and $x' < x$, then $x' \in A$ and if $y \in B$ and $y' > y$, then $y' \in B$;
   - $\forall x \in A, \forall y \in B, x \leq y$;
   - $\forall \varepsilon > 0, \exists x \in A, y \in B$ such that $|x - y| < \varepsilon$.

   These couples are known as *Dedekind cuts*, and they are in a one-to-one relationship with the real numbers.
3. Consider the decimal representation of numbers:

$$\pm \underbrace{\text{finite digits}} \, . \, \underbrace{\text{countable sequence of digits}} \, .$$

   For rational numbers, the part beyond the decimal separator . is finite or periodic, with a pattern that repeats indefinitely. If that part is neither finite nor periodic, then the number is *irrational*, a real number that does not belong to $\mathbb{Q}$.

In what follows, I will denote by $\mathbb{R}_{\geq 0}$ the subset of nonnegative real numbers and by $\mathbb{R}_{>0}$ the subset of positive real numbers. A similar notation will be used for nonnegative $\mathbb{Q}_{\geq 0}$ and positive $\mathbb{Q}_{>0}$ rational numbers. Among the three alternatives in Definition 1.16, the second is probably the less intuitive. A real number x is described in terms of two sets of rational numbers: the set of all rational numbers less than x and the set of all rational numbers greater than x. Despite its abstraction, this is probably the most practical definition when one wants to prove the various properties of real numbers. The third alternative in Definition 1.16 provides an intuitive way of thinking about real numbers and clarifies their relationship to rational ones. Rational numbers do not have a unique decimal representation. For example, 1 and $0.\bar{9}$ are the same number. However, irrational numbers do, as they cannot end with a constant sequence of 9s. The first alternative in Definition 1.16 is the most operative. The expression "compatible with its operations" means that the ordering relation is preserved (or reversed) under addition and multiplication in exactly the same way as we are used to in arithmetic: if $x < y$ then $x + z < y + z$ for any z and $xy > 0$ if $x, y > 0$ or $x, y < 0$.[2] Given $x \in \mathbb{R}$, $[x]$ denotes the *integer part* of x, that is, the largest integer, possibly with sign, that is lower than or equal to x. The order relation on real numbers leads directly to the notion of *interval* (a, b) as the set of points greater than a and lower than b. We will use the notation $[a, b] = \{x \in \mathbb{R} \mid a \leq x \leq b\}$,

---

[2] An axiomatisation of this point is easy to produce. Since nothing is gained, in terms of understanding, from such a tedious exercise, I prefer to omit it.

$[a, b) = \{x \in \mathbb{R} \mid a \le x < b\}$, and $(a, b] = \{x \in \mathbb{R} \mid a < x \le b\}$. Define the *absolute value* $|x|$ of a real number x as $|x| = x$ if $x \ge 0$ and $|x| = -x$ if $x < 0$.

Definition 1.17 (*Increasing and decreasing functions; local maximum and minimum*) Consider a function $f : (a, b) \subseteq \mathbb{R} \rightarrow \mathbb{R}$ and a point $x \in (a, b)$ then

- f is *increasing* in x if $\exists \delta > 0$ such that if $|x - y| < \delta$, then $(f(x) - f(y))(x - y) \ge 0$;
- f is *decreasing* in x if $\exists \delta > 0$ such that if $|x - y| < \delta$, then $(f(x) - f(y))(x - y) \le 0$;
- x is a *local maximum* of f if $\exists \delta > 0$ such that if $|x - y| < \delta$, then $f(y) \le f(x)$;
- x is a *local minimum* of f if $\exists \delta > 0$ such that if $|x - y| < \delta$, then $f(y) \ge f(x)$;
- the function f is *monotonically increasing* or *monotonically decreasing* in the interval (a, b) if it is increasing or decreasing in all points of the interval;
- the "strict" version of the definitions above replaces "$\ge$" and "$\le$" with "$>$" and "$<$".

A function is *monotonic*, or has a *monotonic behaviour*, in an interval if it increases or decreases monotonically in that interval. The definitions of local maximum and minimum are consistent with Definition 1.11 and will be generalised in Definition 2.3. The least upper bound and greatest lower bound properties can be used to prove an important result for a sequence of nested intervals.

Theorem 1.8 (Nested intervals) *Consider the sequence of intervals (I_k) with $k \in \mathbb{N}$ defined as*

$$I_k = [a_k, b_k] = \{x \in \mathbb{R} \mid a_k \le x \le b_k\},$$

where $a_k \le b_k$. If $I_{k+1} \subseteq I_k$, $\forall k$, the set $\bigcap_{k=1}^{\infty} I_k$ is not empty.

Proof Because the intervals are nested, $\forall h, k, a_k \le b_h$. Consider the set $A = \{a_k \mid k \in \mathbb{N}\}$. Since $\forall k, a_k \le b_0$, A is bounded above. Then, for the property of real numbers, there exists $x = \sup A$ such that $x \ge a_k$, $\forall k$. Furthermore, any b_h is an upper bound of $\{a_k\}$, so that $x \le b_k$, $\forall k$. Then $x \in \bigcap_{k=1}^{\infty} I_k$. A similar proof can be derived using the infimum of the set $B = \{b_k \mid k \in \mathbb{N}\}$. $\qquad \square$

There are a couple of useful inequalities that it is important to review. The first is about the absolute value.

Theorem 1.9 (Triangle inequality) *For any two real numbers x and y, $|x| + |y| \ge |x + y|$.*

Proof Just observe that

$$(x + y)^2 = x^2 + y^2 + 2xy \le x^2 + y^2 + 2|x||y| = (|x| + |y|)^2,$$

and remember that if $a^2 = b^2$, then $a = \pm b$. $\qquad \square$

For the second inequality, note that given two positive real numbers x and y and any natural number n,

$$x^n - y^n = (x - y)\left(x^{n-1} + y\,x^{n-2} + y^2\,x^{n-3} + \ldots + y^{n-1}\right).$$

The second parentheses contain n terms, so, if $x > y$, we get the *difference of powers inequality*,

$$n\,(x - y)\,y^{n-1} \le x^n - y^n \le n\,(x - y)\,x^{n-1}.$$

We will use the summation symbol $\sum$ to denote a sum of elements that is difficult to write explicitly. For example, the sum of the first n integer numbers is denoted by $\sum_{k=1}^{n} k$ (see Exercise 1.1). There is a useful rule regarding summations.

Lemma 1.1 (Summation by parts) *Given two generic sets of numbers* $\{a_1, \ldots, a_n\}$ *and* $\{b_1, \ldots, b_{n+1}\}$,

$$\sum_{k=1}^{n} a_k(b_{k+1} - b_k) + \sum_{k=2}^{n} b_k(a_k - a_{k-1}) = a_n b_{n+1} - a_1 b_1.$$

Proof Open all the parentheses and simplify the resulting expression.   □

▶ **Example 1.5** (*Geometric progression*) We derive a simple expression for the *geometric progression*, the sum of a finite number of terms with a fixed ratio. For $a, r \in \mathbb{R}$, define $s_n(a, r) = a + ar + ar^2 + \ldots = \sum_{k=0}^{n} ar^k$. If $r = 0$, then $s_n(a, r) = a$. If $r = 1$, then $s_n(a, r) = na$. If $r \ne 0, 1$, by simple algebraic simplifications, it is easy to see that $rs_n(a, r) - s_n(a, r) = ar^{n+1} - a$. Thus, $s_n(a, r) = a(1 - r^{n+1})/(1 - r)$.

We will use the product symbol $\prod$ to denote a product of elements. For example, the *factorial* of $n \in \mathbb{N}$, is defined as $n! = \prod_{k=1}^{n} k$, while $0! = 1$.

Theorem 1.10 (Archimedean property) *Consider* $y \in \mathbb{R}_{>0}$. *Then* $\forall x \in \mathbb{R}, \exists n \in \mathbb{N}$ *such that* $ny > x$.

Proof Consider the set $E = \{ny \mid n \in \mathbb{N}\}$. Assume that E is bounded above and set $z = \sup E$. Now $z - y < z$, which implies that there exists an element of E, say my for some integer m, such that $my > z - y$, that is, $(m + 1)y > z$, which is absurd for the properties of the supremum. Thus, E is unbounded, and the statement is proved.   □

The Archimedean property is used to prove a fundamental fact about rational numbers.

Theorem 1.11 (Density of rational numbers) *Given two real numbers $x < y$, it is always possible to find a rational number $q \in \mathbb{Q}$ such that $x < q < y$.*

Proof By definition $y - x > 0$, and therefore, for Theorem 1.10, $\exists n \in \mathbb{N}$ such that $n(y - x) > 1$, that is $nx + 1 < ny$. Let m be an integer in the interval $(nx, nx + 1]$, and thus $nx < m \leq nx + 1 < ny$. Dividing the inequalities by n shows that the rational number $q = m/n$ fulfils the statement.   □

Theorem 1.11 is summarised by saying that $\mathbb{Q}$ is *dense* in $\mathbb{R}$. From Definition 1.16 (1), it follows that any strictly increasing function f defined over an interval of rational numbers $(a, b) \subset \mathbb{Q}$ can be directly extended to any real number $x \in (a, b)$ by taking

$$f(x) = \sup_{q \in \mathbb{Q}} \{f(q) \mid a < q < x\} \text{ or } f(x) = \inf_{q \in \mathbb{Q}} \{f(q) \mid x < q < b\}.$$

In this way, a strictly increasing function is defined over the interval of real numbers (a, b). This definition is meaningful because the supremum and infimum are unique and, according to Definition 1.16 (1), the two extensions are identical. If the function is strictly decreasing on the rationals, one replaces the supremum with the infimum, and vice versa, in the above definition, obtaining a strictly decreasing function on the real. In general, since the majority of functions we encounter are piecewise strictly monotonic, we can immediately think of them as real functions, even if we are practically able to compute their value only for rational arguments.

This technique can be applied to obtain a proper definition of the exponential function on the set of real numbers. We need a preliminary result on the existence of the nth root of a positive real number.

Theorem 1.12 (n^{th} root of a positive real number) *For any positive real number x and any natural number n, there exists one positive real number y such that $y^n = x$.*

Proof Consider the set $E = \{z \in \mathbb{R}_{>0} \mid z^n < x\}$. Clearly $x/(1 + x) \in E$ and $(x + 1)$ is an upper bound of E, thus E is not empty and bounded above. Let $y = \sup E > 0$. I want to show that $y^n = x$ and I will prove this by contradiction. Start by assuming $y^n < x$. Then there must be a positive number $0 < h < 1$ such that $(y + h)^n < x$. In fact, from the difference of powers inequality,

$$(y + h)^n - y^h \leq nh(y + h)^{n-1} < nh(y + 1)^{n-1}.$$

Thus, it is sufficient to choose h such that $nh(y + 1)^{n+1} < x - y^n$. But this is a contradiction because there are no numbers greater than y in E. Next, assume that $y^n > x$. In this case, there must be a positive number h such that $(y - h)^n > x$. In fact, from the difference of powers inequality,

$$y^n - (y-h)^h \leq nhy^{n-1}.$$

In this case, it is sufficient to choose h such that $nhy^{n-1} < y^n - x$. This is a contradiction because $y - h$ cannot be an upper bound of E. □

The nth root of x is denoted by $\sqrt[n]{x}$ or $x^{1/n}$. Obviously, $1^{1/n} = 1$ and $0^{1/n} = 0$, $\forall n \in \mathbb{N}$. The function $f(x) = x^{1/n}$ is the inverse function of $g(x) = x^n$. Since the latter is strictly increasing, so is the former. This proves that the y in Theorem 1.12 is unique and that $x^{1/n} > 1$ if $x > 1$ and $x^{1/n} < 1$ if $x < 1$.

For any $x, y > 0$ and any couple of integers n and m, $x^n y^n = (xy)^n$ and $(x^m)^n = (x^n)^m$. Thus, $x^{1/n} y^{1/n} = (xy)^{1/n}$ and $(x^n)^{1/m} = (x^{1/m})^n$. We have extended the definition of power to any positive rational number $q \in \mathbb{Q}_{>0}$, $q = n/m$. This definition can be extended to positive real numbers.

Theorem 1.13 (Power with positive real exponent) *For any $a, b \in \mathbb{R}_{>0}$ the power of base a with exponent b, denoted as a^b, is a positive real number such that $\forall c \in \mathbb{R}_{>0}$:*

1. $1^b = 1$ *and if $b > c$ then $a^b > a^c$ if $a > 1$, $a^b < a^c$ if $a < 1$;*
2. $a^{b+c} = a^b a^c$;
3. $a^{bc} = (a^b)^c$.

Proof We start by considering rational exponents. Any two positive rationals $q_1 > q_2$ admit a representation as the ratio of naturals, $q_1 = n_1/m$ and $q_2 = n_2/m$, with a common denominator m and $n_1 > n_2$. If $a > 1$, then $a^{1/m} > 1$, and $a^{q_1} = (a^{1/m})^{n_1} > (a^{1/m})^{n_2} = a^{q_2}$ so that a^q is a strictly increasing function of the exponent. Analogously, if $a < 1$ one can prove that a^q is strictly decreasing. Note also that $\forall n, m \in \mathbb{N}$, $1^{n/m} = 1$. In general, one has $a^{q_1+q_2} = a^{(n_1+n_2)/m} = (a^{1/m})^{n_1+n_2} = a^{n_1/m} a^{n_2/m} = a^{q_1} a^{q_2}$ and, at the same time, $a^{n_2 q_1} = (a^{q_1})^{n_2}$. Taking the mth root of the last expression, one gets $a^{q_1 q_2} = (a^{q_1})^{q_2}$. Thus, all the properties of the statement are true when the exponents are rational numbers. Now consider the case $a > 1$. Because the function is strictly increasing, $\forall b \in \mathbb{R}_{>0}$ we can define $a^b = \sup_{q \in \mathbb{Q}_{>0}} \{a^q \mid q < b\}$. This function is strictly increasing by construction. Given two positive real numbers b and c,

$$a^b a^c = \sup_{q_1 \in \mathbb{Q}_{>0}} \{a^{q_1} \mid q_1 < b\} \sup_{q_2 \in \mathbb{Q}_{>0}} \{a^{q_2} \mid q_2 < c\},$$

and

$$a^{b+c} = \sup_{q \in \mathbb{Q}_{>0}} \{a^q \mid q < b+c\}.$$

According to Theorem 1.11, for any $q_1 < b$ and $q_2 < c$, $\exists q$ such that $q_1 + q_2 < q < b + c$. Thus $a^{b+c} \geq a^b a^c$. At the same time, for any $q < b+c$, set $\delta = b+c-q$ and let q_1 and q_2 be such that $b - \delta/3 < q_1 < b$ and $c - \delta/3 < q_2 < c$. Then $q_1 < b$, $q_2 < c$, and $q < q_1 + q_2$. Using the property of the power with rational exponent, $a^q < a^{q_1+q_2} < a^{q_1} a^{q_2} \leq a^b a^c$, that implies $a^{b+c} \leq a^b a^c$. Taking both inequalities together, we conclude that $a^b a^c = a^{b+c}$ and property 2 is proved. For property 3, simply note that for any positive rational r, it is

$$\sup_{q \in \mathbb{Q}_{>0}} \{a^q \mid q < rb\} = \sup_{q/r \in \mathbb{Q}_{>0}} \{(a^r)^{q/r} \mid q/r < b\}$$

that is, $a^{rb} = (a^r)^b$. For any positive real c just take the supremum over all rational $r < c$ to get $a^{cb} = (a^c)^b$. The analysis for $a < 1$ is identical, with the only difference that in this case the function a^q is strictly decreasing. Therefore, the infimum should be used rather than the supremum when defining the power with a positive real exponent.   □

The previous theorem implies that $\forall a, b, c \in \mathbb{R}_{>0}$, $(ab)^c = a^c b^c$ and that if $a > c$, then $a^b > c^b$, that is, the power increases with its base. The exponent is extended to the entire set $\mathbb{R}$ by setting $a^0 = 1$ and $a^x = (1/a)^{-x}$ if $x < 0$. This extension preserves the properties defined above. In conclusion, we have learnt that we can raise any positive real number to any real power. This allows us to define two important functions on the real numbers.

Definition 1.18 (*Power function*) For any $a \in \mathbb{R}$, the *power function* from $\mathbb{R}_{>0}$ to $\mathbb{R}_{>0}$ is defined as $f(x) = x^a$.

The power function increases strictly with x if $a > 0$, and decreases strictly with x if $a < 0$. Moreover, in both cases, the power function has an inverse, $f^{-1}(x) = x^{1/a}$, and its image is $\mathbb{R}_{>0}$. If $a = 0$, the function is constant and equal to 1.

Definition 1.19 (*Exponential function*) For any $a \in \mathbb{R}_{>0}$, the *exponential function* from $\mathbb{R}$ to $\mathbb{R}_{>0}$ is defined as $f(x) = a^x$.

The exponential function increases strictly with x if $a > 1$, decreases strictly if $a < 1$, and is constant and equal to 1 if $a = 1$.[3] This function admits an inverse, which will be discussed in Example 5.10. We will use the existence of the exponential function defined on real numbers to derive an important relation between the arithmetic and geometric means of a set of positive numbers.

The n-fold Cartesian product of $\mathbb{R}$ with itself will be denoted as $\mathbb{R}^n$. A bold symbol $\mathbf{x}$ will generically denote an element of $\mathbb{R}^n$ or an n-tuple, with n real *components* $\mathbf{x} = (x_1, \ldots, x_n)$. $\mathbb{R}^n_{\geq 0}$ denotes the set of n-tuples with nonnegative components and $\mathbb{R}^n_{>0}$ the set of n-tuples with positive components. The special n-tuples $\mathbf{0}$ and $\mathbf{1}$ have all components equal to 0 and 1, respectively. With some abuse of notation, given two n-tuples $\mathbf{x}$ and $\mathbf{y}$, I will write $\mathbf{x} \geq \mathbf{y}$ or $\mathbf{x} \gg \mathbf{y}$ if $x_i \geq y_i$ or $x_i > y_i$, respectively, $\forall i = 1, \ldots, n$. Thus, if $\mathbf{x} \in \mathbb{R}^n_{\geq 0}$, $\mathbf{x} \geq \mathbf{0}$ and if $\mathbf{x} \in \mathbb{R}^n_{>0}$, $\mathbf{x} \gg \mathbf{0}$. The expression $\mathbf{x} > \mathbf{y}$ means that $\mathbf{x} \geq \mathbf{y}$ and $\mathbf{x} \neq \mathbf{y}$.

---

[3] An alternative definition of the exponential function is presented in Example 5.10.

Definition 1.20 (*Weighted arithmetic mean*) Let $\mathbf{x} = (x_1, \ldots, x_n) \in \mathbb{R}^n$ be a n-tuple of real numbers and $\mathbf{w} = (w_1, \ldots, w_n) \in \mathbb{R}^n_{>0}$ a n-tuple of positive weights. The *weighted arithmetic mean* of $\mathbf{x}$ with weight $\mathbf{w}$ is defined as

$$a_n(\mathbf{x}, \mathbf{w}) = \frac{\sum_{i=1}^n w_i x_i}{\sum_{i=1}^n w_i}.$$

Definition 1.21 (*Weighted geometric mean*) Let $\mathbf{x} = (x_1, \ldots, x_n) \in \mathbb{R}^n_{\geq 0}$ be an n-tuple of nonnegative real numbers and $\mathbf{w} = (w_1, \ldots, w_n) \in \mathbb{R}^n_{>0}$ an n-tuple of positive weights. The *weighted geometric mean* of $\mathbf{x}$ with weight $\mathbf{w}$ is defined as

$$g_n(\mathbf{x}, \mathbf{w}) = \left(\prod_{i=1}^n x_i^{w_i} \right)^{1/\sum_{i=1}^n w_i}.$$

When all weights are equal, $\forall i, j, w_i = w_j$, the previous definitions reduce to their unweighted regular form. If all components are equal to the same value x, then $a_n(\mathbf{x}, \mathbf{w}) = g_n(\mathbf{x}, \mathbf{w}) = x$. If all weights are multiplied by the same positive constant, the values of a_n and g_n do not change. We can prove that, in general, the arithmetic mean is not lower than the geometric mean.[4]

Theorem 1.14 (AM-GM inequality) *Let* $\mathbf{x} = (x_1, \ldots, x_n)$ *be an n-tuple of non-negative real numbers, and* $\mathbf{w} = (w_1, \ldots, w_n)$ *an n-tuple of positive weights, then* $a_n(\mathbf{x}, \mathbf{w}) \geq g_n(\mathbf{x}, \mathbf{w})$.

Proof If all components of $\mathbf{x}$ are equal, the theorem is obviously true. Thus, we will assume that some components differ. First, we prove the statement when all weights are equal to 1. For generic positive numbers, using the square of the binomial, it is immediately apparent that the statement is true when $n = 2$. From there we will proceed by induction. Assume that the statement is true for n and consider the arithmetic average of $n + 1$ numbers for which one has

$$n a_{n+1}(\mathbf{x}, \mathbf{1}) = \sum_{i=1}^{n-1} x_i + y,$$

with $y = x_n + x_{n+1} - a_{n+1}$. Without loss of generality, we can assume that $x_{n+1} < a_{n+1} < x_n$. In fact, there must be at least one number greater than the average and at least one number lower than the average. Thus, $x_{n+1} < y < x_n$. Since a_{n+1} can be seen as an average of n numbers, applying the theorem for n elements, we obtain that

---

[4] One can easily devise a simpler proof based on the Jensen inequality discussed in Corollary 1.1 and the concavity of the logarithmic function proved in Example 5.9. However, the proof of the latter property without the use of the AM-GM inequality requires the notion of the limit of a sequence that will be covered in Chap. 5.

$$a_{n+1}^n \geq y \prod_{i=1}^{n-1} x_i.$$

Multiplying both sides of the previous equation by a_{n+1} and noting that $0 < (x_n - y)(x_n - a_{n+1}) = a_{n+1}y - x_n x_{n+1}$,

$$a_{n+1}^{n+1} \geq a_{n+1}y \prod_{i=1}^{n-1} x_i > \prod_{i=1}^{n+1} x_i,$$

which proves the assertion.

Now consider the case in which the weights $(w_1, \ldots, w_n)$ are integer numbers. Then the original n elements $(x_1, \ldots, x_n)$ can be replaced with a new set of $\sum_{i=1}^n w_i$ elements, obtained by repeating each element x_i exactly w_i times. The integer weighted averages are just regular unweighted averages on the new set, and the theorem applies.

If the weights are rational numbers, then for each i, $w_i = p_i/q_i$ with $q_i, p_i \in \mathbb{N}$. It can immediately be seen that the weighted averages do not change if each weight w_i is replaced with the integer Qw_i, where $Q = \prod_{i=1}^n q_i$. Thus, we are back to the case of integer weights, and the theorem applies again.

Finally, consider the case of real weights $\mathbf{w}$. Without loss of generality, we can assume that $\forall i = 1, \ldots, n$, $x_1 \leq x_i$, and $\sum_{i=1}^n w_i = 1$, setting $w_1 = 1 - \sum_{i=2}^n w_i$. With this restriction, a_n and g_n become increasing functions of $w_2, \ldots, w_n$ and

$$a_n(\mathbf{x}, \mathbf{w}) = \sup_{(q_2, \ldots, q_n) \in \mathbb{Q}_{>0}^{n-1}} \{a_n(\mathbf{x}, \mathbf{q}) \mid q_i \leq w_i, i = 2, \ldots, n\}$$

$$\geq \sup_{(q_2, \ldots, q_n) \in \mathbb{Q}_{>0}^{n-1}} \{g_n(\mathbf{x}, \mathbf{q}) \mid q_i \leq w_i, i = 2, \ldots, n\} = g_n(\mathbf{x}, \mathbf{w}),$$

where $q_1 = 1 - \sum_{i=2}^n q_i$ and we have used the fact that $\forall \mathbf{q} \in \mathbb{Q}_{\geq 0}^n$, $a_n(\mathbf{x}, \mathbf{q}) \geq g_n(\mathbf{x}, \mathbf{q})$. □

From the previous result, a famous inequality follows.

Theorem 1.15 (The Young inequality) *For any two positive numbers p and q such that $1/p + 1/q = 1$ and for any two nonnegative numbers a and b,*

$$ab \leq \frac{a^p}{p} + \frac{b^q}{q}.$$

Proof Set $x_1 = a^p$, $x_2 = b^q$, $w_1 = 1/p$, and $w_2 = 1/q$ and use Theorem 1.14. □

▶ **Example 1.6** (*Power mean inequality*) Let $\mathbf{x} = (x_1, \ldots, x_n)$ be an n-tuple of positive real numbers and $\mathbf{w} = (w_1, \ldots, w_n)$ an n-tuple of positive weights. For any $z > 0$, define the *weighted power mean*

$$U(z) = \left(\frac{\sum_{i=1}^{n} w_i x_i^z}{\sum_{i=1}^{n} w_i} \right)^{1/z}.$$

We want to prove that if $y > z$, then $U(y) \geq U(z)$ with the equality holding only if all x are equal. By direct substitution in the expression above, it is easy to verify that $\forall j = 1, \ldots, n$,

$$w_j x_j^z U(y)^{-z} = w_j^{1-z/y} \left(\frac{w_j x_j^y \sum_{i=1}^{n} w_i}{\sum_{i=1}^{n} w_i x_i^y} \right)^{z/y}.$$

Recalling that $z/y < 1$, apply the inequality in Theorem 1.14 to the right-hand side to obtain

$$w_j x_j^z U(y)^{-z} \leq \left(1 - \frac{z}{y} \right) w_j + \frac{z}{y} \frac{w_j x_j^y \sum_{i=1}^{n} w_i}{\sum_{i=1}^{n} w_i x_i^y}.$$

Summing over j and rearranging terms,

$$\frac{\sum_{i=1}^{n} w_j x_j^z}{\sum_{j=1}^{n} w_j} \leq U(y)^z,$$

which proves the assertion. Note that the equality in the previous expression requires that $x_j^y = (\sum_{i=1}^{n} w_i x_i^y)/(\sum_{i=1}^{n} w_i)$ for all j, that is, that all x are equal.

1.6   Convexity and Concavity

I will omit any systematic treatment of the topics covered in any standard course of linear algebra. However, as a reference for the reader, I will briefly review some results that will be particularly useful in the next chapters.

Definition 1.22 (*Convex set*) The subset $A \subseteq \mathbb{R}^n$ is *convex* if for any couple of elements $\mathbf{a}, \mathbf{b} \in A$ and any $\lambda \in [0, 1]$ it is $\lambda \mathbf{a} + (1 - \lambda)\mathbf{b} \in A$.

In fact, this definition is not restricted only to pairwise combinations: if A is convex and we consider n elements $\mathbf{v}_1, \ldots, \mathbf{v}_n$ in A, and n nonnegative real numbers $\lambda_1, \ldots, \lambda_n$ such that $\lambda_1 + \lambda_2 + \cdots + \lambda_n = 1$, the linear combination $\lambda_1 \mathbf{v}_1 + \lambda_2 \mathbf{v}_2 + \cdots + \lambda_n \mathbf{v}_n$ is again an element of A. This is easy to prove by induction. The linear combination above is often referred to as *convex combination*. Linear subspaces are an example of convex subsets. On convex subsets, it is possible to define functions with special, and often useful, behaviour.

Definition 1.23 (*Concave and convex functions*) Given a convex subset $A \subseteq \mathbb{R}^n$, the real-valued function $f : A \subseteq \mathbb{R}^n \to \mathbb{R}$ is *concave* if for any couple of elements $\mathbf{x}_1, \mathbf{x}_2 \in A$ and any $\lambda \in (0, 1)$,

$$f((1 - \lambda)\mathbf{x}_1 + \lambda\mathbf{x}_2) \geq (1 - \lambda)f(\mathbf{x}_1) + \lambda f(\mathbf{x}_2).$$

The function f is *strictly concave* if $\forall \lambda \in (0, 1)$,

$$f((1 - \lambda)\mathbf{x}_1 + \lambda\mathbf{x}_2) > (1 - \lambda)f(\mathbf{x}_1) + \lambda f(\mathbf{x}_2).$$

The function is *convex* and *strictly convex* if the previous inequalities hold with $\leq$ instead of $\geq$ and $<$ instead of $>$, respectively.

Note that a function $f(x)$ defined in a convex set A is concave if and only if $-f(x)$ is convex. A linear function is both concave and convex, but neither strictly concave nor strictly convex.

▶ **Example 1.7** (*Convexity and concavity of exponential and power functions*) Using Theorem 1.14, for any positive a, any two distinct numbers x, y, and any $\lambda \in (0, 1)$, it is $a^{\lambda x + (1-\lambda)y} = (a^x)^\lambda (a^y)^{1-\lambda} < \lambda a^x + (1 - \lambda)a^y$. Thus, we can conclude that the exponential function in Definition 1.19 is strictly convex.

Consider $a > 1$. Using the power mean inequality in Example 1.6, for any two distinct positive numbers x, y and any $\lambda \in (0, 1)$, $U(a) = (\lambda x^a + (1 - \lambda)y^a)^{1/a} > U(1) = \lambda x + (1 - \lambda)y$, hence $\lambda x^a + (1 - \lambda)y^a > (\lambda x + (1 - \lambda)y)^a$. In this case, the power function in Definition 1.18 is strictly convex. If $0 < a < 1$ the inequality is reversed, with $U(a) < U(1)$, so that the power function is now strictly concave. For the case with negative exponent $f(x) = x^{-a} = 1/x^a$ with $a > 0$, using Theorem 1.14, $\lambda/x^a + (1-\lambda)/y^a > 1/(x^\lambda y^{1-\lambda})^a > 1/(\lambda x + (1 - \lambda)y)^a$ and we can conclude that the power function in this case is strictly convex.

Concave and convex functions are characterised by a precise ordering of their rates of increase along a straight line.

Theorem 1.16 *Consider a real-valued function f defined on a convex subset $A \subseteq \mathbb{R}^n$ and let $\mathbf{x}_1$, $\mathbf{x}_2$, and $\mathbf{x}_3$ be three aligned elements in A, with $\mathbf{x}_2$ belonging to the segment with end points $\mathbf{x}_1$ and $\mathbf{x}_3$. Then if f is concave,*

$$\frac{f(\mathbf{x}_2) - f(\mathbf{x}_1)}{\|\mathbf{x}_2 - \mathbf{x}_1\|} \geq \frac{f(\mathbf{x}_3) - f(\mathbf{x}_1)}{\|\mathbf{x}_3 - \mathbf{x}_1\|} \geq \frac{f(\mathbf{x}_3) - f(\mathbf{x}_2)}{\|\mathbf{x}_3 - \mathbf{x}_2\|}.$$

The inequalities are strict if the function f is strictly concave. Replace $\geq$ with $\leq$ if the function f is convex and with $<$ if it is strictly convex.

Proof Because x_2 belongs to the segment with end points x_1 and x_3

$$x_2 = \frac{\|x_2 - x_1\|}{\|x_3 - x_1\|}x_3 + \left(1 - \frac{\|x_2 - x_1\|}{\|x_3 - x_1\|}\right)x_1 = \frac{\|x_3 - x_2\|}{\|x_3 - x_1\|}x_1 + \left(1 - \frac{\|x_3 - x_2\|}{\|x_3 - x_1\|}\right)x_3.$$

To prove the assertion, substitute these equations in the inequalities in Definition 1.23 and rearrange terms appropriately.   □

By the previous theorem, if a concave or convex function f is constant at three aligned points, $f(x_1) = f(x_2) = f(x_3)$, then it is constant on the segment. The following result generalises the inequalities in Definition 1.23 to a finite set of elements.

Corollary 1.1 (Jensen's inequality) *Consider a set of points* $(x_1, \ldots, x_n)$ *in a convex subset of* $\mathbb{R}^n$ *and a set of positive real number* $(\lambda_1, \ldots, \lambda_n)$ *such that* $\sum_{i=1}^{n} \lambda_i = 1$. *Then, if* f *is concave,* $f(\sum_{i=1}^{n} \lambda_i x_i) \geq \sum_{i=1}^{n} \lambda_i f(x_i)$ *while if* f *is convex,* $f(\sum_{i=1}^{n} \lambda_i x_i) \leq \sum_{i=1}^{n} \lambda_i f(x_i)$. *If the function is strictly concave or strictly convex, the previous inequalities hold with* $\geq$ *replaced by* $>$ *and* $\leq$ *replaced by* $<$, *respectively.*

Proof This can be easily proved by induction. Because of Definition 1.23, the statement clearly holds for $n = 2$. Assume that it holds for $n - 1$. Then considering a concave function f,

$$f\left((1 - \lambda_n)\sum_{i=1}^{n-1} \frac{\lambda_i}{1 - \lambda_n}x_i + \lambda_n x_n\right) \geq (1 - \lambda_n)f\left(\sum_{i=1}^{n-1} \frac{\lambda_i}{1 - \lambda_n}x_i\right) + \lambda_n f(x_n).$$

But since the statement holds for $n - 1$,

$$f\left(\sum_{i=1}^{n-1} \frac{\lambda_i}{1 - \lambda_n}x_i\right) \geq \sum_{i=1}^{n-1} \frac{\lambda_i}{1 - \lambda_n}f(x_i),$$

and direct substitution proves the assertion for n. The case of a convex function and the strict versions are proved along very similar lines.   □

The extension of the previous inequality to integrals is presented in Corollary 9.5.

The convex cone is a particular kind of convex set defined as the linear combination with nonnegative coefficients of a set of elements.

Definition 1.24 (*Convex cone*) The *convex cone* or *polyhedral cone* C generated by the elements $\{a_1, \ldots, a_k\}$ belonging to $\mathbb{R}^n$ is defined as

$$C = \left\{x = \sum_{i=1}^{k} \lambda_i a_i \mid \lambda_i \geq 0, i = 1, \ldots, k\right\}.$$

The combination with nonnegative coefficients in the above definition is known as a *conical combination*. It is easy to see that the convex cone is closed with respect to the convex combination of its elements. Thus, according to Definition 1.22, it is a convex set. In general, a convex cone is not a subspace. However, if for all $i = 1, \ldots, k$, $-\mathbf{a}_i \in C$, then $C = \mathrm{span}\{\mathbf{a}_1, \ldots, \mathbf{a}_k\}$, that is, C becomes the linear space generated by the elements that define the cone.

Definition 1.25 (*Relative interior*) The *relative interior* C^0 of the convex cone C in Definition 1.24 is defined as

$$C^0 = \left\{ \mathbf{x} = \sum_{i-1}^{k} \lambda_i \mathbf{a}_i \mid \lambda_i > 0, i = 1, \ldots, k \right\}.$$

The relative interior is the set obtained by considering only conical combinations with positive coefficients. A remark is necessary to avoid confusion. In Chap. 2, we introduce the topological notion of interior of a set as those elements of the set that lie "inside" it (cf. Definition 2.2). The relative interior C^0 is, in general, not the topological interior of the set C. In fact, consider the cone generated by a single element $\{\mathbf{a}\} \in \mathbb{R}^n$ with $n > 1$. This set has no interior points in the standard (Euclidean) topology in $\mathbb{R}^n$.

Using the notion of inner product, two special cones related to the convex cone in Definition 1.24 can be defined.

Definition 1.26 (*Dual and polar cone*) Consider the convex cone C generated by the set of elements $\{\mathbf{a}_1, \ldots, \mathbf{a}_k\}$. Its *dual cone* C^+ and *polar cone* C^- are defined as

$$C^+ = \left\{ \mathbf{x} \in \mathbb{R}^n \mid \mathbf{x} \cdot \mathbf{y} \geq 0, \forall \mathbf{y} \in C \right\},$$

and

$$C^- = \left\{ \mathbf{x} \in \mathbb{R}^n \mid \mathbf{x} \cdot \mathbf{y} \leq 0, \forall \mathbf{y} \in C \right\}.$$

Clearly $C^- = -C^+$. That is, all elements of the polar cone can be obtained by considering the opposite of the elements of the dual cone, and vice versa.

Exercises

Exercise 1.1 Prove by induction that the sum of the first n natural numbers $s_n = \sum_{k=1}^{n} k$ is equal to $n(n+1)/2$.

Exercise 1.2 Derive the formula for the sum of the first n even and odd natural numbers.

Exercise 1.3 (*Binomial coefficient*) Prove by induction that for any $a, b \in \mathbb{R}$ and $n \in \mathbb{N}$,

$$(a+b)^n = \sum_{k=0}^{n} \binom{n}{k} a^{n-k} b^k, \quad \binom{n}{k} = \frac{n!}{k!(n-k)!}.$$

Exercise 1.4 With reference to Example 1.4, build the addition and multiplication tables of the field of integers modulus 3.

Exercise 1.5 Consider a bounded set of real numbers $E \subseteq \mathbb{R}$ and let $E^- = \{x \mid -x \in E\}$ be the set of its opposites. Prove that $\inf E = -\sup E^-$.

Exercise 1.6 Prove by induction that for $x > -1$ and any integer n it is $(1+x)^n \geq 1 + nx$. This is called the *Bernoulli inequality*.

Exercise 1.7 The *Peter–Paul inequality* states that $\forall a, b \in \mathbb{R}$ and $\forall \epsilon > 0$, $2ab \leq a^2/\epsilon + \epsilon b^2$. Prove it. The name of the inequality derives from the old saying "Robbing Peter to Pay Paul": if a is Peter's wealth and b is Paul's wealth, this is what one does by increasing ϵ. *Hint: Recall the expression of the square of the binomial.*

Exercise 1.8 Using Theorem 1.13, prove that $\forall a, b, c \in \mathbb{R}_{>0}$, $(ab)^c = a^c b^c$ and that if $a > c$, then $a^b > c^b$.

Exercise 1.9 Using Theorem 1.13, prove that by setting $a^0 = 1$ and $a^x = 1/a^{-x}$ if $x < 0$, then $\forall x, y \in \mathbb{R}$ and $\forall a \in \mathbb{R}_{>0}$ it is $a^{x+y} = a^x a^y$ and $(a^x)^y = a^{xy}$.

Exercise 1.10 Using the AM-GM inequality of Theorem 1.14, prove that for any real $r \in (0, 1)$ and $x > -1$, $(1+x)^r \leq 1 + rx$, which is another form of the *Bernoulli inequality*. *Hint: Consider taking the average of 1 and $1+x$ with appropriate weights.*

Exercise 1.11 Given n positive numbers $\mathbf{x} = (x_1, \ldots, x_n)$ and associated non-negative weights $\mathbf{w} = (w_1, \ldots, w_n)$, the *weighted harmonic mean* is defined as $h_n(\mathbf{x}, \mathbf{x}) = \left(\sum_{i=1}^{n} w_i\right) / \left(\sum_{i=1}^{n} w_i/x_i\right)$. Prove that $g_n(\mathbf{x}, \mathbf{x}) \geq h_n(\mathbf{x}, \mathbf{x})$. *Hint: Use the AM-GM inequality on the inverse of the numbers.*

Exercise 1.12 Using the AM-GM inequality of Theorem 1.14, prove that if a function from $\mathbb{R}$ to $\mathbb{R}$ is convex and admits an inverse, its inverse is concave, and vice versa.

Exercise 1.13 Prove that Definition 1.22 implies that for any n elements $(\mathbf{x}_1, \ldots, \mathbf{x}_n)$ in A and any n nonnegative real numbers $(\lambda_1, \ldots, \lambda_n)$ such that $\sum_{i=1}^{n} \lambda_i = 1$, it is $\sum_{i=1}^{n} \lambda_i \mathbf{x}_n \in A$. *Hint: Prove this by induction.*

Exercise 1.14 Using Theorem 1.16, prove that if the function f is constant at three aligned points, $f(\mathbf{x}_1) = f(\mathbf{x}_2) = f(\mathbf{x}_3)$, then it is constant on the segment containing the three points.

Exercise 1.15 Given a subset $A \subseteq \mathbb{R}^n$ define its *dual cone* as

$$C^+(A) = \{\mathbf{x} \in V \mid \mathbf{x} \cdot \mathbf{y} \geq 0, \forall \mathbf{y} \in A\}.$$

Prove that $C^+(A)$ is a convex set, irrespective of whether A is convex or not.

Exercise 1.16 Consider n elements $\{\mathbf{x}_1, \ldots, \mathbf{x}_n\}$ of $\mathbb{R}^n$ such that $\mathbf{x}_i \cdot \mathbf{x}_j \geq 0, \forall i, j = 1, \ldots, n$. Let C be the cone generated by these elements and C^+ its polar cone. Prove that $C \subseteq C^+$.

Topology

2

2.1 Definition and Basic Properties

This chapter is devoted to the study of the following object.

Definition 2.1 (*Topological space*) Consider a set X and its power set 2^X. Let $T \subseteq 2^X$ be such that

1. $X, \emptyset \in T$;
2. the union of a finite or infinite number of elements of T is in T;
3. the intersection of a finite number of elements of T is in T.

Then T is a *topology* on X, (X, T) is a *topological space*, and the elements of T are called *open sets*.

Points 2 and 3 of Definition 2.1 can be summarised by saying that the set T is closed with respect to finite and infinite unions and with respect to finite intersections. According to the previous definition, on any set X, it is always possible to build the *trivial topology* by taking $T = \{\emptyset, X\}$ and the *discrete topology*, or finest topology, by taking $T = 2^X$.

▶ **Example 2.1** (*Toy topology*) Consider the set $X = \{a, b, c\}$ and the topology $T = \{\emptyset, X, \{a\}, \{a, b\}\} \subset 2^X$. It is immediate to check that the set T is closed under intersection and union, so that the couple (X, T) forms a topological space. In contrast, neither the set $T_1 = \{\emptyset, X, \{a\}, \{b\}\}$ nor the set $T_2 = \{\emptyset, X, \{a, b\}, \{b, c\}\}$ can be used to define a topology on X.

In some sense, the set T is meant to define the degree of proximity of the different elements of X. Generally speaking, the higher the number of open sets in which two elements appear together, the "nearer" they can be considered. In fact, if two elements a and b appear together in every open, nonempty set, a new set $\tilde{X}$ can be

© The Author(s), under exclusive license to Springer Nature Switzerland AG 2023
G. Bottazzi, *Advanced Calculus for Economics and Finance*, Classroom
Companion: Economics, https://doi.org/10.1007/978-3-031-30316-6_2

considered where the two elements are replaced by their union and an equivalent topology $\tilde{T}$ defined in $\tilde{X}$, which is in one-to-one correspondence with T.

The notion of a topological space comes with a number of other related notions. Many of them are collected in the following definition and are likely to be familiar to the reader.

Definition 2.2 Denote with $A^c = \{x \in X | \notin A\}$ the *complement* of A then

1. A is *closed* if A^c is open;
2. A is a *neighbourhood* of x if $A \in T$ and $x \in A$; a neighbourhood of the element x is generically denoted with $N(x)$;
3. x is a *limit point* of A if $\forall N(x)$, $\exists y \in A$ such that $y \in N(x)$ and $y \neq x$; the set of all limit points of A is the *derivative set* of A and is denoted with DA^1;
4. x is an *interior point* of A if $\exists N(x)$ such that $N(x) \subseteq A$; The set of interior points of A is denoted by int A or A^0;
5. x is an *exterior point* of A if $x \in$ int A^c;
6. x is a *boundary point* of A if any neighbourhood of x contains an element of A and an element of A^c; ∂A denotes the boundary of A, that is the set of boundary points of A; clearly $\partial A = \partial A^c$;
7. the *closure* of A is $\bar{A} = A \cup DA$;
8. the set B is *dense* in A if $B \subseteq A$ and $\bar{B} = A$.

Based on the definitions above, it is immediately apparent that if $A \subseteq B$ then $DA \subseteq DB$, $\bar{A} \subseteq \bar{B}$, and int $A \subseteq$ int B. A set composed of a single element, called *singlet*, cannot have its unique element as a limit point. However, this does not mean that it has no limit points (Fig. 2.1).

▶ **Example 2.2** (*Limit points of a singlet*) Consider a set X with at least two elements and the singlet $A = \{a\} \subset X$. Then $T = \{\emptyset, X, A\}$ is a topology on X. In this topology, the only neighbourhood of any point $x \in X \setminus A$ is X itself. The intersection of X with A is the same set A, which, by assumption, does not contain x. Thus $\partial A = DA = X \setminus \{a\} = A^c$.

▶ **Example 2.3** (*Toy topology*) In the topology of Example 2.1, $\{a\}$ is an open set because it belongs to T, $\{b\}$ is a closed set because its complement, $\{a, b\}$, belongs to T, and $\{b\}$ is neither open nor closed. The limit points of the set $A = \{a, b\}$ are b and c. In fact, the intersection with A of all neighbourhoods of these points (X in the case of c and X and $\{a, b\}$ in the case of b) contains a, which is different from b and c. It follows that $\bar{A} = X$. The points of A are internal points, as A is a neighbourhood of both. That is, int $A = A$. The boundary of A is made by the point c. Its only neighbourhood, X, has nonempty intersections with both A and its complement. Thus $\partial A = \{c\}$. Notice that $\partial A = \bar{A}/$ int A.

---

[1] In some texts the derivative set of A is denoted with A'. I find this notation confusing and will never use it in this book.

Fig. 2.1 The abstract set-wise relation between the set A, its interior A^0, its boundary ∂A, and the part of the derivative set that does not belong to the set, $DA \setminus A$

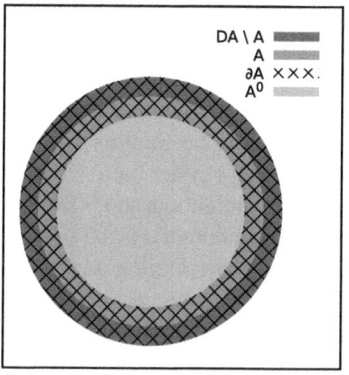

Given a set A and a point $x \in X$, one has $x \in \text{int } A$ or $x \in \text{int } A^c$ or $x \in \partial A = \partial A^c$. If $x \in \partial A$, any $N(x)$ has a nonempty intersection with A and with A^c. So, if $x \in \partial A$, but $x \notin A$, then $x \in DA$. That is $\partial A = \bar{A} \setminus \text{int } A$. The following theorem provides a basic characterisation of open sets.

Theorem 2.1 *A is open if and only if $A = \text{int } A$.*

Proof If A is open, then it is a neighbourhood of each of its points. Then all its points are interior points. Conversely, if $A = \text{int } A$ then A is the union of the interior neighbourhood of all its points and, consequently, it is an open set. □

The interior points of any set A form an open set. In fact, int A is a union of open sets, more precisely, of all open sets interior to A. If A is open, then $\partial A \subseteq A^c$. If A is closed, then $\partial A \subseteq A$, because A^c is open. In general, $A \cup \partial A$ is closed, as its complement is in int A^c, which is open. Also ∂A is closed, since its complement is int $A \cup \text{int } A^c$, which is the union of two open sets. Since $\partial A = \bar{A} \setminus \text{int } A$, if A is open, it is $\partial A = DA \setminus A$. Closed sets can be characterised by the following result.

Theorem 2.2 *A is closed if and only if $DA \subseteq A$.*

Proof Let A be closed and consider an element $x \in DA$. Assume $x \in A^c$. Since A^c is open, x is an internal point of A^c. Thus, there is a neighbourhood of x with no points of A, that is, $x \notin DA$, which is a contradiction. Thus $A^c \cap DA = \emptyset$ and $DA \subseteq A$. Conversely let A be such that $DA \subseteq A$. Consider a point $x \in A^c$. If $x \in \partial A^c$, then any neighbourhood of x has a nonempty intersection with A. This would imply that $x \in DA$, which is not possible because x does not belong to A. Then A^c does not contain any boundary points. This means that all points of A^c are interior, that is, A^c is open and, consequently, A is closed. □

Since the empty set is a subset of any set, it follows that if $DA = \emptyset$, then A is closed.

▶ **Example 2.4** (*Topology of nested sets*) Consider an infinite set X and a strictly nested sequence of subsets $\mathcal{A} = \{A_n\}$ such that $A_n \subset A_{n+1} \subset X$. Assume also that the number of elements of A_1 is greater than one. It is immediate to check that $T = \{\emptyset, X\} \cup \mathcal{A}$ is closed under union and intersection, so that (X, T) is a topological space. First, we want to prove that $\partial A_n = A_n^c$. Since A_n is open, $\partial A_n \subseteq A_n^c$. For any $a \in A_n^c$, either there exists a $k > n$ such that $a \in A_k$ or $a \in \cap_k A_k^c$. In the first case, the neighbourhoods of a are the A_j with $j \geq k$. In the second case, the only neighbourhood of a is the space X itself. In both cases, all neighbourhoods of a have a nonzero intersection with A_n and A_n^c so that $a \in \partial A_n$. Thus, $A_n^c \subseteq \partial A_n$ and the statement follows. Next, we want to prove that $DA_n = X$. For the previous result and because A_n is open, it is $A_n^c = \partial A_n = DA_n \setminus A_n$, which implies $A_n^c \subseteq DA_n$. Take $a \in A_n$. Now take a neighbourhood of a, let it be A_h. Then $A_h \cup A_n$ is equal to A_h or to A_n, depending on which between h or n is the largest, and, in both cases, it contains more than one element of A_n, thus $A_n \subseteq DA_n$.

Definition 2.1 stipulates a specific behaviour of open sets with respect to intersection and union. As the following theorem clarifies, this stipulation also implies a specific behaviour for closed sets. The theorem is based on the set-theoretic property of the complement of the union of sets: $(A \cup B)^c = A^c \cap B^c$.

Theorem 2.3 *The intersection of any number of closed sets is closed. The union of a finite number of closed sets is closed.*

Proof Consider a collection of closed sets $\{A_i\}$. Thus, $\{A_i^c\}$ are open. Now, for the definition of topology, the union of any number of open sets $\cup_i A_i^c$ is open. Thus $(\cup_i A_i^c)^c = \cap_i A_i$ is closed. Analogously, the intersection of a finite number of open sets $\cap_i A_i^c$ is open. Thus, $(\cap_i A_i^c)^c = \cup_i A_i$ is closed. □

The previous theorem suggests that one could define a topology starting from the definition of its closed sets and then introduce open sets as their complements.

In Definition 1.11, we introduced the notion of global maximum and minimum of a function whose image is an ordered set. Using the topological structure of the function domain, we can define the local version of these notions.

Definition 2.3 (*Local maxima and minima*) Consider the topological space (X, T), an ordered set $(Y, \leq)$, and the function $f : A \subseteq X \to Y$. The element $a \in A$ is a *local maximum* of the function f in A if there exists a neighbourhood $N(a)$ of a such that $\forall x \in N(a) \cap A$ it is $f(x) \leq f(a)$. Analogously, the element $a \in A$ is a *local minimum* of the function f in A if there exists a neighbourhood $N(a)$ of a such that $\forall x \in N(a) \cap A$ is $f(a) \geq f(x)$.

A *strict local maximum* or *strict local minimum* is defined using $N(a) \setminus \{a\}$ insetad of $N(a)$ and the strict inequalities $<$ and $>$, respectively. Local maxima and minima are collectively denoted as *extrema*, or *extremal points*, and the values the function takes at these points are called *extremal values*.

▶ **Example 2.5** (*Toy topology*) With reference to Example 2.1, define the function $f : X \rightarrow \mathbb{R}$, with $f(a) = 2$, $f(b) = 3$, and $f(c) = 4$. Then a, b, and c are all local maxima; b and c are strict local maxima; and a is a strict local minimum.

2.2 Base of a Topology

The set T could contain many elements and one can be interested in finding a parsimonious definition of them. The idea is similar to the way things operate in a linear space: all vectors in the space can be generated as linear combinations of a much smaller set of vectors.[2]

Definition 2.4 Consider the topological space (X, T). The set $\mathbb{B} \subseteq T$ is a *base* of the topology if any open set is the union of elements of $\mathbb{B}$.

The set $\mathbb{B}$ generates the topology through the union of its sets. To denote that the base $\mathbb{B}$ generates the topology T, we write $T = \cup \mathbb{B} = \{\cup_\alpha B_\alpha \mid B_\alpha \in \mathbb{B}\}$.

▶ **Example 2.6** (*Base of toy and nested topology*) Sometimes, the notion of base is not particularly useful. In the toy topology of Example 2.1, any base should contain the sets $\{a\}$, $\{a, b\}$, and X. Thus, the base is equivalent to the topology. The same applies for the topology in Example 2.4. If we remove the open set A_h from the topology T, then there is no way of obtaining it back as the union of other open sets. Thus, also in this case, the only base that generates the topology is the topology itself.

How many sets are necessary to build a given topology? The next theorem goes in the direction of answering this question while providing a more useful definition of a base.

Theorem 2.4 *Consider the topological space (X, T). The set $\mathbb{B} \subseteq T$ is a base of the topology T if and only if for any $A \in T$ and any $x \in A$, there exists a $B \in \mathbb{B}$ such that $x \in B \subseteq A$.*

Proof If $\mathbb{B}$ is a base of the topology, then any neighbourhood of x, $N(x) \subseteq A$, which exists because A is open and x is interior, is generated by the union of elements of $\mathbb{B}$, so there should be a $B_x \in \mathbb{B}$ such that $x \in B_x \subseteq A$.

On the contrary, take $A \in T$. By assumption, for any x, there exists a $B_x \in \mathbb{B}$ such that $x \in B_x \subseteq A$. Because $A = \cup_x B_x$, A is the union of elements of $\mathbb{B}$. This is true for any $A \in T$, so $\mathbb{B}$ is a base of the topology. □

---

[2] As already said, the reader should be acquainted with the notion of linear space. If this is not the case, he/she can ignore the sentence before this footnote without any harm.

The previous theorem allows for a redefinition of the notion of neighbourhood based on the elements of the base. For any point $x \in X$ and any neighbourhood $N(x) \in T$, there is another neighbourhood in the base, $B_x \in \mathbb{B}$, contained in $N(x)$. Thus, if we have a base $\mathbb{B}$ and are required to take a neighbourhood of a point, we can safely assume that this neighbourhood belongs to $\mathbb{B}$.

Often the problem is not whether a set is a base for a given topology, that is, if a given T can be obtained via unions of a smaller number of sets. Rather, one is interested in knowing if the elements of a set $\mathbb{B} \subseteq 2^X$ are enough to generate a topology on X. We can derive a theorem that provides sufficient and necessary conditions for this to be the case. It is based on the following basic result in set theory.

Lemma 2.1 *If $A = \cup_j C_j$ and $B = \cup_k C'_k$, then $A \cap B = \cup_{j,k} C_j \cap C'_k$.*

Proof For any $x \in A \cap B$, there exist a C_j and a C'_k such that $x \in C_j$ and $x \in C'_k$, that is $x \in C_j \cap C'_k$, then $A \cap B \subseteq \cup_{j,k} C_j \cap C'_k$.

On the contrary, if there are j and k such that $x \in C_j \cap C'_k$, then $x \in A \cap B$. Thus $\cup_{j,k} C_j \cap C'_k \subseteq A \cap B$.

The statement follows from the previous two observations.   □

Now for the result about the base of a topology.

Theorem 2.5 *Consider a set $X \neq \emptyset$ and $\mathbb{B} \subseteq 2^X$. Then $\mathbb{B}$ is the base of a topology on X if and only if*

1. *the union of elements of $\mathbb{B}$ covers X, that is $X = \cup_\alpha B_\alpha$, $B_\alpha \in \mathbb{B}$, and*
2. *for any couple of sets $B_1, B_2 \in \mathbb{B}$, if $x \in B_1 \cap B_2$ there exists $B_3 \in \mathbb{B}$ such that $x \in B_3 \subseteq B_1 \cap B_2$.*

Proof First, assume that $\mathbb{B}$ generates a topology T on X. Because $X \in T$, X is the union of elements in $\mathbb{B}$ and point 1 is satisfied. Moreover, for the properties of the topology, because B_1 and B_2 in the statement are open, $B_1 \cap B_2$ is open as well. Thus, any $x \in B_1 \cap B_2$ has a neighbourhood $N(x)$ contained in the intersection and, for Theorem 2.4, there exists an element of the base B_x that contains x and is contained in the neighbourhood, $x \in B_x \subseteq N(x) \subseteq B_1 \cap B_2$. This proves point 2.

On the contrary, assume that 1 and 2 are valid and consider T as the set generated by all the unions of sets in $\mathbb{B}$. Now $X \in T$ according to hypothesis 1 and $\emptyset \in T$ because it is the union of zero elements of $\mathbb{B}$. We have to prove that the set T generated by the unions of the elements of $\mathbb{B}$ is closed with respect to the finite or infinite union and with respect to the finite intersection. Consider a set (finite or infinite) $\{A_\alpha\}$ of elements in T. It is clear that since any A_α is the union of sets of $\mathbb{B}$, also $\cup_\alpha A_\alpha$ is the union of sets of $\mathbb{B}$ and $\cup_\alpha A_\alpha \in T$. So T is closed under the (finite or infinite) union of sets. Now take a finite sequence $\{A_j\}$ of sets in T and consider $x \in \cap_{j=1}^J A_j$.

Since for each j it is $A_j = \cup_{\alpha_j} B_{\alpha_j}$,

$$\cap_{j=1}^{J} A_j = \cap_{j=1}^{J} \cup_{\alpha_j} B_{\alpha_j} = \cup_{\alpha_1,\dots,\alpha_J} B_{\alpha_1} \cap B_{\alpha_2}\dots \cap B_{\alpha_J}.$$

Then there exists a set of indexes $(\alpha_1, \alpha_2, \dots, \alpha_J)$ such that $x \in B_{\alpha_1} \cap B_{\alpha_2}\dots \cap B_{\alpha_J}$. This implies that there is a $B_x \in \mathbb{B}$ such that $x \in B_x \subseteq \cap_{j=1}^{J} B_{\alpha_j}$. We can thus write $\cap_{j=1}^{J} A_j = \cup_x B_x$ so that $\cap_{j=1}^{J} A_j \in T$ and T is closed under intersection. Therefore, T is a topology on X. $\qquad\square$

▶ **Example 2.7** (*A topology on* $\mathbb{N}$) Let $\mathbb{N} = \{1, 2, 3, \dots\}$ be the set of natural numbers and $\mathbb{O} \subset \mathbb{N}$ the set of odd numbers. Consider the collection of sets $\mathbb{B} = \{\{n, n+1\} \mid n \in \mathbb{O}\} \cup \{\{n\} \mid n \in \mathbb{O}\} \subseteq 2^{\mathbb{N}}$. First, the whole space $\mathbb{N}$ can be written as union of elements of $\mathbb{B}$. Second, the intersection of two elements of $\mathbb{B}$ is either the empty set or an element of $\mathbb{B}$. Thus, the requirements of Theorem 2.5 are satisfied and we can conclude that $\mathbb{B}$ is the base of a topology T on $\mathbb{N}$. Exercise 2.7 uses this topology.

In general, the base of a topology is not unique (see Exercise 2.6). Since multiple bases are possible, it is useful to have a criterion to know if two bases $\mathbb{B}_1$ and $\mathbb{B}_2$ are equivalent, that is if they generate the same topology.

Theorem 2.6 *Consider a set $X \neq \emptyset$, and let $\mathbb{B}_1$ and $\mathbb{B}_2$ be the bases of two topologies T_1 and T_2. Then the two topologies are the same if and only if*

- $\forall B \in \mathbb{B}_1$ *and* $\forall x \in B$, $\exists B' \in \mathbb{B}_2$ *such that* $x \in B' \subseteq B$
- $\forall B' \in \mathbb{B}_2$ *and* $\forall x \in B'$, $\exists B \in \mathbb{B}_1$ *such that* $x \in B \subseteq B'$

Proof Assume that the two topologies are the same. Then $\forall B \in \mathbb{B}_1$ is an union of elements of $\mathbb{B}_2$, because $B \in T_1 = T_2$. At the same time, $\forall B' \in \mathbb{B}_2$ is an union of elements of $\mathbb{B}_1$, so that the two conditions immediately follow.
Now assume that the two conditions are valid and take $A \in T_1$. Then $A = \cup_\alpha B_\alpha$ with $B_\alpha \in \mathbb{B}_1$ and $\forall x \in A$, $\exists B_{\alpha(x)}$ such that $x \in B_{\alpha(x)}$. By hypothesis, $\exists B'_x \in \mathbb{B}_2$ such that $x \in B'_x \subseteq B_{\alpha(x)}$. As a consequence, it is $A = \cup_{x \in A} B'_x$ and $A \in T_2$. It follows that $T_1 \subseteq T_2$. Analogously, it can be shown that $T_2 \subseteq T_1$ and the statement is proved. $\qquad\square$

2.2.1 Countability

Until now, no assumptions have been made about the number of elements in T and $\mathbb{B}$. If the elements of T are finite, little is gained by introducing a base $\mathbb{B}$, which will, in turn, have a finite number of elements (see Example 2.6). The possibility of using a base becomes interesting when the number of elements of T is infinite, in particular uncountably infinite. In this case, we can be interested in describing the topology using a countable subset of open sets.

Definition 2.5 (*First-countable topological space*) The topological space (X, T) is *first-countable* if $\forall x \in X$ there exists a countable set of open sets $\mathcal{U}(x) = \{U_n(x)\}$ such that for any neighbourhood $N(x)$ of x, $x \in U_j(x) \subseteq N(x)$ for some $U_j(x) \in \mathcal{U}(x)$.

This means that the behaviour of the topology near any point x can be described using only a countable set of neighbourhoods. A stronger version of the same idea is introduced by the following definition.

Definition 2.6 (*Second-countable topological space*) The topological space (X, T) is *second-countable* if it admits a countable base.

Obviously, a second-countable space is also first-countable. The opposite is not true. For example, the discrete topology defined over a set X is always first-countable, as for any element $x \in X$, the singlet $\{x\}$ constitutes a countable (finite) local base. However, if the number of elements of X is uncountable, the discrete topology is not second-countable.

2.2.2 Euclidean Topology

The next definition introduces an important topology on the set of real numbers naturally induced by its order relation.

Definition 2.7 (*Euclidean topology on* $\mathbb{R}$) Let $\mathbb{B} = \{I_{(a,b)} \mid a, b \in \mathbb{R}, a < b\}$ be the collection of all open intervals $I_{(a,b)} = \{x \in \mathbb{R} \mid a < x < b\}$. The set $\mathbb{B}$ is the base of the *Euclidean topology* on $\mathbb{R}$.

The fact that $\mathbb{B}$ is the base of a topology follows directly from Theorem 2.5. In fact, any real number belongs to at least one element of $\mathbb{B}$ and if $x \in I_{(a,b)} \cap I_{(a',b')}$ then $x \in I_{(a'',b'')}$ where $a'' = \max\{a, a'\}$ and $b'' = \min\{b, b'\}$. This is the topology to which every reader should be used. The open sets of this topology are the unions of, possibly infinitely many, open intervals $I_{(a,b)}$. The reason for the name "Euclidean" will be clarified in Example 3.4. The definition of a topology in the ordered set $\mathbb{R}$ provides a topological characterisation of the supremum and infimum.

Theorem 2.7 (Supremum and infimum are in the boundary) *Consider a subset of real numbers $A \subseteq \mathbb{R}$. If $\sup A$ or $\inf A$ exist, they belong to ∂A.*

Proof Let A be bounded above. Then, for the property of real numbers, there exists $x = \sup A$. Any neighbourhood (in the base) of x, that is, any open interval (a, b) such that $a < x < b$, contains points of A, because otherwise there would be an upper bound lower than x, and points of A^c, because otherwise x would not be an upper bound. This means that $x \in \partial A$. In the same way, if A is bounded below, then it is inf $A \in \partial A$. □

Alternatively, the Euclidean topology can be defined using the absolute value.

▶ **Example 2.8** (*Euclidean topology induced by the absolute value*) Consider the set $\mathbb{R}$ and the set of all symmetric intervals $B(x, r) = (x - r, x + r)$ with $x, r \in \mathbb{R}$ and $r > 0$. First of all, notice that all $B(x, r)$ are open sets according to Definition 2.7. Also, for any open interval $I_{(a,b)}$ and any $x \in I_{(a,b)}$, there exists a r such that $(x - r, x + r) \subseteq I_{(a,b)}$, that is $B(x, r) \subseteq I_{(a,b)}$. Thus, according to Theorem 2.6, the Euclidean topology can be generated by the set of all symmetric intervals.

Symmetric intervals are directly related to the absolute value $|.|$ on $\mathbb{R}$ by the simple observation that $B(x, r) = \{y \in \mathbb{R} \mid |y - x| < r\}$. In this way, the absolute value induces a topology on the set of real numbers and this topology is precisely the Euclidean topology of Definition 2.7. These symmetric intervals are a special case of "open balls" of radius r and centre x introduced in Theorem 3.5. They hint to a more general relation between norms (such as the absolute value) and topology that will be studied in Chap. 4.

Two important properties of the Euclidean topology are established in the following. Both follow from the fact that the set of rational numbers $\mathbb{Q}$ is countable (see Theorem 1.7) and dense in $\mathbb{R}$ (see Theorem 1.11).

Theorem 2.8 *The Euclidean topology is second-countable.*

Proof Consider the set of open intervals with rational boundaries, $\mathbb{B}' = \{I_{(q,r)} \mid q, r \in \mathbb{Q}, q < r\}$. Since $\mathbb{Q}$ is dense in $\mathbb{R}$, if two intervals with real boundaries $I_{(a,b)}$ and $I_{(c,d)}$ overlap, then any point in their intersection $x \in I_{(a,b)} \cap I_{(c,d)}$ is contained in an open interval with rational boundaries $x \in I_{(q,r)} \subseteq I_{(a,b)} \cap I_{(c,d)}$. The opposite is also easy to see. Thus, according to Theorem 2.6, $\mathbb{B}'$ and $\mathbb{B}$ in Definition 2.7 define the same topology. However, the elements in $\mathbb{B}'$ are an infinite subset of $\mathbb{Q} \times \mathbb{Q}$; thus, according to Theorem 1.5, they are countable. □

Theorem 2.9 *Any open set in the Euclidean topology can be written as the countable union of disjoint intervals.*

Proof Let $A \subseteq \mathbb{R}$ be open. Take $x \in A$ and let I_x be the largest interval such that $x \in I_x \subseteq A$. Next take $y \in A \setminus I_x$ and let I_y be the largest interval such that $y \in I_y \subseteq A \setminus I_x$. Clearly $I_x \cap I_y = \emptyset$. Repeat this procedure finding a covering $A = \cup_\alpha I_\alpha$. These are all disjoint intervals. In each interval, there is at least one different rational number; thus, they cannot be more than countable. □

2.3 Cover and Compactness

Compact sets play an important role in calculus and optimisation theory. This section discusses their most general definition in terms of a generic topology. The starting point is the notion of cover.

Definition 2.8 (*Cover*) A *cover* of a set A is a collection of open sets $\{C_\alpha\} \subseteq T$ such that $A \subseteq \cup_\alpha C_\alpha$. Given a cover $\{C_\alpha\}$, a subset $\{C'_\alpha\} \subseteq \{C_\alpha\}$ which is still a cover of A is a *subcover* of $\{C_\alpha\}$.

Then we have the following definition.

Definition 2.9 $K \subseteq X$ is *compact* if it is possible to extract a *finite* subcover from *any* cover of K.

Obviously, any set admits a finite cover, that is, a cover made of a finite number of sets. Consider the whole space X: it covers any set. However, a set that is not compact will have at least one infinite cover that does not admit a finite subcover. The empty set is compact by definition.

▶ **Example 2.9** (*Compact sets in trivial and discrete topologies*) In the trivial topology (X, T) with $T = \{\emptyset, X\}$, any set is compact. The only cover of any set that is nonempty is X, and it is finite. In general, in any topology with a finite number of open sets, any set is compact. In the discrete topology $(X, 2^X)$, any set made of a finite number of elements is compact, but any set made of an infinite number of elements is not compact. Indeed the cover made by the union of the singlets of all the elements of the set does not admit a finite subcover.

The collection of compact sets in a topology is closed with respect to union, as the following theorem clarifies.

Theorem 2.10 *The union of two compact sets is compact.*

Proof Let K_1, K_2 be compact. Consider a cover $\{C_\alpha\}$ of $K_1 \cup K_2$. Then $\{C_\alpha\}$ is a cover of both K_1 and K_2. Now, let $\{C_{\alpha_1}\}$ be a finite subcover of K_1 and $\{C_{\alpha_2}\}$ a finite subcover of K_2. Thus $\{C_{\alpha_1}\} \cup \{C_{\alpha_2}\}$ is a finite subcover of $K_1 \cup K_2$. □

In any topological space, the properties of compactness and closure are related to one another.

Theorem 2.11 *A closed subset of a compact set is compact.*

Proof Let K be compact and $C \subseteq K$ be closed. Let $\{V_\alpha\}$ be a cover of C, since C is closed, $\{V_\alpha, C^c\}$ is a cover of K. Since K is compact, there exists a finite subcover, name it $\{V'_j, C^c\}$. Then $\{V'_j\}$ is a finite subcover of C. □

The previous theorem does not mean that, in general, compact sets are closed. Counterexamples are provided in Example 2.9 and Exercise 2.13. This is, however, true in a special class of topologies that we will analyse in the next section.

▶ **Example 2.10** (*Compact sets in left- and right-order topologies*) Two special topologies are available on the set of real numbers (or any totally ordered set). For any $a \in \mathbb{R}$, define $I_a^- = \{x \in \mathbb{R} \mid x < a\}$ and $I_a^+ = \{x \in \mathbb{R} \mid x > a\}$. The *left-order topology* $(\mathbb{R}, <)$ is the topology generated by the base $\{I_a^- \mid a \in \mathbb{R}\}$, while the *right-order* topology $(\mathbb{R}, >)$ is generated by the base $\{I_a^+ \mid a \in \mathbb{R}\}$. For any $a > b$, $I_a^- \cup I_b^- = I_b^-$ and $I_a^- \cap I_b^- = I_a^-$. Analogously, $I_a^+ \cup I_b^+ = I_a^+$ and $I_a^+ \cap I_b^+ = I_b^+$. Thus, the left- and right-order topologies are nested topologies; see Example 2.4. The elements of their bases and the whole space $\mathbb{R}$ are the only open sets of these topologies.

A set $A \subset \mathbb{R}$ is compact in the left-order topology if and only if it has a maximum. In fact, if $\bar{a} = \sup A$ and $\bar{a} \in A$, then any cover should contain a set I_a^- such that $a > \bar{a}$ and, consequently, I_a^- is a finite subcover. Conversely, if $\bar{a} \notin A$, the cover $\cup_{n=1}^\infty I_{\bar{a}-1/n}^-$ does not admit any finite subcover. Along the same lines, it is easy to show that a set is compact in the right-order topology if and only if it has a minimum.

2.3.1 Hausdorff Spaces

In this section, we focus on special topological spaces that take their name from Felix Hausdorff, a German mathematician who was active in the late nineteenth and early twentieth centuries.

Definition 2.10 (*Hausdorff topology*) The topological space (X, T) is of *Hausdorff* if $\forall x, y \in X, x \neq y$ there exist $N(x)$ and $N(y)$ such that $N(x) \cap N(y) = \emptyset$.

The discrete topology is of Hausdorff, whereas the trivial topology is not if there are at least two elements in the space.

▶ **Example 2.11** (*The toy topology is not Hausdorff*) Consider the toy topology in Example 2.1. There are no neighbourhoods of b, or of c, which do not contain a. Thus, we can conclude that this topology is not of Hausdorff type.

The basic idea of Hausdorff spaces is that you can separate, topologically speaking, any pair of elements. This property is very natural and is connected with the idea of proximity that we derive from the physical world: irrespective of how close two distinct points are in space, there is always some space between them. For our purposes, the practical effect of the Hausdorff property is twofold: on the one hand, it strengthens the relationship between compactness and closure, and on the other hand, it enriches the definition of limit point.

Theorem 2.12 *In a Hausdorff space, if K is compact, it is closed.*

Proof Consider a point $y \in K^c$. Due to Hausdorff, $\forall x \in K$ there exists a neighbourhood $N(x)$ of x and a neighbourhood $N_x(y)$ of y such that $N(x) \cap N_x(y) = \emptyset$. Now $\cup_x N(x)$ is a cover of K. Let $\{N(x_i)\}, i = 1, \ldots, I$ be a finite subcover and consider

$$A = \cup_{i=1}^I N(x_i) \quad \text{and} \quad B = \cap_{i=1}^I N_{x_i}(y).$$

Note that B is a neighbourhood of y and $B \cap A = \emptyset$ by construction. Since $K \subseteq A$, $B \cap K = \emptyset$, and hence $y \in \text{int } K^c$. As this is true for any $y \in K^c$, it follows that K^c is an open set, so that K is closed. □

A direct consequence of the previous theorem is that, in Hausdorff's space, the compactness works much more smoothly with the set-wise intersection and the collection of compact sets is also closed under intersection.

Corollary 2.1 *In a Hausdorff space, the intersection of a closed set with a compact set is compact.*

Proof If C is closed and K is compact, $C \cap K$ is closed because it is the intersection of two closed sets. Thus, according to Theorem 2.11, it is a closed subset of a compact set. Hence, it is compact. □

Corollary 2.2 *In a Hausdorff space, the intersection of compact sets is compact.*

Proof Let K_1, K_2 be compact, then they are closed and thus $K_1 \cap K_2$ is closed. Since this is a closed subset of a compact set it is compact. □

Another important property derived from Theorem 2.12 concerns the intersection of non-disjoint compact sets.

Theorem 2.13 *Let (K_n) be a sequence of compact sets in a Hausdorff space. If for any finite subsequence $(K_{n_1}, \ldots, K_{n_L})$, $\cap_{j=1}^L K_{n_j} \neq \emptyset$, then $\cap_n K_n \neq \emptyset$.*

Proof We prove this by contradiction. Consider K_1. Assume that no point of K_1 belongs to all $\{K_n\}$, that is $K_1 \cap (\cap_2^n K_n) = \emptyset$. Because of Hausdorff, the sets K_n^c are open. Consequently, the sequence $\{K_{n \geq 2}^c\}$ is a cover of K_1, i.e. $K_1 \subseteq \cup_{n \geq 2} K_n^c$. Now let $\{K_{n_1}^c, \ldots, K_{n_L}^c\}$ be the finite subcover of K_1. Since $K_1 \subseteq K_{n_1}^c \cup \ldots K_{n_L}^c$, $K_1 \cap (K_{n_1} \cap \ldots K_{n_L}) = \emptyset$, which contradicts the hypothesis. Thus, any compact set has a nonempty intersection with the intersection of all other sets and the statement follows. □

In other terms, a countable sequence of non-disjoint and nonempty compact sets identifies a nonempty set of points. This result will be exploited in Sect. 5.2 to prove the convergence of Cauchy sequences in metric spaces through the following simple corollary.[3]

Corollary 2.3 *Let (K_n) be a sequence of compact sets in a Hausdorff space, such that $K_n \neq \emptyset$ for any n and $K_{n+1} \subseteq K_n$. Then $\cap_n K_n \neq \emptyset$.*

As mentioned before, the second advantage of working in a Hausdorff space is to obtain a richer definition of limit points.

Theorem 2.14 *In a Hausdorff space, if $x \in DA$, then any neighbourhood of x contains an infinite number of elements of A.*

Proof Assume $\exists N(x)$ such that only a finite number $\{y_1, ..., y_L\}$ of elements of A different from x belongs to it. Let $N_j(x)$, with $j = 1, ..., L$, be a neighbourhood of x that does not contain y_j. Consider $\cap_{j=1}^{L} N_j(x) \cap N(x) = N^*(x)$. Now $N^*(x)$ is a neighbourhood of x that does not contain any element of A apart from x, so x cannot be a limit point of A, which contradicts the hypothesis. $\qquad\qquad\square$

The previous theorem helps in clarifying the relationship between limit points and compact sets in any Hausdorff space.

Theorem 2.15 *In an Hausdorff space consider a compact set K. Any infinite subset $A \subseteq K$ has at least one limit point ($DA \neq \emptyset$) and all the limit points are inside K ($DA \subseteq K$).*

Proof Assume that $DA \cap K = \emptyset$, so that A has no limit points in K. Then $\forall x \in K$, $\exists N(x)$ neighbourhood of x, which contains a finite number of elements of A. The union of these neighbourhoods covers the set K, $K \subseteq \cup_x N(x)$. Because K is compact, there exists a finite subcover $\{N(x_1), ..., N(x_L)\}$ such that $K \subseteq \cup_{j=1}^{L} N(x_j)$. But since each $N(x_j)$ contains only a finite number of elements of A, also $\cup_{j=1}^{L} N(x_j)$ contains a finite number of elements of A, which is absurd because A is infinite and $A \subseteq \cup_{j=1}^{L} N(x_j)$.

The second part of the theorem is obvious because, in a Hausdorff space, if K is compact, it is closed and thus contains all its limit points. $\qquad\qquad\square$

In a generic Hausdorff space, the inverse of Theorem 2.15 is not true. That is, even if any infinite subset of a set A has at least one limit point, and all limit points are in A, we cannot conclude that A is compact. To do so, we need the further assumption of second-countability.

---

[3] This corollary could be seen as a generalisation of Theorem 1.8, if we only knew that the closed intervals of real numbers were compact. This is actually the case in Euclidean topology; see Theorem 2.18.

Theorem 2.16 (Bolzano–Weierstrass) *Let (X, T) be a Hausdorff, second-countable topological space. If $K \subseteq X$ is such that, for any infinite subset $E \subseteq K$, it is $DE \neq \emptyset$ and $DE \subseteq K$, then K is compact.*

Proof Proving the theorem is equivalent to prove that if K is not compact, then there exists a set $A \subseteq K$, with an infinite number of points, such that $DA = \emptyset$ or $DA \not\subseteq K$.

If K is not compact there exists a cover $\{V_\alpha\}$, of which no finite subcover can be found. However, because the space is second-countable, it is always possible to find a countable subcover. To see it, let $\{B_n\}$ be the elements of the countable base of the space. For each B_n, if there exists a V_{α_n} such that $B_n \subseteq V_{\alpha_n}$, set $W_{\alpha_n} = V_{\alpha_n}$, otherwise set $W_{\alpha_n} = \emptyset$. The set $\{W_{\alpha_n}\}$ is at most countable and it is a cover of K. Indeed, take any $x' \in K$. Then x' belongs to some V'_α and, for the definition of base, there is some $B_{n'}$ such that $x' \in B_{n'} \subseteq V'_\alpha$.

From the hypothesis, it is not possible to extract a finite subcover from $\cup_n W_{\alpha_n}$. Consider $x_n \in K$ and $x_n \notin \cup_{j=1}^n W_n$ for each value of n. Consider the set $E = \{x_1, ..., x_j, ...\}$. This set is infinite, because for any finite set of points $A \subseteq K$, there is an m such that $A \subseteq \cup_{j=1}^m W_j$, but this is absurd for E, because $x_{m'} \notin \cup_{j=1}^m W_j$ if $m' \geq m$. Now, if $DE = \emptyset$, the theorem is proved. Otherwise, let $x^* \in DE$ and assume $x^* \in K$. Then $\exists W_{n^*}$ such that $x^* \in W_{n^*}$. Since the space is Hausdorff, this implies that there exists an infinite number of points of E in W_{n^*}. But this is absurd, because for $\forall k > n^*$, it is $x_k \notin W_{n^*}$.   □

2.3.2 Compactness in Euclidean Topology

Theorem 2.8 already established that the Euclidean topology is second-countable. The following result is also easy to prove.

Theorem 2.17 *The Euclidean topology on $\mathbb{R}$ is of Hausdorff.*

Proof Take two real numbers $x, x' \in \mathbb{R}$, $x < x'$. Then there is always a third real number x'' such that $x < x'' < x'$. Consider the open intervals $(x - 1, 0.75\, x + 0.25\, x'')$ and $(0.75\, x' + 0.25\, x'', x' + 1)$ which are, respectively, a neighbourhood of the smallest and largest point. The intersection of the two intervals is empty.   □

Therefore, Theorem 2.16 applies to the Euclidean topology.

Corollary 2.4 (Bolzano–Weierstrass in $\mathbb{R}$) *In the Euclidean topology on $\mathbb{R}$ a set K is compact if and only if any infinite subset of K has at least one limit point and all its limit points belong to K.*

A further straightforward consideration is that, in the Euclidean topology, any compact set is bounded. To see it, note that any set can be covered by elements of the base.

This cover admits a finite subcover, that is a cover by a finite number of intervals. Because any interval is bounded, so is the set. In fact, using Theorem 2.16, we can prove more.

Theorem 2.18 (Heine–Borel theorem on real numbers) *In the Euclidean topology on R, any bounded closed set is compact.*

Proof First, notice that any closed and bounded set is contained in a closed interval $[a, b] = \{a \leq x \leq b | x \in R\}$. If we can show that $[a, b]$ is compact, then we prove the theorem because the original set is just a closed subset of a compact set.

To prove that the closed interval $[a, b]$ is compact, we will prove that any infinite set of points in the interval has at least one limit point (see Theorem 2.16). Since the interval is closed, we already know that all its limit points belong to it. Take any infinite subset $A = \{x_n\} \subset [a, b]$. Split the closed interval into two by considering the mid point $c = (b - a)/2$. Then at least one of the two intervals $[a, c]$ and $[c, b]$ contains an infinite number of elements of A, otherwise A would be finite, contradicting the hypothesis. Let I_1 be this interval. Repeat the splitting procedure on I_1, and call I_2 one of its "halves" that contains an infinite number of points of A. In this way, a nested sequence of closed intervals $\{I_k\}$ is built. The length of the interval I_k is $(b-a)/2^k$. According to Theorem 1.8, there exists a $x^* \in \cap_k I_k$. We can prove that $x^* \in DA$. Indeed consider the neighbourhood $N(x^*) = (x^* - \delta, x^* + \delta)$ for some $\delta > 0$. For sufficiently small δ, it is $N(x^*) \subseteq [a, b]$. Notice that if k is sufficiently large, then $(b - a)/2^k < \delta$, which implies that $I_k \subseteq N(x^*)$. Thus, an infinite number of elements of A is in $N(x)$, proving the statement. □

It is important to stress that the previous theorem has been specifically derived for the Euclidean topology.[4] As the next example shows, the fact that a bounded closed set is a compact set is not generally true for any Hausdorff second-countable space.

▶ **Example 2.12** (*Bounded closed sets are not compact in* $\mathbb{Q}$) Let $\mathbb{B}'$ be the set of open intervals of rational numbers $I_{(a,b)} = \{x \in \mathbb{Q} | a < x < b\}$ with $a, b \in \mathbb{Q}$. It is immediate to see that $\mathbb{B}'$ is a base of a topology defined on $\mathbb{Q}$. This topology is second-countable (see the proof of Theorem 2.8). Given two distinct rational numbers $x, x' \in \mathbb{Q}$, $x \neq x'$, there is always a third rational number x'' such that $x < x'' < x'$. Consequently, it is immediate to show that the topology $(\mathbb{Q}, \cup \mathbb{B}')$ is of Hausdorff type.

Now in $(\mathbb{Q}, \cup \mathbb{B}')$ consider the set of rational numbers $E = \{x \in Q, 2 < x^2 < 3\}$. Define $\sqrt{m}|_n$ the n-digit truncation of the square root of m and $\sqrt{m}|^n = \sqrt{m}|_n + 10^{-n}$ the n-digit rational number obtained rounding up the n-digit of $\sqrt{m}$. It is $\sqrt{3}|_n < \sqrt{3}$ and $\sqrt{2}|^n \geq \sqrt{2}$. Notice that $\{x \in Q | x^2 > 3\} = \cup_{n=1}^{\infty} \{x \in Q | x > \sqrt{3}|_n\}$ and $\{x \in Q | x^2 < 2\} = \cup_{n=1}^{\infty} \{x \in Q | x < \sqrt{2}|_n\}$ are open. Thus $E^c = \{x \in Q | x^2 > 3\} \cup \{x \in Q | x^2 < 2\}$ is open too and then E is closed. To determine whether the

---

[4] In Chap. 4 we will see that it is also valid in $\mathbb{R}^n$ when a norm topology is adopted.

set E is compact or not, consider the sequence $S = \{\sqrt{3}|_n\}$. S is countable and it is $S \subset E$. However, S does not have any limit point in E (or in $\mathbb{Q}$). We can conclude that the set E is not compact. Indeed $\cup_{n=1}^{\infty}(\sqrt{2}|^n, \sqrt{3}|_n)$ is an infinite cover of E, but any finite subcover does not cover it.

Since E is closed but not compact, any set that contains E, for instance the interval $[0, 4]$, cannot be compact (see Theorem 2.11).

2.3.3 The Extended Real Number System

Although the real line, that is, the entire space $\mathbb{R}$, is not a compact set in the Euclidean topology, it is sometimes useful to make it compact. There are several possible procedures that can be adopted to do it, depending on what one is looking for.

Definition 2.11 (*Extended real numbers*) Consider the extended set $\bar{\mathbb{R}} = \mathbb{R} \cup \{+\infty, -\infty\}$. The point $+\infty$ is greater than any other real number and belongs to any set that is unbounded above. The point $-\infty$ is lower than any other real number and belongs to any set that is unbounded below. We assume that $\forall x \in R, x + \infty = +\infty$, $x - \infty = -\infty, x/\pm\infty = 0$, and $x \cdot (\pm\infty) = \pm\infty$ if $x > 0$, while $x \cdot (\pm\infty) = \mp\infty$ if $x < 0$.

The expressions $0 \cdot (\pm\infty)$ and $\infty - \infty$ remain undetermined. The intervals of type $(a, +\infty) = \{x \in \mathbb{R}|x > a\}$ are a basis for the neighbourhoods of $+\infty$ and those of type $(-\infty, a) = \{x \in \mathbb{R}|x > a\}$ are a basis for the neighbourhoods of $-\infty$. With this extension, the set $\bar{\mathbb{R}}$ becomes compact: any infinite set which is bounded has a limit point by Theorem 2.18. If the infinite set is not bounded, then $+\infty$, $-\infty$, or both are limit points.

The procedure above is often called *two-points compactification* of the real line. There is also a *one-point compactification* of $\mathbb{R}$ in which the two points $+\infty$ and $-\infty$ are identified with the same point. In this way, a one-to-one relationship is built between the element of $\mathbb{R} \cup \{\infty\}$ and the points of the circle, as shown in Fig. 2.2.

2.4 Connectedness

This section introduces the topological notion of connectedness. The idea is to distinguish topological spaces that can be considered as made up of a single "piece", from spaces that are, conversely, the union of separate "pieces". We are naturally acquainted with a notion of separate sets based on the geometric intuition of the physical world, from which the metric properties of points in space are derived. As we will see, a purely topological treatment is possible.

Definition 2.12 A topological space (X, T) is *connected* if the only open and closed sets are $\emptyset$ and X.

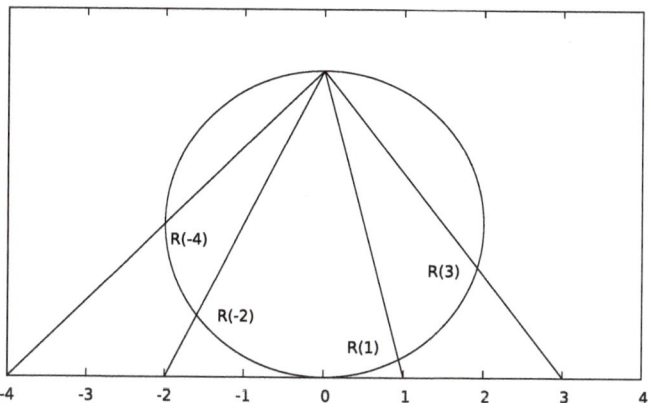

Fig. 2.2 The one-point compactification of the real line is a one-to-one relation between the real numbers and the points of a circle. The relation is defined by drawing a straight line from the uppermost part of the circle to the real line. The unique intersection $R(x)$ of the line with the circle is the point in relation with $x \in \mathbb{R}$

The discrete topology $T = 2^X$ is not connected, while the trivial topology $T = \{X, \emptyset\}$ is connected.

Theorem 2.19 *The Euclidean space* $(\mathbb{R}, |\cdot|)$ *is connected.*

Proof We will prove that no sets apart $\mathbb{R}$ (and $\emptyset$) are closed and open at the same time. Consider an $A \subset \mathbb{R}$ and, without loss of generality, assume that there exists a $x \in A^c$ such that some element of A is below x. Define $B = \{z \in A | z < x\} \subseteq A$. The set B is nonempty and bounded above so we can define $z^* = \sup B$. Note that $z^* \in \partial A$. In fact, $\forall \epsilon > 0$, there must be some element of A in $(z - \epsilon, z)$, or $z - \epsilon$ would be an upper bound, and no elements of A in $(z, z+\epsilon)$, or the supremum would be higher than z. Now note that if A is open, then $z^* \in A^c$, because open sets do not contain their boundary, while if A is closed, then $z^* \in A$, because its complement would be open. These two things are mutually exclusive, and we conclude that A cannot be closed and open at the same time. □

The notion of connectedness can be extended from the entire space to specific sets using the following definition.

Definition 2.13 Let (X, T) be a topological space and $Y \subseteq X$. Define $T_Y = \{Y \cap A \mid A \in T\}$. Then (Y, T_Y) is a *topological subspace* of (X, T).

The set T_Y is closed under countable union and finite intersection, thus the topological subspace is a topological space. Using this notion, we can say that a subset $Y \subseteq X$ of a topological space (X, T) is connected if the topological subspace (Y, T_Y) is a connected space. The property of being connected depends on both the set Y and the topology T.

▶ **Example 2.13** (*Nested topology*) The topology in Example 2.4 is connected. The open sets of the topology are nested: $A_n \subset A_{n+1} \subset X$. Thus all nonempty open sets must contain A_1, and any A_n^c does not. Thus, there are no nonempty, open sets, apart X, whose complement is open.

▶ **Example 2.14** (*Rational numbers*) The space $(\mathbb{Q}, \cup \mathbb{B}')$ in Example 2.12 is a subspace of $(\mathbb{R}, \cup \mathbb{B})$ in Definition (2.7). In $(\mathbb{Q}, \cup \mathbb{B}')$ the set $E = \{x \in Q, 2 < x^2 < 3\}$ is both open and closed. Thus, $(\mathbb{Q}, \cup \mathbb{B}')$ is not connected. The difference from Theorem 2.19 is that the set of rational numbers does not have the least upper bound property of Definition 1.10.

The property of topological connectedness is further illustrated by the following useful result.

Theorem 2.20 *The set Y is not connected if and only if $\exists A, A' \in T$ such that*

1. $Y = (Y \cap A) \cup (Y \cap A')$;
2. $Y \cap A' \cap A = \emptyset$ *(that is, $Y \cap A$ and $Y \cap A'$ are disjoint).*

Proof Assume that Y is not connected. Then there is a $B \in T_Y$ such that $Y \setminus B \in T_Y$. Let $A, A' \in T$ be such that $B = Y \cap A$ and $B' = Y \cap A'$. Then $(Y \cap A) \cup (Y \cap A') = B \cup (Y \setminus B) = Y$ and $Y \cap A' \cap A = (Y \cap A) \cap (Y \cap A') = B \cap (Y \setminus B) = \emptyset$.

On the contrary, assume that 1 and 2 are true and set $B = Y \cap A$ and $B' = Y \cap A'$. By definition, it is $B, B' \in T_Y$. Since $B \cup B' = Y$ and $B \cap B' = \emptyset$, it is $B' = Y \setminus B$ so B is open and closed in T_Y.   □

The previous theorem expresses the idea that the set Y is made up of two disconnected pieces, $Y \cap A$ and $Y \cap A'$. In the Euclidean topology on $\mathbb{R}$, bounded and unbounded intervals are the only connected sets.

2.5   Limit of Functions and Continuity

This section will use the notation on images and preimages of sets introduced at the end of Sect. 1.1. If both the domain and the image of a function are subsets of topological spaces, then we can describe the local behaviour of the function using the following definition.

Definition 2.14 (*Limit of a function*) Let $f : X \to Y$ be a function between two topological spaces (X, T_X) and (Y, T_Y). We say that y_0 is a *limit* of f as $x \to x_0$ and write $\lim_{x \to x_0} f(x) = y_0$ if $\forall N(y_0)$, there exists a $N(x_0)$ such that $f(N(x_0) \backslash \{x_0\}) \subseteq N(y_0)$.

The neighbourhood of a point, less the point itself, $N(x_0) \backslash \{x_0\}$, is sometimes called *punctured neighbourhood*. The previous definition is based on the idea that the topology captures the notion of proximity. In this sense, we can say that the limit of the function f in x_0 describes or approximates the value of f around this point. The images of the points in a neighbourhood of x_0, that is, close to it, are mapped by the function to a neighbourhood of y_0. The value of the function in x_0 itself is not relevant. The actual implication of Definition 2.14 depends on where the point x_0 is located. If x_0 is an interior point of $\mathbb{D}_f$, then there are neighbourhoods $N(x_0)$ that will be fully mapped by the function f within $N(y_0)$. In contrast, if x_0 belongs to the boundary of the domain, $\partial \mathbb{D}_f$, then only part of the neighbourhood, the part within the domain, will be mapped by f. If x_0 is not a limit point in the function domain f, the definition is not particularly informative. In this case, there is a neighbourhood $N(x_0)$ such that $f(N(x_0) \backslash \{x_0\}) = \emptyset$. Since the empty set is a subset of any set, this implies that any $y \in Y$ is a limit of the function. Thus, in general, one applies Definition 2.14 to the case $x_0 \in D\mathbb{D}_f$. In any case, Definition 2.14 does not exclude that the function might have, in a point $x \in D\mathbb{D}_f$, more than one limit. However, it is immediately clear that if the image space (Y, T_Y) is Hausdorff, then if the limit of the function at a point exists, it is unique. The definition of the limit of a function is related to the notion of continuity.

Definition 2.15 (*Continuity in a point*) The function $f : X \to Y$ is continuous in x_0 if $\lim_{x \to x_0} f(x) = f(x_0)$.

For this definition to be meaningful, both $x_0 \in \mathbb{D}_f$ and $x_0 \in D\mathbb{D}_f$. It is important to understand that the property of being continuous at a given point does not depend exclusively on the function f, but also on the topologies considered in the domain space X and the image space Y. For instance, continuity stops being an interesting property when the discrete topology is considered.

▶ **Example 2.15** (*Continuity in discrete topology*) Let $f : X \to Y$ be a function between two topological spaces and assume that T_X and T_Y are discrete topologies. Then for any $x_0 \in X$, let $f_0 = f(x_0)$ and A_{f_0} be the set of points that the function maps to f_0 (there might be more than one point if the function is not injective). The set A_{f_0} is a neighbourhood of x_0, because any set that contains x_0 is a neighbourhood of x_0 in a discrete topology. Then, for any neighbourhood $N(f_0) \subseteq Y$, $f(A_{f_0}) = \{f_0\} \in N(f_0)$, and we can conclude that the function is continuous in x_0. In conclusion, when the discrete topology is considered, all functions are continuous at all points.

In general, when analysing the continuity of a function, the topological space is the subspace defined by the function's domain. So in what follows, when we look at a topological space (X, T_X), in general we are working with $(\mathbb{D}_f, T_{\mathbb{D}_f})$

(see Definition 2.13). With this caveat in mind, consider the following important alternative definition of continuity.

Definition 2.16 (*Global Continuity*) Let (X, T_X) and (Y, T_Y) be two topological spaces. A function $f : X \to Y$ is a *continuous function* or a *continuous map* if $\forall A \in T_Y, f^{-1}(A) \in T_X$.

The equivalence between Definition 2.16 and Definition 2.15 is the subject of the following two theorems.

Theorem 2.21 *If $f : X \to Y$ is continuous according to Definition 2.16, then $\forall x_0 \in X, \lim_{x \to x_0} f(x) = f(x_0)$.*

Proof For any $x_0 \in X$ and any $N(f(x_0))$, simply consider $N(x_0) = f^{-1}(N(f(x_0)))$. This set is a neighbourhood of x_0, it is open by hypothesis and $f(N(x_0)) \subseteq N(f(x_0))$. □

The opposite is also true.

Theorem 2.22 *Let $f : A \subseteq X \to Y$ be continuous on all the points $a \in A$ according to Definition 2.15, then A is open and f is globally continuous on the subspace (A, T_{X_A}).*

Proof Consider any open subset of the image $B \subseteq f(A)$ and $B \in T_Y$. We have to prove that $f^{-1}(B)$ is open.

Take an element $x \in f^{-1}(B)$. Then $f(x) \in B$. Since the function is locally continuous in x, there exists a neighbourhood $N(x)$ such that $f(N(x)) \subseteq B$. This implies that $N(x) \subseteq f^{-1}(B)$ and, consequently, that x is an interior point of $f^{-1}(B)$.

Since the above statement is true for any $x \in f^{-1}(B)$, we can conclude that all points of $f^{-1}(B)$ are interior points and, thus, that $f^{-1}(B)$ is open (see Theorem 2.1). The same is true for $A = f^{-1}(f(A))$. □

In summary, the local and global continuity conditions are basically the same.

▶ **Example 2.16** (*Continuity of the constant function*) Let $f : A \subseteq X \to Y$ be the constant function $f(x) = c$ for any $x \in A$. The image $\mathbb{I}_f = \{c\}$ being a single point, it is endowed with the trivial topology $T' = \{\emptyset, \{c\}\}$. In this topology $\{c\}$ is an open set. At the same time, if one considers the topology restricted to the domain of the function T_{X_A}, then A becomes the whole set and it is open. Thus, according to Definition 2.16, the preimage of any open set is open and we can conclude that the constant function f is continuous.

Conversely, consider $f : X \to Y$ and a set $A \subset X$. Let $f(x) = c_1$ if $x \in A$ and $f(x) = c_2$ if $x \in A^c$, with $c_2 \neq c_1$. If, in T_Y, there exists a neighbourhood of c_2 which does not contain c_1, then the topology restricted to the image of the function contains the singlets $\{c_1\}$ and $\{c_2\}$. Thus, if the space X is connected, the function is not continuous, because $A = f^{-1}(c_1)$ and $A^c = f^{-1}(c_2)$ cannot be both open.

The previous example can be generalised to say that a function defined over a connected topology which takes a finite number of values in a Hausdorff space is not continuous. Interestingly, continuity can also be defined using closed sets.

Theorem 2.23 *The function $f : X \to Y$ is continuous if and only if for every closed set $C \subseteq Y$, $f^{-1}(C)$ is closed.*

Proof For any set $A \subseteq Y$, because f is defined on whole X, $f^{-1}(A^c) = (f^{-1}(A))^c$, i.e. the preimage of the complement is the complement of the preimage. The statement trivially follows.   □

The fact that the openness and closure conditions in Definition 2.16 and in Theorem 2.23, respectively, are imposed on the preimage is essential. A function between topological spaces is an *open map* if, under this function, the image of an open set is open, that is, if the function maps open sets to open sets. Analogously, a function is a *closed map* if it maps closed sets to closed sets. In general, a continuous function is neither open nor closed.

▶ **Example 2.17** (*Open and closed functions*) To see that continuous functions are not required to be open or closed, consider the two functions $f(x) = x^2$ and $g(x) = 2^{-x}$. As discussed in Example 1.7, these functions are, respectively, concave and convex. Thus, according to Theorem 2.31 below, they are continuous in the Euclidean topology. The function f maps the open set $(-1, 1)$ in the set $[0, 1)$, which is neither closed nor open. Thus, f is not open. However, it is easy to show that it is closed. Conversely, g maps the closed semi-line $[0, +\infty)$ to $(0, 1]$, which is neither closed nor open. Thus, the function g is not closed.

At the same time, a closed and/or open function is not required to be continuous. Consider the *floor function* from $\mathbb{R}$ (with the Euclidean topology) to the integers $\mathbb{Z}$ (with the discrete topology), which maps a real number to its integer part. This is a particular *step function* and it is both open and closed. However, it is not continuous: the preimage of the open set (in the $\mathbb{Z}$ topology) $\{2\}$ is the set $[2, 3) \subseteq \mathbb{R}$, which is not open.

It is straightforward to see that the composition of continuous functions is continuous.

Theorem 2.24 *Consider the topological spaces (X, T_X), (Y, T_Y), and (Z, T_Z). If $f : X \to Y$ and $g : Y \to Z$ are continuous functions, then $g \circ f$ is a continuous function.*

Proof Take $B \in T_Z$, then $g^{-1}(B) \in T_Y$ and $f^{-1}(g^{-1}(B)) = (g \circ f)^{-1}(B) \in T_X$ so that the composition is continuous.   □

In general, if a continuous function exists between two topological spaces, these spaces possess similar properties. The first thing they have in common is connectedness.

Theorem 2.25 *Let $f : X \to Y$ be a surjective (onto) continuous function between two topological spaces (X, T_X) and (Y, T_Y). Then if (X, T_X) is connected, (Y, T_Y) is connected.*

Proof Suppose that (Y, T_Y) is not connected. Then there exists a $B \subseteq Y$ that is both open and closed. Since f is surjective $f^{-1}(B) \neq \emptyset$ and, according to Definition 2.16 and Theorem 2.23, it is both open and closed, contradicting the hypothesis. $\square$

The hypothesis of surjectivity is essential for the previous theorem, as the following example shows.

▶ **Example 2.18** (*A continuous not surjective function*) Consider a set with two elements $Y = \{a, b\}$ and the finest topology $T_Y = 2^Y$. This is clearly not a connected topology. Now consider a generic connected topology (X, T_X) and the constant function $f : X \to Y$, $f(x) = a$. This function is continuous but not surjective: the preimage of any set in T_Y is either X or $\emptyset$, thus the function is continuous but the connectedness property is not preserved.

Note that given (X, T_X), we can always take $(f(X), T_{f(X)})$ as the image topology. In this way, the function becomes surjective by construction. With this specification, we can say that a continuous function maps connected sets to connected sets. Continuous functions also preserve compactness.

Theorem 2.26 *If $f : X \to Y$ is continuous and X is compact, then $f(X)$ is compact.*

Proof Let $\{V_\alpha\}$ be an open cover of $f(X)$, then $X \subseteq \cup_\alpha f^{-1}(V_\alpha)$. Since X is compact, there exists a finite subcover $\{f^{-1}(V_1), \ldots, f^{-1}(V_n)\}$ such that $X \subseteq \cup_{i=1}^n f^{-1}(V_i)$. It follows that $f(X) \subseteq \cup_{i=1}^n V_i$, that is, $\{V_1, \ldots, V_n\}$ is a finite subcover of $\{V_\alpha\}$. $\square$

Compactness is also useful if one is interested in defining the inverse of a one-to-one function.

Theorem 2.27 *If $f : X \to Y$ is continuous and one-to-one, i.e. it is both surjective and injective, X is compact, and Y is Hausdorff, then $f^{-1} : Y \to X$ is continuous.*

Proof We have to show that if $A \subseteq X$ is open, then $f(A)$ is open. Consider A^c. Since A^c is closed and is a subset of a compact set $A^c \subseteq X$, A^c is compact. It follows that also $f(A^c)$ is compact, and because Y is Hausdorff, it is closed. It follows that $f(A^c)^c$ is open. For the one-to-one condition, the latter set is equal to $f(A)$. $\square$

The theorem above states that a continuous one-to-one function between two compact sets is open. The one-to-one condition in the previous theorem is essential for two reasons. It guarantees that $f(A^c)^c = f(A)$ and that f^{-1} is a function (for the latter, however, injectivity would be enough).

Continuous functions with continuous inverse can be used to induce an equivalence relation between topologies. These functions are given a specific name.

Definition 2.17 (*Homeomorphism*) Let (X, T_X) and (Y, T_Y) be two topological spaces. If $f : X \to Y$ is continuous and one-to-one, and at the same time f^{-1} is continuous, then f is called a *homeomorphism*.

Two homeomorphic spaces, that is, two spaces linked by a homeomorphism, are topologically identical: a one-to-one relation between open sets can be established using the function f. This relation is clearly symmetric and reflexive. Transitivity is guaranteed by the fact that the composition of continuous functions is continuous. Thus, homeomorphisms can be used to build equivalence classes across topologies by saying that two topological spaces are equivalent if a homeomorphism exists between them.

2.5.1 Continuity in Euclidean Topology

The following result is widely used in the following chapters.

Theorem 2.28 *Consider a real-valued function $f : \mathbb{R} \to \mathbb{R}$. In the Euclidean topology, if the function is monotonic, and with a connected domain and image, then it is continuous.*

Proof The domain and the image of the function, being connected, are open or closed, bounded or unbounded, intervals. Assume that the function is increasing and consider any closed interval $[a, b]$ in the image of f. For any $y \in [a, b]$, $\exists x \in f^{-1}([a, b])$ such that $f(x) = y$. Since the function is increasing, $f^{-1}(a) \le x \le f^{-1}(b)$. Thus $f^{-1}([a, b]) \subseteq [f^{-1}(a), f^{-1}(b)]$. Moreover, the function is defined on any $x \in [f^{-1}(a), f^{-1}(b)]$, and because is increasing, $a \le f(x) \le b$. Thus, $[f^{-1}(a), f^{-1}(b)] \subseteq f^{-1}([a, b])$. The two inclusions taken together imply $f^{-1}([a, b]) = [f^{-1}(a), f^{-1}(b)]$. In other terms, the preimage of all closed intervals in the image are closed intervals in the domain. Since closed intervals are a base of all closed sets, the statement follows from Theorem 2.23. If the function f is decreasing, just apply the previous analysis to $-f$. $\qquad\square$

Theorem 2.28 guarantees that the power function in Definition 1.18 is continuous. Since, given a functional relationship between two spaces, one is actually considering the subspace $(\mathbb{D}_f, T_{\mathbb{D}_f})$ in Definition 2.16, the functional relationship can be continuous on some sets and their associated topological subspaces and not continuous on other sets and their associated topological subspaces.

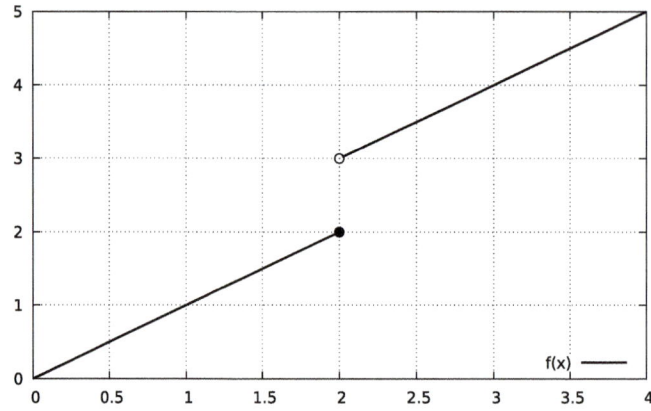

Fig. 2.3 The function of Example 2.19

▶ **Example 2.19** (*A jumping function*) Consider the function

$$f(x) = \begin{cases} x & x \le 2, \\ 1 + x & x > 2. \end{cases}$$

Now the set (0, 3) has preimage (0, 2] which is not open. Analogously, the preimage of [3, 5] is (2, 4], which is not close. Thus, the function f is not continuous (Fig. 2.3). In contrast, the function defined over $\mathbb{R}\backslash\{2\}$

$$f(x) = \begin{cases} x & x < 2, \\ 1 + x & x > 2, \end{cases}$$

is continuous. Note that in this case the domain subtopology is not connected as $A = \{x \in \mathbb{R} \mid x < 2\}$ is both closed and open.

Extending Example 2.18, notice that a continuous function defined over the Euclidean topology in $\mathbb{R}$ cannot map an interval (a, b) to a finite number of points greater than one. This is because the interval is a connected set, while the image set is not connected. Thus, a continuous function on an interval is either a constant function, so that its image is a single real number, or it takes an infinite number of values.

Theorem 2.29 *Consider a continuous function* $f : A \subseteq \mathbb{R} \to \mathbb{R}$ *defined over a connected subset A. If f takes the values a and b > a, then it takes all the values between.*

Proof Assume that the value $c \in (a, b)$ is not taken. Then the image of the function is contained in two disjoint sets $(-\infty, c)$ and $(c, +\infty)$ and, according to Theorem 2.20, it is not connected. But this is impossible for Theorem 2.25. □

Theorem 2.26 guarantees that a continuous real function defined over a compact set in Hausdorff spaces has both maximum and minimum, as clarified in the following result.

Theorem 2.30 (Weierstrass extreme value theorem) *Consider a topological space* (X, T) *and a continuous function* $f : X \rightarrow \mathbb{R}$. *If* $K \subseteq X$ *is a compact set, then the function* f *has a maximum and a minimum in* K.

Proof According to Theorem 2.26 $f(K)$ is compact. Thus, according to Theorem 2.18, it is closed and bounded. Consequently, $f(K)$ has both a supremum and an infimum. By Theorem 2.7, the supremum and infimum belong to the boundary of the set. This in turn belongs to the set because it is closed. This means that there are two points $x', x'' \in X$ such that $f(x') = \sup f(K)$ and $f(x'') = \inf f(K)$, which proves the assertion. □

▶ **Example 2.20** (*Upper and lower semicontinuity*) Consider a real-valued function $f : X \rightarrow \mathbb{R}$ defined on a generic topological space (X, T). The function f is *upper semicontinuous* if it is continuous when the image space $\mathbb{R}$ is equipped with the left-order topology (see Example 2.10). An upper semicontinuous function has a maximum in any compact set. In fact, a compact set $K \subseteq X$ is mapped by an upper semicontinuous function f in a set $f(K)$ that is compact in the left-order topology, that is, it has a maximum.

Analogously, the function f is *lower semicontinuous* if it is continuous when the image space $\mathbb{R}$ is equipped with the right-order topology. Thus, it has a minimum in any compact set $K \subseteq X$.

Alternatively, upper and lower semicontinuity can be defined by saying that f is upper (lower) semicontinuous in $x \in X$, if for any $y > f(x)$ $(y < f(x))$ there exists a neighbourhood $N(x)$ such that $f(N(x)) \subseteq (-\infty, y)$ $(f(N(x)) \subseteq (y, +\infty))$.

The next result proves the continuity of many real functions, including the power and exponential functions defined in Sect. 1.5.

Theorem 2.31 (Continuity of concave and convex functions) *If* f *is a concave or convex function defined over an open interval* $(a, b) \in \mathbb{R}$ *then* f *is continuous in* (a, b).

Proof Assume f to be concave. If f is constant, we know that it is continuous. So assume that it is not constant. Let $x \in (a, b)$ and $\forall \epsilon > 0$ consider the symmetric neighbourhood $N(f(x)) = (f(x) - \epsilon, f(x) + \epsilon)$. Define $M = \max\{|f(x) - f(a)|/(x - a), |f(b) - f(x)|/(b - x)\}$. Since f is not constant,

$M > 0$. Set $\delta = \epsilon/M$. If $z \in (x, x + \delta)$ and $z < b$, by Theorem 1.16,

$$M \geq \frac{f(x) - f(a)}{x - a} \geq \frac{f(z) - f(x)}{z - x} \geq \frac{f(b) - f(x)}{b - x} \geq -M,$$

that is, $|f(z) - f(x)| \leq (z - x) M < \delta M < \epsilon$. Analogously, if $z \in (x - \delta, x)$ and $z > b$,

$$M \geq \frac{f(x) - f(a)}{x - a} \geq \frac{f(x) - f(z)}{x - z} \geq \frac{f(b) - f(x)}{b - x} \geq -M,$$

that is, again, $|f(z) - f(x)| \leq (x - z) M < \delta M < \epsilon$. Thus, for any $z \in (a, b) \cap (x - \delta, x + \delta)/\{x\}$, $f(z) \in N(f(x))$ that proves the assertion. If f is convex, the proof is identical.   □

Exercises

Exercise 2.1 Let T_1 and T_2 be two topologies on X. Prove that $T_1 \cap T_2 \subseteq 2^X$ is a topology.

Exercise 2.2 Consider a topological space (X, T), a subset $A \subseteq X$, and a point $x \in X$. Prove that $x \in \text{int } A$ or $x \in \text{int } A^c$ or $x \in \partial A = \partial A^c$.

Exercise 2.3 Prove that if the singlet $\{x_0\}$ is a closed set and $N(x_0)$ is a neighbourhood of x_0, then the punctured neighbourhood $N(x_0) \setminus \{x_0\}$ is an open set.

Exercise 2.4 (*Cofinite topology*) Consider a set X with an infinite number of elements and let $T \subset 2^X$ be made by the empty set, the whole X, and all sets whose complement contains a finite number of elements. Prove that (X, T) is a topological space. It is known as the *cofinite topology*.

Exercise 2.5 With reference to Example 2.4, let $X = \mathbb{N}$ and define $A_k = \{1, 2, \ldots, k+1\}$. Consider the identity function $f : \mathbb{N} \to \mathbb{N}$, $f(x) = x$. Find the local maxima and minima of f in the nested topology.

Exercise 2.6 In contrast to the bases in linear spaces, topological bases are in general redundant. Consider the topological base introduced in Definition 2.7. Show that if a particular interval (a, b) is removed from the set $\mathbb{B}$, the set remains a base for the same topology.

Exercise 2.7 In the topological space $(\mathbb{N}, T)$ defined in Example 2.7, consider the set $A = \{2, 3, 4\}$ and find its interior, its boundary, and its derivative set. Repeat the analysis for the set $A' = \{3, 4, 5\}$.

Exercise 2.8 (*Euclidean topology on* $\mathbb{R}^2$) Consider the set of ordered couples of real numbers $\mathbb{R} \times \mathbb{R} = \mathbb{R}^2$. Consider two points $\mathbf{x}, \mathbf{y} \in \mathbb{R}^2$, $\mathbf{x} = (x_1, x_2)$ and $\mathbf{y} = (y_1, y_2)$, such that $x_1 < y_1$ and $x_2 < y_2$ and define the interval $I(\mathbf{x}, \mathbf{y}) = \{\mathbf{yx} \in \mathbb{R}^2 \mid x_1 < z_1 < y_1, x_2 < z_2 < y_2\}$. Prove that the collection of all these intervals $\mathbb{B} = \{I\}$ is the base of a topology, named *Euclidean topology* on $\mathbb{R}^2$.

Exercise 2.9 Given three numbers $a, b, c \in \mathbb{R}$, formally define

$$T_{a,b,c} = \{(x, y) \in \mathbb{R}^2 \mid x > a, y > b, x + y < c\}.$$

Consider $\mathbb{B}_1 = \{T_{a,b,c} \mid a, b, c \in \mathbb{R}\}$, $\mathbb{B}_2 = \{T_{a,a,c} \mid a, c \in \mathbb{R}\}$, $\mathbb{B}_3 = \{T_{a,b,c} \mid a, b, c \subset \mathbb{Z}\}$. For each of these sets, discuss whether it is the base for a topology in $\mathbb{R}^2$. If the answer is positive, determine whether the generated topology is equivalent to the topology in Exercise 2.8.

Exercise 2.10 Given two real numbers a and b such that $b > a$ define the interval $\bar{J}_{a,b} = \{x \in \mathbb{R} \mid a \le x \le b\} = [a, b]$. Prove that the set of all these intervals, $\mathbb{B} = \{\bar{J}_{a,b}\}$, cannot be the base of a topology on $\mathbb{R}$. *Hint: Check the second part of Theorem 2.5.*

Exercise 2.11 (*Lower and upper limit topology*) Given two real numbers a and b such that $b > a$ define the interval $J_{a,b} = \{x \in \mathbb{R} \mid a \le x < b\} = [a, b)$. Prove that the set of all these intervals, $\mathbb{B} = \{J_{a,b}\}$, is the base of a topology on $\mathbb{R}$. This is called *lower limit topology*, denoted by T_l. Is T_l equivalent to the Euclidean topology? The *upper limit topology* T_u is generated by the base $\mathbb{B}' = \{K_{a,b}\}$, where $K_{a,b} = \{x \in \mathbb{R} \mid a < x \le b\} = (a, b]$. Is T_u equivalent to T_l?

Exercise 2.12 Consider two real numbers a and b such that $b > a$. Prove that in $(\mathbb{R}, T_l)$ of Exercise 2.11 the set $\{x \in \mathbb{R} \mid a < x < b\}$ is open and the set $\{x \in \mathbb{R} \mid a \le x \le b\}$ is closed.

Exercise 2.13 Prove that in the cofinite topology of Exercise 2.4 all sets are compact.

Exercise 2.14 With reference to Example 2.10 prove that the left- and right-order topologies are second-countable, connected but not Hausdorff.

Exercise 2.15 Prove that $(\mathbb{R}, T_l)$ of Exercise 2.11 is a Hausdorff space.

Exercise 2.16 Prove that the topology in Exercise 2.8 is Hausdorff and second-countable.

Exercise 2.17 Consider the statement "given any point x and any compact set C such that $x \notin C$, we can find two open sets A and B such that $x \in A$, $C \subseteq B$, and $A \cap B = \emptyset$". Is this statement true in any topological space? Is it true in any Hausdorff space?

Exercise 2.18 Prove that the set T_Y in Definition 2.13 is closed with respect to infinite union and finite intersection.

Exercise 2.19 Prove that in the lower limit topology $(\mathbb{R}, T_l)$ of Exercise 2.11, the intervals $J_{a,b}$ are not connected.

Exercise 2.20 Consider a topological space (X, T). Let $Y \subseteq X$ be open, $Y \in T$, and (Y, T_Y) be the topological subspace on Y. Prove that if $A \subseteq X$ is open (close) in T, then $A \cap Y$ is open (close) in T_Y.

Exercise 2.21 Consider a function $f : X \to Y$ between two topological spaces (X, T_X) and (Y, T_Y) and let x_0 be a limit point of the domain of f. Prove that if T_Y is Hausdorff, then if $\lim_{x \to x_0} f$ exists, it is unique.

Exercise 2.22 Considering the usual topology on $\mathbb{R}$, prove that $f(x) = x^2$ is a closed map and $f(x) = 2^{-x}$ is an open map.

Exercise 2.23 Consider a continuous function f from the topological space (X, T_X) to the topological space (Y, T_Y). Let $X' \subset X$ be dense in X and let (Y, T_y) be Hausdorff. Prove that if f is constant in X', then it is constant in X.

Exercise 2.24 Consider a continuous function f from the topological space (X, T_X) to the topological space (Y, T_Y). Let $A' \subset Y$ be open and $A = f^{-1}(A')$. Prove that $\bar{A} \subseteq f^{-1}(\bar{A}')$. In other terms, the closure of the preimage is contained in the preimage of the closure. *Hint: Use Theorem 2.23.*

Exercise 2.25 Consider a continuous function f from a topological space (X, T) to $(\mathbb{R}, |\cdot|)$. Let $X' \subset X$ be dense in X. Prove that if $f(x) \leq 0, \forall x \in X'$, then it is $f(x) \leq 0$ for each $x \in X$.

Exercise 2.26 Build a continuous function f from a topological space (X, T) to $(\mathbb{R}, |\cdot|)$ such that $f(x) > 0$ for each $x \in X'$ where X' is dense in X, but $f(x) \leq 0$ for some x in X. *Hint: The choice of the topological space is left to you.*

Exercise 2.27 Consider the Euclidean topology in $\mathbb{R}^2$ defined in Exercise 2.8, and the sets $A = \{(x, y) \mid x^2 + y^2 < 2\}$, $B = \{(x, 0) \mid 2 \leq x \leq 3\}$, and $C = \{(x, y) \mid 2 < x < 3, 0 < y < 1, x, y \in \mathbb{Q}\}$. Discuss which sets among A, B, C, $A \cup B$, $B \cup C$, $\bar{A} \cup C$ are compact and / or connected.

Exercise 2.28 Consider the function $f : \mathbb{R} \to \mathbb{R}^2$ defined as $f(t) = (t/(1 + t^2), t^2/(1 + t^2))$ and its image $A = f(\mathbb{R})$. In the Euclidean topology introduced in Exercise 2.8, prove that A is connected and $\bar{A}$ is compact. Find $\bar{A} \setminus A$.

Metric Spaces

3

3.1 Definition and Basic Properties

In this chapter, we are mainly concerned with a special class of functions.

Definition 3.1 *(metric space)* Consider a set X and a real nonnegative function $d : X \times X \to \mathbb{R}_{\geq 0}$ such that $\forall x, y, z \in X$ it is:

- $d(x, y) = 0$ if and only if $x = y$ (*positive definiteness*);
- $d(x, y) = d(y, x)$ (*symmetry*);
- $d(x, y) + d(y, z) \geq d(x, z)$ (*triangle inequality*).

Then d is a *distance function* or *metric* and (X, d) is a *metric space*.

When considering a subset $Y \subseteq X$ and the restriction of d to $Y \times Y$, the structure (Y, d) is again a metric space. Together with the idea of distance comes the idea of the "size" of a set.

Definition 3.2 *(Diameter)* Consider a metric space (X, d) and a subset $A \subseteq X$. If the set $\{d(x, y) \mid \forall x, y \in A\}$ is bounded above then the set A is *bounded* and the *diameter* of A is $\operatorname{diam}(A) = \sup\{d(x, y) \mid \forall x, y \in A\}$.

A set that is not bounded is said to be *unbounded*.

Definition 3.3 *(Bounded function)* Let (X, d) be a metric space and A a generic set. A function $f : A \to X$ is a *bounded function* if the set $\{d(f(a), f(b)) \mid \forall a, b \in A\}$ is bounded, that is if the image $f(A)$ is a bounded subset of X.

Using the distance function, it is also possible to define the distance of a point from a set.

© The Author(s), under exclusive license to Springer Nature Switzerland AG 2023
G. Bottazzi, *Advanced Calculus for Economics and Finance*, Classroom
Companion: Economics, https://doi.org/10.1007/978-3-031-30316-6_3

Definition 3.4 *(Distance of a point from a set)* Let $x \in X$ and consider a nonempty subset $A \subseteq X$, then the distance of x from A is defined as

$$d(x, A) = \inf_{y} \{d(x, y) \mid y \in A\}.$$

The definition is always meaningful because the set considered is bounded below by zero. If $x \in A$, then $d(x, A) = 0$. As a first example, we can build the "usual" metric space on the set of real numbers using the absolute value.

▶ **Example 3.1** *(Euclidean metric on $\mathbb{R}$)* Consider the function $d : \mathbb{R} \times \mathbb{R} \to \mathbb{R}_{\geq 0}$ defined as $d(x, y) = |x - y|$. Using the property of the absolute value and the triangle inequality Theorem 1.9, it is immediate to show that d fulfils the properties of a distance function in Definition 3.1. In particular,

$$d(x, y) + d(y, z) = |x - y| + |y - z| \geq |x - y + y - z| \geq |x - z| = d(x, z).$$

Thus, $(\mathbb{R}, |.|)$ is a metric space, usually known as the *Euclidean metric space* of real numbers. The distance function is just the Euclidean "length" between two points on a line. In this space, the metric notion of diameter introduced in Definition 3.2 is that derived from the order relation. In fact, given any subset $A \subset \mathbb{R}$, $\text{diam}(A) = \sup(A) - \inf(A)$. Consistently with the discussion in Sect. 1.2, a bounded set is just a set with an upper and lower bound.

Exercise 3.6 provides other examples of distance functions on $\mathbb{R}$. However, metric spaces can be very different from those that our geometric intuition might suggest.

▶ **Example 3.2** *(Hamming distance)* Consider a finite set A. This set is an alphabet and its elements are symbols. Let A^n be the n times Cartesian product of A with itself. The elements of A^n are the n-tuples of symbols $\sigma = (s_1, s_2, \ldots, s_n)$ with $s_i \in A$ for $i = 1, \ldots, n$. The *Kronecker delta function* is defined on $A \times A$ as $\delta_{s,s'} = 1$ if $s = s'$ and $\delta_{s,s'} = 0$ if $s \neq s'$, with $s, s' \in A$. Given two n-tuples σ and σ' define

$$d(\sigma, \sigma') = \sum_{i=1}^{n} (1 - \delta_{s_i, s_i'}).$$

Note that $d(\sigma, \sigma') \geq 0$ and $d(\sigma, \sigma') = 0$ if and only if $\sigma = \sigma'$. For any three n-tuples σ, σ' and σ'', $d(\sigma, \sigma') + d(\sigma', \sigma'') \geq d(\sigma, \sigma'')$ so that the function d is a distance on the set A. It is called *Hamming distance* from the name of the American mathematician who first introduced it in computer science in 1950.

Next, we have the first example of a *functional space*, a space whose points are, in fact, functions.

▶ **Example 3.3** *(Metric space of bounded functions)* Consider a set A and let $B(A)$ be the set of bounded functions f from A to $\mathbb{R}$. An element of $B(A)$ is for instance the function that maps all the elements of A to a single real number (the constant function). Consider two functions f and g in $B(A)$ and the function $d_\infty : B(A) \times B(A) \to \mathbb{R}_{\geq 0}$ defined as

$$d_\infty(f, g) = \sup_{a \in A} \{|f(a) - g(a)|\} \quad \text{for any} \quad f, g \in B(A).$$

Since both g and f are bounded, and since the supremum exists for any bounded set, the previous function is well defined. To each couple of functions, the function d_∞ assigns a nonnegative real number. Using the property of the supremum, it is easy to show that the function d satisfies all the requirements of Definition 3.1. Thus, $(B(A), d_\infty)$ is a metric space. The distance d_∞ is known as the *Chebyshev distance*, after the Russian mathematician Pafnuty Chebyshev (1821–1894). The reason for its peculiar "infinity" symbol will be made clear in Example 4.2.2. The use of $\mathbb{R}$ as the image space of the functions aids familiarity but is not required. The same analysis can be applied to functions with image in a generic metric space, see Exercise 3.10.

3.2 Metric and Topology

Any metric induces a first-countable (see Definition 2.10) and Hausdorff (see Definition 2.10) topology. To see how, we start by defining the open balls.

Definition 3.5 *(Open ball)* Given a metric space (X, d), the *open ball* of radius $r > 0$ centred in x is defined as $B(x, r) = \{y \in X \mid d(x, y) < r\}$.

The set of open balls is a base of a topology.

Theorem 3.1 (Topology induced by the distance) *Given the metric space (X, d), consider the collection of all open balls $\mathbb{B} = \{B(x, r) \mid x \in X, r \in \mathbb{R}_{>0}\}$. Then $\mathbb{B}$ is a base of a Hausdorff first-countable topology on X.*

Proof First, note that the union of elements of $\mathbb{B}$ covers the entire space, as $X \subseteq \bigcup_{x \in X} B(x, 1)$. Next consider $y \in B(x, r) \cap B(x', r') \neq \emptyset$ and define $d(x, y) = \eta < r$, $d(y, x') = \eta' < r'$. Take $r'' = \min\{r - \eta, r' - \eta'\}$. Now $B(y, r'') \subseteq B(x, r) \cap B(x', r')$. Indeed, if $d(y, z) < r''$ then

$$d(x, z) \leq d(x, y) + d(y, z) \leq \eta + r'' \leq r,$$

and, at the same time,

$$d(x', z) \leq d(x', y) + d(y, z) \leq \eta' + r'' \leq r'.$$

Thus, for Theorem 2.5 the set $\mathbb{B}$ generates a topology. Now consider any x, $y \in X$ and define $d(x, y) = r$. Then $B(x, r/3) \cap B(y, r/3) = \emptyset$ so that the space is Hausdorff. To see that the topology is first-countable for any point x, consider the collection of balls $B(x, r)$ with $r \in \mathbb{Q}^+$. This is a countable collection that, due to the density of rationals (see Theorem 1.11), clearly satisfies Definition 2.5. □

In the following, we will implicitly assume that any metric space (X, d) is also a topological space with the topology induced by the distance function. The distance function is always continuous with respect to the topology it induces on the set. In fact, for any $x' \in X$, consider the function $f : X \to \mathbb{R}_{\geq 0}$ defined as $f(x) = d(x', x)$. The counter image of any open interval (a, b) under this function is the set $B(x', b) \setminus \bar{B}(x', b)$, where $\bar{B}$ is the closure of the ball B. This is an open set, so the function f is continuous. Because the distance-induced topology is Hausdorff, compact sets are closed (see Theorem 2.12). In addition, we have the following.

Theorem 3.2 *In a metric (topological) space, a compact set is bounded.*

Proof Let $K \subseteq X$ be compact and consider $\cup_{x \in X} B(x, 1)$ as its cover. There exists a finite subcover $K \subseteq \cup_{j=1}^{J} B(x_j, 1)$. Define $\eta_j = d(x_1, x_j)$ for $h = 1, \ldots, J$ and set $\eta = \sum_{j=1}^{J} \eta_j$. Then $\cup_{j=1}^{J} B(x_j, 1) \subseteq B(x_1, \eta + 1)$. Indeed for any $z \in \cup_{j=1}^{J} B(x_j, 1)$, there exists a k such that $z \in B(x_k, 1)$, thus $d(z, x_1) \leq d(z, x_k) + d(x_k, x_1) < 1 + \eta_k < 1 + \eta$. Thus diam$(K) \leq \eta + 1$. □

The topology induced by the distance function has some other useful properties. First, note that if $x \in A$, then $d(x, A) = 0$. Generally, we have the following.

Theorem 3.3 *Let $x \in X$ and consider a subset $A \subseteq X$. Then $d(x, A) = 0$ if and only if $x \in \bar{A}$.*

Proof If $d(x, A) = 0$, then for any open ball of x, $B(x, \delta)$ there exists an element $a \in A$ such that $d(x, a) < \delta$, that is $a \in B(x, \delta)$. One possibility is that $a = x$, so that $x \in A$. Otherwise, if x does not belong to A, x is a limit point of A.

Suppose instead that $x \in \bar{A}$. Then for any $r > 0$ there exists an $y \in A$ such that $y \in B(x, r)$ (y can be equal to x or not), that is, $d(x, y) < r$. This implies that $d(x, A) = 0$. □

Second, adding limit points to a set does not increase its diameter.

Theorem 3.4 *Let $A \subseteq X$ be a subset of a metric space (X, d), then* diam$(\bar{A}) =$ diam(A).

Proof First of all, because $A \subseteq \bar{A}$, we know that $\text{diam}(\bar{A}) \geq \text{diam}(A)$. Now $\forall \epsilon > 0$ and $\forall p, q \in \bar{A}, \exists p', q' \in A$ such that $d(p', p) < \epsilon/2$ and $d(q', q) < \epsilon/2$. Therefore,

$$d(p, q) \leq d(p, p') + d(p', q') + d(q', q) < \epsilon + \text{diam}(A).$$

Thus, for any $\epsilon > 0$, $\text{diam}(A) \leq \text{diam}(\bar{A}) < \text{diam}(A) + \epsilon$, which proves the statement. $\square$

For example, in the Euclidean metric of Example 3.1, the open interval (a, b) has the same diameter as the interval $[a, b]$, and in the metric space of bounded functions of Example 3.3, given any bounded function g, the set of all bounded functions f such that $d_\infty(f, g) < 1$ has the same diameter as the set of all bounded functions f such that $d_\infty(f, g) \leq 1$, that is, 1.

▶ **Example 3.4** *(From Euclidean metric to Euclidean topology)* The trick performed in Theorem 3.1 is the generalisation to any metric space of what we did in Example 2.2.2 for the real line. This also explains why in Definition 2.7 we called the topology "Euclidean". Notice that the Euclidean topology on the real numbers has more properties than the topology generically induced by a distance function in a metric space. For instance, in Theorem 2.8, we prove that the topology $(\mathbb{R}, |.|)$ is also second-countable. This allowed us to prove the fundamental Bolzano–Weierstrass Theorem 2.16 and Heine–Borel Theorem 2.18 for real numbers. So it seems that the space $(\mathbb{R}, |.|)$ is somehow special among the metric spaces. In fact, in Chap. 4 we will see that this is the simplest example of a special class of metric spaces, the *normed spaces*, and we will discuss how to generalise to them many of the special results we obtained for the set of real numbers.

3.3 Uniform Continuity

The topological notion of continuous functions has been generally defined in Sect. 2.5. We met both a local definition based on the notion of limit in Definition 2.15, and a global definition based on the counter image of open sets in Definition 2.16. With metric spaces, we can introduce a stricter definition of continuity.

Definition 3.6 *(Uniform continuity)* Consider two metric spaces (X, d_X) and (Y, d_Y). A map $f : X \to Y$ is *uniformly continuous* on X if $\forall \epsilon > 0, \exists \delta_\epsilon > 0$ such that if $d_x(x, x') < \delta_\epsilon$ then $d_y(f(x), f(x')) < \epsilon$.

The fact that this definition is stricter than the original definition of continuity is immediately verified. If the function f is uniformly continuous in X, it is also continuous with respect to the topology induced by the metric. The opposite is generally not true.

▶ **Example 3.5** (*Continuous but not uniformly continuous*) Consider the function $f(x) = 1/x$ on the interval $(0, 1)$. The image of this open interval is made of all real numbers greater than 1. For any $x_2 > x_1 > 1$, the inverse image of the open interval (x_1, x_2) is the open interval $(1/x_2, 1/x_1)$. Since the open intervals are a base of the topology, we can conclude that the inverse image of any open set is an open set and, consequently, the function f is continuous in $(0, 1)$.

Regarding uniform continuity, choose $\epsilon \in (0, 1)$. Then for any $\delta \in (0, 1)$ there exists x_1 and x_2 in $(0, 1)$ such that $d(x_1, x_2) < \delta$ but $d(f(x_1), f(x_2)) > \epsilon$. To see it, simply consider the positive points $x_1 = \sqrt{\delta} - \epsilon\delta$ and $x_2 = \sqrt{\delta}$. The smaller the values of δ, the closer the points considered to zero. However, they always exist. Thus, f is not uniformly continuous in $(0, 1)$.

If one considers functions defined over compact sets, then continuity and uniform continuity turn out to be the same thing.

Theorem 3.5 (Heine–Cantor) *Let f be a map from a compact metric space X to a metric space Y. Then, if f is continuous, it is uniformly continuous.*

Proof Consider a fixed $\epsilon > 0$. Since the function is continuous, for any point $x \in X$, there exists a δ_x such that for any $x' \in X$, if $d_X(x, x') < \delta_x$, then $d_Y(f(x), f(x')) < \epsilon/2$. Clearly, $\cup_x B(x, \delta_x/2)$ covers X and because X is compact, from this cover it is possible to extract a finite subcover $\cup_{k=1}^{N} B(x_k, \delta_{x_k}/2)$.

Let $\delta_\epsilon = \min\{\delta_{x_1}, \ldots, \delta_{x_N}\}/2 > 0$. Consider two points $q, q' \in X$ such that $d_X(q, q') < \delta_\epsilon$. We will show that the distance of their images is less than ϵ. Among the different sets that compose the finite cover, there exists one that contains q. That is, there exists a k such that $q \in B(x_k, \delta_{x_k}/2)$. The distance of q' from x_k is less than δ_{x_k}. In fact,

$$d_X(q', x_k) \leq d_X(q', q) + d_X(q, x_k) < \delta_\epsilon + \frac{\delta_{x_k}}{2} \leq \delta_{x_k}.$$

We have shown that q and q' both belongs to $B(x_k, \delta_{x_k})$. Thus, by construction $d_Y(f(q), f(q')) \leq d_Y(f(q), f(x_k)) + d_Y(f(q'), f(x_k)) < \epsilon$.   □

The distance from a given point is itself a uniformly continuous function from (X, d) to $(\mathbb{R}_{\geq 0}, |.|)$. To see it, fix a point $z \in X$ and consider the function $f : X \to \mathbb{R}$, $f(x) = d(x, z)$ defined on the whole X. For the triangle inequality, $|f(x) - f(y)| = |d(x, z) - d(y, z)| \leq d(x, y)$ so that for any ϵ, if $d(x, y) < \epsilon$, $|f(x) - f(y)| < \epsilon$.

3.4 Lipschitz Continuity

A type of function that is commonly encountered in applications takes its name after the German mathematician Rudolf Lipschitz (1832–1903).

Definition 3.7 *(Lipschitz continuity)* Consider two metric spaces (X, d_X) and (Y, d_Y). A function $f : X \to Y$ is *Lipschitz continuous* on X if $\exists k > 0$ such that $\forall x, x' \in X, d_Y(f(x), f(x')) < k d_X(x, x')$.

If $k < 1$, then the function f a *contraction* or *contraction map*. A function is *locally Lipschitz continuous* in $x \in X$ if it is Lipschitz continuous when restricted to a neighbourhood of x. If f is Lipschitz continuous, $\forall \epsilon > 0$, if $d_x(x, x') < \epsilon/k$, then $d_y(f(x), f(x')) < \epsilon$, thus we can conclude the following.

Corollary 3.1 *If the function f is Lipschitz continuous, then it is uniformly continuous.*

In general, the opposite is not true.

▶ **Example 3.6** *(Uniformly but not Lipschitz continuous)* Consider the function $f(x) = \sqrt{x}$ in the Euclidean metric. The function is uniformly continuous in $\mathbb{R}_{\geq 0}$. In fact, $\forall \epsilon > 0$, if $|x - x'| < \epsilon^2$,

$$|f(x) - f(x')| \leq \sqrt{|f(x) - f(x')|(\sqrt{x} + \sqrt{x'})} = \sqrt{|x - x'|} < \epsilon.$$

However, the function is not Lipschitz continuous in $\mathbb{R}_{\geq 0}$. To see it, note that $|\sqrt{x} - \sqrt{x'}|/|x - x'| = 1/(\sqrt{x} + \sqrt{x'})$ and there is no $k > 0$ such that $1/(\sqrt{x} + \sqrt{x'}) < k$ for any $x, x' \in \mathbb{R}_{\geq 0}$. If restricted to $(1, +\infty)$, the function f becomes Lipschitz continuous with $k = 1/2$.

From Definition 3.2 it follows that for a Lipschitz continuous function $f : E \subseteq X \to Y$, it is $\mathrm{diam}(f(E)) \leq K \, \mathrm{diam}(E)$, then we have the following.

Corollary 3.2 *A Lipschitz continuous function on a bounded set is bounded.*

Exercises

Exercise 3.1 *(Quadrangular inequality)* Let (X, d) be a metric space. Show that for every $x, y, z, w \in X$ it holds that $|d(x, y) - d(z, w)| \leq d(x, z) + d(y, w)$.

Exercise 3.2 Consider a function $f : \mathbb{R}_{\geq 0} \to \mathbb{R}_{\geq 0}$ that is strictly increasing and *sub additive*, that is, $f(x + y) \leq f(x) + f(y)$. Assume further that $f(0) = 0$. If the function f satisfies the properties above, is $d(x, y) = f(|x - y|)$ a distance on $\mathbb{R}$? If so, prove it. If not, provide a counterexample.

Exercise 3.3 Prove that in the Euclidean metric space $(\mathbb{R}, |.|)$ of Example 3.1 a set $A \subset \mathbb{R}$ is bounded if and only if there exists a positive real number M such that $|a| < M$ for any $a \in A$.

Exercise 3.4 Consider two functions f and g from a generic set X to the Euclidean metric space $(\mathbb{R}, |.|)$ of Example 3.1. Prove that if they are bounded, then their sum $f + g$ and their product fg are bounded. *Hint: Use the triangle inequality and the properties of the absolute value.*

Exercise 3.5 Consider the function $d_1(\mathbf{x}, \mathbf{y}) = |x_1 - y_1| + |x_2 - y_2|$ defined over $\mathbb{R}^2 \times \mathbb{R}^2$ where the indices denote the components of the vectors. Prove that it is a distance. Draw the circle of radius 1 and centre the origin in the metric space $(\mathbb{R}^2, d_1)$ (the set of point with distance 1 from the origin). Prove that it is contained in the closure of the Euclidean ball of radius 1 and centre in the origin.

Exercise 3.6 Determine which functions from $\mathbb{R}^2$ to $\mathbb{R}$ in the following list can be used to define a metric on $\mathbb{R}$:

$$d_1(x, y) = (x - y)^2, \quad d_2(x, y) = \sqrt{|x - y|}, \quad d_3(x, y) = |x^2 - y^2|,$$

$$d_4(x, y) = |x - 2y|, \quad d_5(x, y) = \frac{|x - y|}{1 + |x - y|}, \quad d_6(x, y) = |e^x - e^y|.$$

Exercise 3.7 Let X be an infinite set and consider the following function $d : X \times X$ to R, $d(a, b) = 1$ if $a = b$ and $d(a, b) = 0$ otherwise. Prove that it is a distance. In the induced topology, which sets are open? Which ones are closed? Which are compact?

Exercise 3.8 Prove that the function d_∞ in Example 3.3 is a distance.

Exercise 3.9 Consider the set $A = \{1, 2, 3\}$. Using the Hamming distance of Example 3.2 in A^3, find the elements of the open ball of radius one and centre $(1, 2, 3)$ and those of the open ball of radius 0.5 and centre $(3, 2, 1)$.

Exercise 3.10 Extending Example 3.3, let A be a generic set, (V, d) a metric space, and $B(A)$ the set of bounded functions $f : A \to V$. Consider the function $d_\infty : B(A) \times B(A) \to \mathbb{R}_{\geq 0}$ defined as $d_\infty(f, g) = \sup_{a \in A}\{d(f(a), g(a))\}$. Prove that $(B(A), d_\infty)$ is a metric space. *Hint: Use the properties of the supremum.*

Exercise 3.11 Prove that if A is bounded, then given any point x_0, there exists an open ball $B(x_0, r)$ such that $A \subseteq B(x_0, r)$.

Exercise 3.12 Prove that a uniformly continuous function is continuous. *Hint: Use the local definition of continuity in a point.*

Exercise 3.13 Explicitly prove that the function $y = x^2$ is uniformly continuous in $[0, 1]$ (do not use Theorem 3.5). Is it uniformly continuous in $(0, 1)$? Is it uniformly continuous in $\mathbb{Q} \cap (0, 1)$?

Exercise 3.14 With reference to Example 3.1, take two functions f and g from a metric space (X, d) to $(\mathbb{R}, |.|)$. Prove that if f and g are uniformly continuous, then their sum $f + g$ is uniformly continuous. *Hint: Use the triangle inequality.*

Exercise 3.15 With reference to Example 3.1, consider two functions f and g from a metric space (X, d) to $(\mathbb{R}, |.|)$. Prove that if they are bounded and uniformly continuous, then their product $f\,g$ is uniformly continuous. *Hint: Use the fact that they are bounded.*

Exercise 3.16 With reference to Example 3.1, consider a function f from a metric space (X, d) to $(\mathbb{R}, |.|)$ which is bounded away from zero, that is, $\inf_{x \in X} |f(x)| > 0$. Prove that if f is uniformly continuous, then $1/f$ is uniformly continuous.

Normed Spaces

4

4.1 Definition and Basic Properties

The Euclidean notion of the length of a linear segment can be generalised to any linear space by the introduction of a norm.

Definition 4.1 (*Normed space*) Consider a linear space V on $\mathbb{R}^n$. A *norm* is a function $\rho : V \to \mathbb{R}_{\geq 0}$ such that

- $\rho(\mathbf{x}) = 0$ only if $\mathbf{x} = 0$ (*separate points*);
- $\rho(\mathbf{x} + \mathbf{y}) \leq \rho(\mathbf{x}) + \rho(\mathbf{y})$ (*triangle inequality*);
- $\rho(a\mathbf{x}) = |a|\rho(\mathbf{x})$ (*positive homogeneity*).

The pair (V, ρ) is said to be a *normed space*.

The simplest example of a normed space is the set of real numbers together with the absolute value $(\mathbb{R}, |.|)$. Often, the norm is denoted with a couple of double vertical lines $\|.\|$ and the normed space $(V, \|.\|)$. In this book, I reserve this symbol for the Euclidean norm introduced in Sect. 4.2.1 and denote the generic norm by ρ. The first important thing is to clarify the relationship between normed spaces and metric spaces.

Lemma 4.1 (*Norm and distance*) *Any normed space* (V, ρ) *is a metric space with the distance defined as* $d_\rho(\mathbf{x}, \mathbf{y}) = \rho(\mathbf{x} - \mathbf{y})$, $\forall \mathbf{x}, \mathbf{y} \in V$.

Proof The function d_ρ is, by definition, nonnegative and fulfils the first two properties of Definition 3.1. From the triangle inequality in Definition 4.1

$$\rho(\mathbf{x} - \mathbf{z}) = \rho(\mathbf{x} - \mathbf{y} + \mathbf{y} - \mathbf{z}) \leq \rho(\mathbf{x} - \mathbf{y}) + \rho(\mathbf{y} - \mathbf{z})$$

and the third property of Definition 3.1 follows. □

© The Author(s), under exclusive license to Springer Nature Switzerland AG 2023
G. Bottazzi, *Advanced Calculus for Economics and Finance*, Classroom
Companion: Economics, https://doi.org/10.1007/978-3-031-30316-6_4

From now on, when we refer to (V, ρ) as a metric space, we will implicitly assume that the definition of distance used is the one implied by the norm. The norm can be used to introduce the notion of a bounded set in any linear space.

Definition 4.2 A subset of a normed space $A \subseteq V$ is *bounded* if $\{\rho(\mathbf{x}) \mid \mathbf{x} \in A\}$ is a bounded subset of $\mathbb{R}$.

This definition is analogous to Definition 3.2 but is sometimes more practical to use. If the linear space V has a finite dimension n, one can build a one-to-one linear function $f : V \rightarrow \mathbb{R}^n$ by selecting a base and mapping any vector $\mathbf{x} \in V$ into the n-tuple of its coordinates. Since the function f is linear, it is easy to show that the function $\rho \circ f^{-1}$ is a norm on $\mathbb{R}^n$. In fact, let $\mathbf{x}, \mathbf{y} \in \mathbb{R}^n$, then $\rho(f^{-1}(\mathbf{x}+\mathbf{y})) = \rho(f^{-1}(\mathbf{x})+f^{-1}(\mathbf{y})) \leq \rho(f^{-1}(\mathbf{x}))+\rho(f^{-1}(\mathbf{y}))$. Thus, any finite-dimensional normed space defined over $\mathbb{R}$ is isomorphic to some normed space $(\mathbb{R}^n, \rho \circ f^{-1})$. Similarly, any norm ρ' defined on $\mathbb{R}^n$ induces a norm on a vector space of dimension n, $(V, \rho' \circ f)$.

Because the normed space (V, ρ) is a metric space, we can naturally induce a topology on it using the distance function implied by the norm (see Lemma 4.1). The topology on (V, ρ) is generated by the set of open balls $B(\mathbf{x}, r) = \{\mathbf{v} \in V \mid \rho(\mathbf{v} - \mathbf{x}) < r\}$, $\forall \mathbf{x} \in V$ and radius $r > 0$. This topology is Hausdorff and first-countable (see Theorem 3.1) and in this topology all compact sets are bounded (see Theorem 3.2).

▶ **Example 4.1** *(Continuity of the norm)* Any norm is a continuous function in the topology it defines. Consider a normed space (V, ρ) together with the topology induced by the norm. Take the open interval (a, b). The inverse image of this interval contains all elements $\mathbf{x} \in V$ such that $a < \rho(\mathbf{x}) < b$. This set is just the intersection of $B(\mathbf{0}, b)$, which is an open set, with the complement of the closure of $B(\mathbf{0}, a)$, which is another open set. Thus, the inverse image of any open interval (a, b) is an open set. Since the open intervals form a base of the topology in $\mathbb{R}$, we have shown that the inverse image of any open set is an open set and therefore ρ is continuous.

In a convex subspace of a normed space, convex and concave real-valued functions have special properties that are often exploited in applications.

Theorem 4.1 (Uniqueness of extrema) *Let (V, ρ) be a normed space and A be a convex subset of V. Consider a concave (convex) function $f : A \rightarrow \mathbb{R}$. Then, if f has a strict local maximum (minimum) in A, it is unique.*

Proof Let us start with the concave case. Assume that $\mathbf{x}_1$ and $\mathbf{x}_2$ are strict local maxima of f in A, and that $f(\mathbf{x}_1) \geq f(\mathbf{x}_2)$. Then for $\mathbf{z} = (1 - \lambda)\mathbf{x}_2 + \lambda\mathbf{x}_1$ with $\lambda \in [0, 1]$, $f(\mathbf{z}) \geq (1 - \lambda)f(\mathbf{x}_2)+\lambda f(\mathbf{x}_1) \geq f(\mathbf{x}_2)$. Since in any neighbourhood of $\mathbf{x}_2$ there is at least one point of this kind, $\mathbf{x}_2$ cannot be a strict maximum. A similar reasoning applies to the convex case. □

Since global maxima (minima) are also local maxima (minima), the previous analysis directly extends from the latter to the former. Note that the previous considerations cannot be extended to extremal points that are not strict. In fact, any constant function on A is concave (and convex), and all points of A are local maxima and minima. However, the result extends to non-strict maxima and minima if the functions considered are strictly concave or strictly convex, respectively.

4.2 Example of Normed Spaces

In the following, we review three widely used normed spaces: the Euclidean norm on real vectors and its generalisation, the p-norm, and the operator norm on linear maps (or matrices). In all cases, the procedure will be the same: we introduce a function from the linear space in question to $\mathbb{R}_{\geq 0}$ and show that it satisfies all the requirements of Definition 4.1.

4.2.1 Euclidean Norm in $\mathbb{R}^n$

The nth power of the real set $\mathbb{R}^n = \mathbb{R} \times \mathbb{R} \ldots \times \mathbb{R}$ defines the most natural linear space. In this space, the vectors are identified with ordered n-tuples of real numbers $\mathbf{x} = (x_1, \ldots, x_n)$. The *Euclidean norm* of a vector is defined as $\|\mathbf{x}\| = \sqrt{\sum_{j=1}^{n} x_j^2}$. We want to prove that this is actually a norm. The following results are useful.

Theorem 4.2 (Cauchy–Schwarz inequality) *Let* $\mathbf{x}, \mathbf{y} \in \mathbb{R}^n$ *be two n-tuples of real numbers, then*

$$\|\mathbf{x}\| \|\mathbf{y}\| \geq \left| \sum_{i=1}^{n} x_i y_i \right| .$$

Proof Consider the double summation $\sum_{i=1}^{n} \sum_{j=1}^{n} (x_i y_j - x_j y_i)^2 \geq 0$. Using the binomial expansion of the squares, it reduces to

$$\sum_{i=1}^{n} \sum_{j=1}^{n} x_i^2 y_j^2 + x_j^2 y_i^2 \geq 2 \sum_{i=1}^{n} \sum_{j=1}^{n} x_i x_j y_i y_j.$$

The left-hand side is equal to $2 \|\mathbf{x}\|^2 \|\mathbf{y}\|^2$ while the right-hand side is $2(\sum_{i=1}^{n} x_i y_i)^2$. Simplifying out the factor 2 and taking the square roots of the two nonnegative expressions prove the assertion. $\qquad\square$

Using the Cauchy–Schwarz inequality, we can prove the *triangle inequality* required by Definition 4.1.

Theorem 4.3 (Triangle Inequality) *Let* $\mathbf{x}, \mathbf{y} \in \mathbb{R}^n$ *be two n-tuples of real numbers,* *then* $\|\mathbf{x} + \mathbf{y}\| \leq \|\mathbf{x}\| + \|\mathbf{y}\|$.

Proof Using the inequality in Theorem 4.2,

$$\|\mathbf{x} + \mathbf{y}\|^2 = \|\mathbf{x}\|^2 + \|\mathbf{y}\|^2 + 2\sum_{i=1}^{n} x_i y_i \leq \|\mathbf{x}\|^2 + \|\mathbf{y}\|^2 + 2|\sum_{i=1}^{n} x_i y_i|$$

$$\leq \|\mathbf{x}\|^2 + \|\mathbf{y}\|^2 + 2\|\mathbf{x}\|\|\mathbf{y}\| = (\|\mathbf{x}\| + \|\mathbf{y}\|)^2,$$

and the statement is proved.   □

The other two properties of separate points and positive homogeneity are immediate. We can conclude that the Euclidean norm is actually a norm. The space $(\mathbb{R}^n, \|.\|)$ is the Euclidean normed space of dimension n.[1] The topology of this space is the Euclidean topology (see Example 3.4). Note that the Cauchy–Schwarz inequality also implies the following inequality.

Corollary 4.1 *Let* $\mathbf{x}, \mathbf{y} \in \mathbb{R}^n$ *be two n-tuples of real numbers, then* $\|\mathbf{x} + \mathbf{y}\| \geq |\|\mathbf{x}\| - \|\mathbf{y}\||$.

Proof Take the squares of the two sides to obtain

$$\sum_{i=1}^{n} (x_i + y_i)^2 \geq \sum_{i=1}^{n} x_i^2 + \sum_{i=1}^{n} y_i^2 - 2\|\mathbf{x}\|\|\mathbf{y}\|.$$

After simplifications, it reduces to $\sum_{i=1}^{n} x_i y_i \leq \|\mathbf{x}\|\|\mathbf{y}\|$, which follows from Theorem 4.2.   □

▶ **Example 4.2** (*Space of polynomials*) Consider the set of polynomials of all orders n, $V = \{p(x) = \sum_{i=0}^{n} c_i x^i\}$. The sum of two polynomials is a polynomial and the product of a polynomial for a constant is again a polynomial. Thus, V represents a linear space on $\mathbb{R}$. The natural basis for this space is the infinite countable sequence $(1, x, x^2, x^3, \ldots)$. In this space, consider the function $\rho : V \to \mathbb{R}_{\geq 0}$ defined as

$$\rho\left(\sum_{i=0}^{n} c_i x^i\right) = \sqrt{\sum_{i=0}^{n} \frac{c_i^2}{2^{i+1}}}.$$

---

[1] A similar definition applies to complex normed spaces. In this case, the linear space is defined on $\mathbb{C}$ and the absolute value is replaced by the modulus of the complex number. The complex normed space of dimension n is denoted by $(\mathbb{C}^n, \|.\|)$.

It is immediate to see that $\rho = 0$ only when the function is applied to the zero polynomial (the polynomial having all coefficients set to zero) and that $\rho(a\, p(x)) = |a|\, \rho(p(x))$. Consider two polynomials $p_a(x)$ and $p_b(x)$, then

$$\rho(p_a(x) + p_b(x)) = \sqrt{\sum_{i=0}^{n}(c_{a,i} + c_{b,i})^2/2^{i+1}} \leq$$

$$\sqrt{\sum_{i=0}^{n} c_{a,i}^2/2^{i+1}} + \sqrt{\sum_{i=0}^{n} c_{b,i}^2/2^{i+1}} = \rho(p_a(x)) + \rho(p_b(x)),$$

where n is the order of the largest monomial and the c's for any lacking monomial are set to zero. The inequality comes from a direct application of the triangle inequality in Theorem 4.3 to vectors $(c_{a,i}/\sqrt{2^{i+1}})$ and $(c_{b,i}/\sqrt{2^{i+1}})$. Thus, we can conclude that (V, ρ) is an infinite-dimensional normed space.

4.2.2 p-Norm in $\mathbb{R}^n$

The *p-norm* of a vector $\mathbf{x} \in \mathbb{R}^n$, denoted as $\|\mathbf{x}\|_p$, is defined by

$$\|\mathbf{x}\|_p = \left(\sum_{i=1}^{n} |x_i|^p\right)^{1/p}, \quad \forall p \geq 1.$$

In the case of $p = 2$, it reduces to the Euclidean norm discussed in Sect. 4.2.1. The expression above satisfies the criteria for separate points and positive homogeneity required by Definition 4.1. To prove that it is a norm, one has to show that it also satisfies the triangle inequality. For this purpose, we need a preliminary result which can be easily proved using the Young inequality from Theorem 1.15.

Theorem 4.4 (Holder's inequality for sums) *For any two positive numbers p and q such that $1/p + 1/q = 1$ and $\forall \mathbf{x}, \mathbf{y} \in \mathbb{R}^n$, $\sum_{i=1}^{n} |x_i\, y_i| \leq \|\mathbf{x}\|_p \|\mathbf{y}\|_q$.*

Proof Define $\mathbf{x}' = \mathbf{x}/\|\mathbf{x}\|_p$ and $\mathbf{y}' = \mathbf{y}/\|\mathbf{y}\|_q$. For the Young inequality, Theorem 1.15, applied to each couple x_i' and y_i', and summing on all components,

$$\sum_{i=1}^{n} |x_i'\, y_i'| \leq \frac{1}{p}\sum_{i=1}^{n} |x_i'|^p + \frac{1}{q}\sum_{i=1}^{n} |y_i'|^q = \frac{\|\mathbf{x}'\|_p^p}{p} + \frac{\|\mathbf{y}'\|_q^q}{q}.$$

By construction, the norms on the right-hand side are both equal to 1, such that $\sum_{i=1}^{n} |x_i'\, y_i'| \leq 1$. Substituting the definitions of $\mathbf{x}'$ and $\mathbf{y}'$, the statement is proved. □

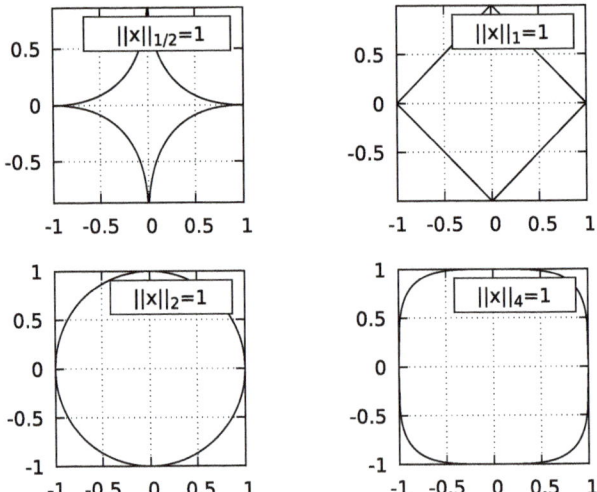

Fig. 4.1 Unit radius circle in $\mathbb{R}^2$ for the p-norm expression with different values of p. The case $p = 1/2$ is not a norm

Using the previous result, one can prove the triangle inequality for the p-norm, which takes its name from the Lithuanian mathematician Hermann Minkowski (1864–1909) (Fig. 4.1).

Theorem 4.5 (Minkowski inequality) *For any two vectors* $\mathbf{x}, \mathbf{y} \in \mathbb{R}^n$ *and any* $p \geq 1$, $\|\mathbf{x} + \mathbf{y}\|_p \leq \|\mathbf{x}\|_p + \|\mathbf{y}\|_p$.

Proof For $p = 1$, the theorem follows directly from the properties of the absolute value. Consider $p > 1$. By taking the p power of the p-norm, one has

$$\|\mathbf{x} + \mathbf{y}\|_p^p = \sum_{i=1}^n |x_i + y_i|^p \leq \sum_{i=1}^n (|x_i| + |y_i|) \, |x_i + y_i|^{p-1}.$$

At the same time, using Holder's inequality in Theorem 4.4 with exponents p and $p/(p - 1)$,

$$\sum_{i=1}^n |x_i| |x_i + y_i|^{p-1} \leq \left(\sum_{i=1}^n |x_i|^p \right)^{\frac{1}{p}} \left(\sum_{i=1}^n |x_i + y_i|^{(p-1)\frac{p}{p-1}} \right)^{\frac{p-1}{p}}$$

$$\left(\sum_{i=1}^n |x_i|^p \right)^{\frac{1}{p}} \left(\sum_{i=1}^n |x_i + y_i|^p \right)^{\frac{1}{p}(p-1)} = \|\mathbf{x}\|_p \|\mathbf{x} + \mathbf{y}\|_p^{p-1}.$$

Summing the last inequality and the same inequality with x_i and y_i exchanged, one gets

$$\sum_i (|x_i| + |y_i|) |x_i + y_i|^{p-1} \le (\|\mathbf{x}\|_p + \|\mathbf{y}\|_p) \|\mathbf{x} + \mathbf{y}\|_p^{p-1}$$

which, using the first inequality above, proves the assertion.   □

▶ **Example 4.3** *(Normed space of bounded functions)* Consider the set of bounded functions $B(A)$ defined in Example 3.3. This set is closed under sum of functions and multiplication by a constant, thus it is a linear space. Now consider the function $\|.\|_\infty : B(A) \to \mathbb{R}_{\ge 0}$ defined as

$$\|f\|_\infty = \sup_{a \in A} \{|f(a)|\}.$$

Because f is bounded and because the supremum exists for any bounded set, the previous function is well defined. It is easy to show that it satisfies all the requirements of Definition 4.1. Thus, $(B(A), \|.\|_\infty)$ is a normed space. The norm $\|.\|_\infty$ is generally known as the *uniform norm* or *sup-norm*. This norm is what was used in Example 3.3 to define a distance on $B(A)$. If the set A is compact, the supremum in the definition of the uniform norm can be replaced with the maximum; see Theorem 2.30. The symbol ∞ in the notation for the uniform norm comes from its relation with the p-norm; see Exercise 4.8.

4.2.3 Operator Norm

Linear functions connecting two normed spaces are themselves a normed space. To see how, we start by introducing the notion of a bounded operator. Consider two normed spaces (V, ρ) and (V', ρ').

Definition 4.3 *(Bounded operator)* A function or *operator* $f : V \to V'$ is bounded if $\exists c > 0$ such that $\rho'(f(\mathbf{x})) \le c\rho(\mathbf{x})$ for all $\mathbf{x} \in V$.

The constant c is a sort of upper bound on how much the operator f can increase the norm of a vector in V. A bounded operator sends any bounded set of V to a bounded set of V'. In fact, if $A \subset V$ is bounded and $a = \sup\{\rho(\mathbf{x}) \mid \mathbf{x} \in A\}$, then $\sup\{\rho'(\mathbf{x}') \mid \mathbf{x}' \in f(A)\} \le ca$. In general, which operators are bounded and which are not depends on the norms one chooses on the two spaces V and V'. Now, let $L(V, V')$ be the space of linear bounded operators from V to V'. This is clearly a linear space. In this space, consider the following.

Definition 4.4 *(Operator norm)* Let $f \in L(V, V')$ and define

$$\|f\|_{op} = \sup\{\rho'(f(\mathbf{x})) \mid \rho(\mathbf{x}) = 1\},$$

where ρ and ρ' are the norms for V and V', respectively.

In other terms, the norm of a bounded linear operator is the maximal norm of the images of all vectors with a norm equal to one. Alternative definitions are given in Exercises 4.14 and 4.15. It is easy to see that the function in Definition 4.4 is a norm. Clearly $\|f\|_{op} = 0$ if and only if f is the zero map and $\|a\,f\|_{op} = |a|\,\|f\|_{op}$. The triangle inequality follows directly from the properties of ρ'. Indeed, given $f, g \in L(V, V')$, $\rho'(f(\mathbf{x}) + g(\mathbf{x})) \leq \rho'(f(\mathbf{x})) + \rho'(g(\mathbf{x}))$ and for the property of the supremum

$$\sup \{\rho'(f(\mathbf{x})) + \rho'(g(\mathbf{x})) \mid \rho(\mathbf{x}) = 1\} \leq$$
$$\sup \{\rho'(f(\mathbf{x})) \mid \rho(\mathbf{x}) = 1\} + \sup \{\rho'(g(\mathbf{x})) \mid \rho(\mathbf{x}) = 1\},$$

so that $\|f + g\|_{op} \leq \|f\|_{op} + \|g\|_{op}$.

Given the definition of norm operator, it is immediate to see that for all $\mathbf{x} \in V$, $\rho'(f(\mathbf{x})) \leq \|f\|_{op}\,\rho(\mathbf{x})$. In fact, $\|f\|_{op}$ is the lowest among all possible values c that satisfy Definition 4.3 for a given f. Thus, given $f \in L(V, V')$ and $g \in L(V', V'')$, $\forall \mathbf{x} \in V$, $\rho''(g(f(\mathbf{x}))) \leq \|g\|_{op}\rho'(f(\mathbf{x})) \leq \|g\|_{op}\|f\|_{op}\rho(\mathbf{x})$. Thus, the operator norm is submultiplicative with respect to the composition of functions $\|g \circ f\|_{op} \leq \|g\|_{op}\|f\|_{op}$. In particular, because the operator norm of the identity is 1, if the function f has the inverse, $\|f^{-1}\|_{op} \geq 1/\|f\|_{op}$.

If the space V' has dimension m and the space V has dimension n, then all linear maps $f : V \to V'$ are bounded and the space $L(V, V')$ is simply the space of $m \times n$ real matrices $M_{m,n}$. The matrix $A \in M_{m,n}$ is associated with the mapping from $\mathbb{R}^n$ to $\mathbb{R}^m$, $\mathbf{x} \to A\,\mathbf{x}$, where the last expression denotes the multiplication of rows by columns. Thus, by the definition of the operator norm, for any matrix A and any vector $\mathbf{x} \neq 0$, we have $\|A\,\mathbf{x}\| \leq \|A\|_{op}\,\|\mathbf{x}\|$. The operator norm cannot be lower than the largest eigenvalue of A. If the matrix A can be diagonalized, they are equal. The next result clarifies that the space of bounded linear operators is also the space of continuous linear operators. This implies that any one-to-one bounded linear function between two normed spaces is a homeomorphism.

Theorem 4.6 *Any linear operator between two normed spaces is bounded if and only if it is continuous.*

Proof Let $f : V \to V'$ be a linear function from the normed space (V, ρ) to the normed space (V', ρ'). If f is the zero map, the theorem directly follows from Example 2.16. We will assume that for some $\mathbf{x} \in V$, $f(\mathbf{x}) \neq \mathbf{0}$.

Assume that f is bounded and $c > 0$ is an upper bound according to Definition 4.3. Take any open set $A' \subseteq f(V)$. For any $\mathbf{y} \in A'$ let $\epsilon(\mathbf{y}) > 0$ be such that $B(\mathbf{y}, \epsilon(\mathbf{y})) \subseteq A'$ and consider a point $\mathbf{x_y} \in f^{-1}(\mathbf{y})$. Note that, by definition, $\forall \mathbf{z} \in V$ such that $\rho(\mathbf{x_y} - \mathbf{z}) < \epsilon(\mathbf{y})/c$, $\rho'(f(\mathbf{x_y}) - f(\mathbf{z})) < \epsilon(\mathbf{y})$, thus $f(B(\mathbf{x_y}, \epsilon(\mathbf{y})/c)) \subseteq B(\mathbf{y}, \epsilon(\mathbf{y}))$. It follows that $f^{-1}(A') = \cup_{\mathbf{y} \in A'} B(\mathbf{x_y}, \epsilon(\mathbf{y})/c)$. Thus, $f^{-1}(A')$ is open, which proves that f is continuous.

Assume, conversely, that f is continuous. Let $\mathbf{0}$ be the zero vector. Since f is linear, $\mathbf{0} \in f^{-1}(\mathbf{0})$ and $\exists \delta$ such that $f(B(\mathbf{0}, \delta)) \subseteq B(\mathbf{0}, 1)$ and also $\bar{B}(\mathbf{0}, \delta) \subseteq f^{-1}(\bar{B}(\mathbf{0}, 1))$ (see Exercise 2.24). In other terms, $\rho'(f(\mathbf{x})) \le 1$ if $\rho(\mathbf{x}) \le \delta$. Since for any vector $\mathbf{z} \in V, \mathbf{z} \ne 0, \rho(\mathbf{z}\delta/\rho(\mathbf{z})) = \delta$, we also have $\rho'(f(\mathbf{z}\delta/\rho(\mathbf{z}))) \le 1$. This implies that

$$\rho'(f(\mathbf{z})) = \frac{\rho(\mathbf{z})}{\delta} \rho'\left(f\left(\frac{\mathbf{z}\delta}{\rho(\mathbf{z})}\right)\right) \le \frac{\rho(\mathbf{z})}{\delta},$$

and $1/\delta$ is the constant c of Definition 4.3. $\qquad\square$

4.3 Finite-Dimensional Normed Spaces

In this section, we review some fundamental results of normed spaces with finite dimension. They will play an essential role in the study of real functions of one or many variables.

Theorem 4.7 *Any norm topology on $\mathbb{R}^n$ is second-countable.*

Proof The norm topology is generated from the base of open balls $\mathbb{B}$; see Definition 3.5. Since $\mathbb{Q}$ is dense in $\mathbb{R}$, using Theorem 2.6, it is straightforward to show that the same topology can be generated by the set $\mathbb{B}'$ of open balls having rational radius and a centre with rational components. The set $\mathbb{B}'$ is one-to-one with a subset of $\mathbb{Q}^{n+1}$. Therefore, it is countable. $\qquad\square$

We can extend to finite-dimensional normed spaces a useful result that we have already proved in Theorem 2.9 for the Euclidean topology on $\mathbb{R}$.

Theorem 4.8 *Any open set in $(\mathbb{R}^n, \rho)$ can be written as the countable union of disjoint open balls.*

Proof Let $A \subseteq \mathbb{R}^n$ be open. Take $\mathbf{x} \in A$ and let $B(\mathbf{x}, \rho_x)$ be the largest ball such that $B(\mathbf{x}, \rho_x) \subseteq A$. Next take $\mathbf{y} \in A \setminus B(\mathbf{x}, \rho_x)$ and let $B(\mathbf{y}, \rho_y)$ be the largest ball such that $B(\mathbf{y}, \rho_y) \subseteq A \setminus B(\mathbf{x}, \rho_x)$. Clearly $B(\mathbf{x}, \rho_x) \cap B(\mathbf{y}, \rho_y) = \emptyset$. Repeat this procedure to find a cover $A = \cup_\alpha B(\mathbf{x}_\alpha, \rho_\alpha)$. These balls are all disjoint. In each ball there is a different element of $\mathbb{Q}^n$. Because this set is countable, the number of balls cannot be more than countable. $\qquad\square$

Since any finite-dimensional normed space $(\mathbb{R}^n, \rho)$ is Hausdorff (Theorem 3.1) and second-countable (Theorem 4.7) according to Theorem 2.16, we have the following.

Corollary 4.2 (*BolzanoWeierstrass*) *In the topology induced by the norm on finite-dimensional real linear spaces* $(\mathbb{R}^n, \rho)$, *a set K is compact if and only if any infinite subset of K has at least one limit point and all its limit points belong to K.*

In a generic Hausdorff space, we cannot conclude that a bounded and closed set is compact (see Example 2.12). Indeed, for this to be true, one needs an additional property. To discuss this property, we introduce a special kind of open sets.

Definition 4.5 (*n-cell*) In $\mathbb{R}^n$ an *n-cell* I is defined by n couples (a_j, b_j) with $a_j < b_j$, $j = 1, \ldots, n$, as $I = \{\mathbf{x} \in \mathbb{R}^n \mid a_j \leq x_j \leq b_j\}$ where x_j denotes the jth component of the element $\mathbf{x} \in \mathbb{R}^n$.

A 1-cell is a closed interval in $\mathbb{R}$ and a 2-cell is a rectangle in $\mathbb{R}^2$, including its boundaries. Alternatively, we can write the n-cell as a Cartesian product $I = \times_{k=1}^n [a_k, b_k]$. The special property that characterises the n-cell is the following.

Lemma 4.2 *Consider a nested sequence of n-cells in $\mathbb{R}^n$, that is a sequence (I^k) such that $\forall k$, $I^{k+1} \subseteq I^k$. Then $\cap_k I^k$ is not empty.*

Proof The jth components of the n-cells in the set (I^k) form a sequence of nested intervals $[a_j^k, b_j^k]$ whose intersection, according to Theorem 1.8, is not empty. Let $x_j^* \in \cap_k [a_j^k, b_j^k]$. Repeat the same procedure for any j. Then the vector $\mathbf{x}^*$ with components $(x_1^*, \ldots, x_n^*)$ belongs to $\cap_k I^k$. □

The previous result, which generalises Theorem 1.8, is due to the existence of supremum and infimum for real bounded sets. It has an important consequence.

Theorem 4.9 *Any n-cell I of the Euclidean normed space $(\mathbb{R}^n, \|\cdot\|)$ is compact.*

Proof We prove the statement by contradiction. Consider an n-cell I and assume there exists a cover $C \subseteq \cup_\alpha V_\alpha$ from which no finite subcover can be extracted. Now divide the cell I into 2^n smaller cells, considering the midpoint along any dimension $c_j' = (a_j + b_j)/2$. Among all the cells $\times_{1=j}^n ([a_j', b_j'])$ with $[a_j', b_j'] = [a_j, c_j']$ or $[a_j', b_j'] = [c_j', b_j]$, at least one is not covered by a finite subcover, or otherwise I would be. Let this cell be I^2. Repeat the splitting procedure again for I^2 to find an n-cell $I^3 \subset I^2$ which is not covered by a finite subcover. Iterating this procedure, we obtain a sequence of nested n-cells (I^k) such that $I^{k+1} \subset I^k$. All these sets have the property that no finite subset of $\{V_\alpha\}$ is sufficient to cover them. However, according to Lemma 4.2, $\exists \mathbf{x}^* \in \cap_k I^k$ and a α^* such that $x^* \in V_{\alpha^*}$. Since V_{α^*} is open, $\exists \rho^* > 0$ such that $B(x^*, \rho^*) \subseteq V_{\alpha^*}$.

The length of the jth side of I^k is $(b_j - a_j)/2^k$. Define $\delta = \sup_j \{b_j - a_j\}$ so that $\forall \mathbf{x}, \mathbf{y} \in I^k$, $\|\mathbf{x} - \mathbf{y}\| \leq \sum_j |x_j - y_j| \leq n\delta/2^k$. If $h > \log_2 \delta n/\rho^*$, then $n\delta/2^h < \rho^*$, so that if $\mathbf{x} \in I^h$, $\mathbf{x} \in B(x^*, \rho^*)$. In other terms, $I^h \subseteq B(x^*, \rho^*)$. But this is a contradiction, because then a single element of the cover, V_{α^*}, would be

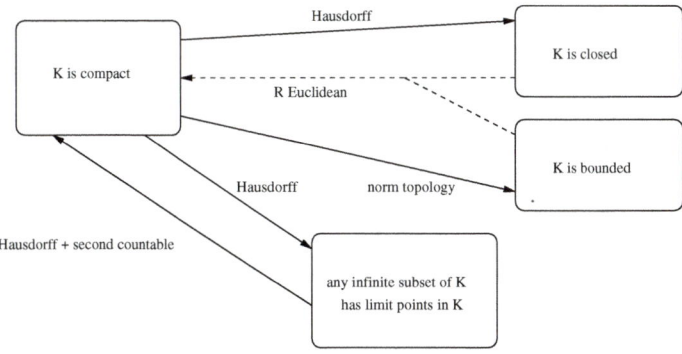

Fig. 4.2 Relationship between compact, closed, and bounded sets. The direction of the arrow identifies logical implication. The dotted line relations are valid for real numbers $\mathbb{R}$ and real linear normed spaces $(\mathbb{R}^n, \rho)$

sufficient to cover I^h. Thus, there are no covers from which a finite subcover cannot be extracted, which proves the statement. □

Theorem 4.9 directly leads to the following result.[2]

Theorem 4.10 (Heine–Borel) *In* $(\mathbb{R}^n, || \cdot ||)$, *a set is compact if and only if it is bounded and closed.*

Proof Assume that the set K is bounded and closed. Since it is bounded, there exists a n-cell I such that $K \subseteq I$. Then K is a closed subset of a compact set, and by Corollary 2.1, it is compact.

Suppose instead that K is compact. Then, because the Euclidean topology on $\mathbb{R}^n$ is Hausdorff, by Theorem 2.12 the set K is closed and by Theorem 3.2 it is bounded. □

Theorems 4.9 and 4.10 were proved only for a specific norm, the Euclidean norm. In the next section, we will see that these results can be generalised to any norm on $\mathbb{R}^n$ (Fig. 4.2).

► **Example 4.4** *(Homeomorphism of convex sets in $\mathbb{R}^n$)* We will prove that any compact convex set in $\mathbb{R}^n$ with a nonempty interior is homeomorphic (see Definition 2.17) to the closed ball of unit radius $\bar{B}_n(\mathbf{0}, 1)$. Let $A \subset \mathbb{R}^n$ be compact and convex and

---

[2] A more general version of the Heine–Borel theorem would require the notion of product topology and the Tychonoff theorem. The advantage of using the Tychonoff theorem is that the result can be directly extended to infinite-dimensional normed spaces. The disadvantage is that the basic treatment of the subject would require several additional definitions. In any case, the derivation below, which is valid *only* for finite-dimensional spaces, is sufficient for our purposes.

assume that its interior is not empty. Without loss of generality, we can assume that $\mathbf{0} \in A^0$. Then $\exists r > 0$ such that $\mathbf{0} \in B_n(\mathbf{0}, r) \subset A^0$. For any $\mathbf{x} \in \mathbb{R}^n$ define $\eta(\mathbf{x}) = \inf\{t > 0 \mid \mathbf{x}/t \in A\}$. Note that $\eta(\mathbf{0}) = 0$ and $\eta(c\,\mathbf{x}) = c\,\eta(\mathbf{x})$. Because A is bounded, $\eta(\mathbf{x})$ is strictly positive $\forall \mathbf{x} \neq \mathbf{0}$ and, moreover, $\eta(\mathbf{x}) < \|\mathbf{x}\|/r$. For any $l > \eta(\mathbf{x})$, $\mathbf{x}/l \in A^0$ and $\mathbf{x}/\eta(\mathbf{x}) \in \partial A$.

Due to the convexity of A, any combination $\lambda\,\mathbf{x}/\eta(\mathbf{x}) + (1-\lambda)\,\mathbf{y}/\eta(\mathbf{y})$ for $\lambda \in [0, 1]$ belongs to A. In particular, if one sets $\lambda = \eta(\mathbf{x})/(\eta(\mathbf{x})+\eta(\mathbf{y}))$, $(\mathbf{x}+\mathbf{y})/(\eta(\mathbf{x})+\eta(\mathbf{y})) \in A$ so that $\eta(\mathbf{x} + \mathbf{y}) \leq \eta(\mathbf{x}) + \eta(\mathbf{y})$. This implies that the function η is convex and, hence, according to Theorem 2.31, continuous.

Now consider the function from the set A to the unit closed ball $\phi : A \to \bar{B}_n(\mathbf{0}, 1)$ defined as

$$\phi(\mathbf{x}) = \begin{cases} \mathbf{x}\,\eta(\mathbf{x})/\|\mathbf{x}\| & \text{if } \mathbf{x} \neq \mathbf{0}, \\ 0 & \text{if } \mathbf{x} = \mathbf{0}. \end{cases}$$

If $\mathbf{x} \in \partial A$, then $\eta(\mathbf{x}) = 1$ and $\phi(\mathbf{x}) \in \partial \bar{B}_n(\mathbf{0}, 1)$. If $\mathbf{x} \in A^0$, then $\eta(\mathbf{x}) < 1$ and $\phi(\mathbf{x}) \in B_n(\mathbf{0}, 1)$. For $\mathbf{x} \neq \mathbf{0}$ the function ϕ is the composition of continuous functions. When $\mathbf{x} \to 0$, the quantity $\eta(\mathbf{x})/\|\mathbf{x}\| = \eta(\mathbf{x}/\|\mathbf{x}\|)$ remains bounded so that $\lim_{\mathbf{x}\to 0} \phi(\mathbf{x}) = \mathbf{0}$. We can conclude that the function ϕ is continuous.

Now consider the function $\phi' : \bar{B}_n(\mathbf{0}, 1) \to A$ defined as

$$\phi'(\mathbf{x}) = \begin{cases} \mathbf{x}\,\|\mathbf{x}\|/\eta(\mathbf{x}) & \text{if } \mathbf{x} \neq \mathbf{0}, \\ 0 & \text{if } \mathbf{x} = \mathbf{0}. \end{cases}$$

Again, we have $\lim_{\mathbf{x}\to 0} \phi'(\mathbf{x}) = \mathbf{0}$, so that the function ϕ' is continuous. Since $\phi' \circ \phi$ is the identity in A, this proves the assertion. The function ϕ maps A^0 to B_n. Therefore, we have also shown that any open bounded convex set in $\mathbb{R}^n$ is homeomorphic to the open ball with unit radius.

4.3.1 Equivalence of Norms in $\mathbb{R}^n$

Any function that satisfies the properties of Definition 4.1 can be used to define a norm in a linear space V. The question arises whether different norms might induce radically different properties on the space itself. For instance, could it be that a set is bounded according to a given norm and not bounded with respect to another? We will see that as long as the space V has a finite dimension, this is not the case. To proceed, we need an operational definition of the equivalence between norms.

Definition 4.6 Consider a linear space V and two norms $\rho : V \to \mathbb{R}$ and $\rho' : V \to \mathbb{R}$. The two norms are *equivalent* if there exist two positive constants c and C such that $\forall \mathbf{x} \in V$, $c\rho(\mathbf{x}) \leq \rho'(\mathbf{x}) \leq C\rho(\mathbf{x})$.

Note that if c and C exist, then $\rho'(\mathbf{x})/C \leq \rho(\mathbf{x}) \leq \rho'(\mathbf{x})/c$, so the equivalence of the norms is a symmetric relation. One can easily prove that it is also transitive, so that it can be used to define an equivalence relation on the space of all possible norms. It turns out that, on a finite-dimensional linear space, all norms are equivalent.

Theorem 4.11 (Equivalence of norms) *All norms defined on a finite-dimensional real linear space V are equivalent.*

Proof We will prove the theorem by showing that all norms are equivalent to the Euclidean norm defined over the components with respect to a given basis. Let n be the dimension of V and $(\mathbf{b}_1, \ldots, \mathbf{b}_n)$ a basis. For any $\mathbf{v} \in V$, there is a unique n-tuple of components $\mathbf{x} = (x_1, \ldots, x_n)$ such that $\mathbf{v} = \sum_{i=1}^{n} x_i \, \mathbf{b}_i$. Define the function $\rho_E : V \to \mathbb{R}_{\geq 0}$ as $\rho_E(\mathbf{v}) = \|\mathbf{x}\|$. This function is simply the Euclidean norm defined over the space of components. It is immediate to see that it defines a norm on V. We will show that any other norm ρ on V is equivalent to ρ_E. For the triangle and Cauchy–Schwartz inequalities (see Theorem 4.2),

$$\rho(\mathbf{v}) = \rho \left(\sum_{i=1}^{n} x_i \, \mathbf{b}_i \right) \leq \sum_{i=1}^{n} |x_i| \rho(\mathbf{b}_i) \leq \|\mathbf{x}\| \, \|\mathbf{b}\|,$$

where $\mathbf{b} = (\rho'(\mathbf{b}_1), \ldots, \rho'(\mathbf{b}_n))$. Thus, setting $C = \|\mathbf{b}\|$, $\rho(\mathbf{v}) \leq C \, \rho_E(\mathbf{v})$ for any $\mathbf{v}$.

Consider now the set $S = \{\mathbf{x} \mid \|\mathbf{x}\| = 1\} \subseteq \mathbb{R}^n$ and define the function $f(\mathbf{x}) = \rho \left(\sum_{i=1}^{n} x_i \, \mathbf{b}_i \right)$ from S to $\mathbb{R}_{\geq 0}$. The set S is just the boundary of the Euclidean open ball with centre $\mathbf{0}$ and radius 1. Thus, it is a closed set. Since it is also bounded, Theorem 4.10 implies that it is compact. The function f is the composition of two functions: the first function maps the n-tuple of real numbers into a vector of V, and the second function is just the norm ρ on V. The first function is bounded as $\rho(\sum_{i=1}^{n} x_i \, \mathbf{b}_i) \leq \|\mathbf{b}\| \|\mathbf{x}\|$; hence, being linear, it is continuous by Theorem 4.6. The second function is continuous by definition (see Example 4.1). Thus, the function f, being the composition of two continuous functions, is continuous and it reaches its minimum in S at a point $\mathbf{x}' \in S$ associated to a vector $\mathbf{v}' \in V$, with $\mathbf{v}' = \sum_{i=1}^{n} x_i' \, \mathbf{b}_i$. Let $\rho(\mathbf{v}') = c$. Thus, $\forall \mathbf{v} \in V$,

$$\rho(\mathbf{v}) = \rho \left(\sum_{i=1}^{n} x_i \, \mathbf{b}_i \right) = \|\mathbf{x}\| \rho \left(\sum_{i=1}^{n} \frac{x_i}{\|\mathbf{x}\|} \mathbf{b}_i \right) \geq c \, \|\mathbf{x}\| = c \, \rho_e(\mathbf{v}),$$

and the statement is proved. □

The procedure adopted in Theorem 3.1 to generate a topology gives precisely the same result when using distance functions derived from two equivalent norms (see Exercise 4.20). This is straightforward to prove using Theorem 2.6. Thus, we can conclude that all norms induce precisely the same topology on a finite-dimensional normed space.[3] For example, the Euclidean topology on $\mathbb{R}^n$ can be generated using n-cells (hyperrectangles) as a base; see Exercise 4.13. A direct consequence of Theorem 4.11 is the generalisation of Theorem 4.10.

Corollary 4.3 *(Heine–Borel) In any finite-dimensional normed space (V, ρ), a set is compact if and only if it is bounded and closed.*

This immediately implies the following.

Theorem 4.12 (Nearest point) *Let $\mathbf{x} \in \mathbb{R}^n$ and consider a subset $A \subseteq \mathbb{R}^n$ such that $\mathbf{x} \notin A$. Then if A is closed, $\exists \mathbf{y} \in A$ such that $d(\mathbf{x}, \mathbf{y}) = d(\mathbf{x}, A)$. $\mathbf{y}$ is the nearest point of the set A to the point x.*

Proof Pick an element $\mathbf{a} \in A$. If $d(\mathbf{x}, \mathbf{a}) = d(\mathbf{x}, A)$, we are done. Otherwise, consider the closed ball $\bar{B}(\mathbf{x}, d(\mathbf{x}, \mathbf{a}))$. Since the closed ball is closed and bounded, it is compact. Thus, by Corollary 2.1, its intersection with A is compact, so that the function $f(\mathbf{z}) = d(\mathbf{z}, \mathbf{x})$ has a minimum $\mathbf{y}$ in this intersection (see Theorem 2.30). Since no points of A can be closer to $\mathbf{x}$ than the point $\mathbf{y}$, the statement is proved. $\square$

The statement above cannot be generalised to any metric space. For example, the set of rational numbers $I = \{x \in \mathbb{Q} \mid 3 < x^2 < 5\}$ is closed and its distance from the origin is $\sqrt{3}$, but any point in the set is more distant from the origin than that. More generally, Corollary 4.3 has been proven only for finite-dimensional normed spaces. When infinite-dimensional spaces are considered, one has to check this property on a case-by-case basis.

Exercises

Exercise 4.1 Prove that in any normed space, Definition 4.2 of bounded sets is equivalent to Definition 3.2.

Exercise 4.2 In the normed space $(\mathbb{R}^n, \rho)$ with a generic norm ρ, prove that the boundary of the open ball $B(\mathbf{x}, r)$ is made of all points $\mathbf{y}$ such that $\rho(\mathbf{x} - \mathbf{y}) = r$.

Exercise 4.3 Given a normed space (V, ρ), let $\bar{\mathbb{B}}$ be the set of all closed balls $\bar{B}(\mathbf{x}, r) = \{v \in V \mid \rho(v - x) \le r\}$ with $r > 0$ and their interior (the open balls). Prove that $\mathbb{B}$ is the base of a topology. Is the resulting topological space a Hausdorff space? Characterise compact sets and connected subspaces in such a topological space. *Hint: The topology generated by $\mathbb{B}$ is peculiar but not unheard of.*

---

[3] The statement is easily extendable to finite-dimensional normed spaces defined over the field of complex numbers. In the proof above, the Euclidean norm is replaced with the norm defined using the complex conjugate $\|\mathbf{x}\| = \sqrt{\sum_{h=1}^{n} x_h x_h^*}$.

Exercise 4.4 Prove that Theorem 4.1 is valid for maxima and minima if the function f is strictly concave or strictly convex.

Exercise 4.5 In $(\mathbb{R}^n, ||.||)$, prove that any open ball $B(\mathbf{x}, r)$ and its closure $\bar{B}(\mathbf{x}, r)$ are convex sets, while the boundary of the ball is not convex.

Exercise 4.6 Let V be a linear space of dimension n and consider a basis $(\mathbf{b}_1, \ldots, \mathbf{b}_n)$. For any $\mathbf{v} \in V$ with $\mathbf{v} = \sum_{i=1}^{N} x_i \, \mathbf{b}_i$, define the function $\rho_E(\mathbf{v}) = ||\mathbf{x}||$. Prove that this function is a norm. *Hint: Use the linear properties of the representation of vectors on a basis.*

Exercise 4.7 (*open n-cell*) Prove that in $(\mathbb{R}^n, ||.||)$, for any ball $B(\mathbf{x}, \rho)$ there exists an open n-cell $I = \times_{i=1}^{n}(a_i, b_i)$ with $a_i < b_i$, $\forall i$, such that $\mathbf{x} \in I \subset B(\mathbf{x}, \rho)$. Analogously, prove that for any open n-cell and any $\mathbf{x} \in I$, there exists an open ball $B(\mathbf{x}, \rho)$ such that $B(\mathbf{x}, \rho) \subset I$. Discuss how this result implies that open n-cells are a base of $(\mathbb{R}^n, ||.||)$.

Exercise 4.8 (*Uniform norm*) Define the function $||\mathbf{x}||_\infty = \max\{|x_1|, \ldots, |x_n|\}$, $\forall \mathbf{x} \in \mathbb{R}^n$. Prove that $||.||_\infty$ is a norm.

Exercise 4.9 Prove that $|| \cdot ||_\infty$ defined in Example 4.3 is a norm.

Exercise 4.10 Extending Example 4.3 let A be a generic set, (V, ρ) a normed space, and $B(A)$ the set of bounded functions $f : A \to V$, that is, functions such that $||f||_\infty = \sup_{a \in A}\{\rho(f(x))\} < +\infty$. Prove that $(B(A), ||.||_\infty)$ is a normed space.

Exercise 4.11 Consider the p-norm of Sect. 4.2.2 in $\mathbb{R}^2$ and let $B_p(\mathbf{x}, \rho)$ be the associated ball of centre $\mathbf{x}$ and radius $\rho > 0$. Prove that $B_2(\mathbf{0}, \rho) \subset B_3(\mathbf{0}, \rho)$.

Exercise 4.12 Given $q > p > 1$, prove that $||\mathbf{x}||_p \geq ||\mathbf{x}||_q$. *Hint: Define $y_j = |x_j|/||\mathbf{x}||_q$ and notice that $y_j^p \geq y_j^q$.*

Exercise 4.13 (*n-cell*) With reference to Exercise 4.8, consider the normed space $(\mathbb{R}^n, ||.||_\infty)$. Prove that in this space the ball $B(\mathbf{x}, \delta)$ is in fact the open n-cell $\times_{i=1}^{n}(x_i - \delta, x_i + \delta)$.

Exercise 4.14 Starting from Definition 4.4 prove that for all $\mathbf{x} \in V$, $\rho'(f(\mathbf{x})) \leq ||f||_{op} \, \rho(f(\mathbf{x}))$.

Exercise 4.15 Prove that Definition 4.4 can be replaced with the equivalent definition $||f||_{op} = \sup\{\rho'(f(\mathbf{x})) \mid \rho(\mathbf{x}) \leq 1\}$.

Exercise 4.16 Prove that Definition 4.4 can be replaced with the equivalent definition $||f||_{op} = \sup\{\rho'(f(\mathbf{x}))/\rho(\mathbf{x})) \mid \mathbf{x} \neq 0\}$.

Exercise 4.17 Compute the operator norm of the matrix

$$A = \begin{pmatrix} 1 & 1 \\ 0 & 1 \end{pmatrix}$$

and compare it with the matrix eigenvalues. *Hint: Remember that if $x^2 + y^2 = 1$, then you can substitute $x = \cos\phi$ and $y = \sin\phi$ with $\phi \in [0, 2\pi)$.*

Exercise 4.18 (*Dual norm*) Consider a normed space $(\mathbb{R}^n, \rho)$ and the function $\rho^* : \mathbb{R}^n \to \mathbb{R}_{\geq 0}$, $\rho^*(\mathbf{x}) = \sup\{\sum_{i=1}^n x_i z_i \mid \rho(\mathbf{z}) \leq 1\}$. Prove that ρ^* is a norm. In fact, it is named the *dual norm* of ρ.

Exercise 4.19 Prove that the equivalence of norms in Definition 4.6 is transitive.

Exercise 4.20 Prove that two equivalent norms on a linear space V induce on V the same topology.

Sequences and Series

5

5.1 Sequences in Topological Spaces

A sequence is an ordered collection of elements of a set. Each element is assigned a progressive natural number that represents the relative position of the element in the sequence.

Definition 5.1 *(Sequence)* A *sequence* (x_n) is a map from $\mathbb{N}$ to a set X:

$$f : \mathbb{N} \to X,$$
$$n \to x_n.$$

A sequence is generically denoted by (x_n) or $(x)_n$. The functional dependence can be written in parentheses, for example, $(2n + 1)$ or $(x_n = 2n + 1)$ represents the sequence of odd natural numbers $(3, 5, 7, \ldots)$. Alternatively, we can write $(2n)_{n>3}$ for the sequence of even natural numbers greater than 6, $(8, 10, 12, \ldots)$. The elements of the sequence define the subset $\{x_n\} \subseteq X$. This is the image of the function f in Definition 5.1. It can be finite or countable. For example, if $(x_n = (-1)^n)$, then $\{x_n\} = \{-1, 1\}$. In general, we assume that X has a topological structure.

Definition 5.2 *(Limit of a sequence)* Consider the topological space (X, T) and let $(x_n) \subseteq X$ be a sequence of elements of X. The element $y \in X$ is a *limit* of the sequence (x_n) if $\forall N(y) \in T, \exists n \in \mathbb{N}$ such that $\forall m > n, x_m \in N(y)$.

If y is a limit of the sequence, the sequence *converges* to y. Similarly, a sequence with a limit is said to be *convergent*. We denote the limit of a sequence by writing $\lim_{n \to \infty} x_n = y$ or, for short, $x_n \to y$. Note that Definition 5.2 does not imply or assume that the limit exists or is unique.

▶ **Example 5.1** (*Sequence in the toy topology*) Consider the topology defined in Example 2.1. The sequence $(x_n = a)$ has two limits, a and b. The point b is also the limit of the sequence defined as

$$x_n = \begin{cases} a & \text{if } n \text{ is odd,} \\ b & \text{if } n \text{ is even.} \end{cases}$$

The sequence $(x_n = c)$ has c as its unique limit.

We will say that a given property applies *eventually* or *asymptotically* to the terms of a sequence if there exists an integer n such that it applies to all terms with an index greater than n.

Definition 5.3 (*Constant sequence*) A sequence is a *constant sequence* if it is eventually constant, that is, if it takes a constant value $x_n = x$ starting from some index n.

Since we are exclusively interested in the behaviour of the sequence when n is large, the previous definition is the most appropriate. The range $\{x_n\}$ of a constant sequence contains a finite number of elements. The notion of limit of a sequence differs from the notion of limit point of a set introduced in Chap. 2. In a constant sequence, it may eventually be $x_n = x$, but x can or cannot be a limit point of the set $\{x_n\}$. If a sequence is not constant, then $x_n \to x$ implies that $x \in D\{x_n\}$, that is, the limit of the sequence is a limit point of the range of the sequence.

5.1.1 Subsequences

Starting from a sequence (x_n), we can imagine building a new sequence picking elements with progressively higher indices. Each element of the new sequence is assigned a new increasing index. The new sequence can be denoted as (x_{m_n}) to say that the nth element of the new sequence was the m_nth element of the original sequence. For example, (x_{2n+1}) is the subsequence formed by the elements of the original sequence (x_n) with an odd index. Of course, it must be the case that $\forall n$, $m_{n+1} > m_n$. This procedure is formalised in the following definition.

Definition 5.4 (*Subsequence*) Given a sequence (x_n) consider a strictly increasing function on the natural numbers

$$f : \mathbb{N} \to \mathbb{N},$$
$$i \to n_i,$$

with $n_i > n_j$ if $i > j$. The composition of this function with the function that defines the sequence identifies a new sequence (x_{n_i}) that is a *subsequence* of the original sequence.

The range of the subsequence is a subset of the range of the original sequence. Moreover, if the original sequence has a limit, any subsequence will converge to it.

Theorem 5.1 *Any subsequence of a converging sequence is converging to the limit of the sequence.*

Proof Let $x_n \to y$. Then $\forall N(y) \in T, \exists n \in \mathbb{N}$ such that $\forall m > n, x_m \in N(y)$. Using the notation in Definition 5.4, for any subsequence (x_{n_i}), $\exists i$ such that $n_i > n$. Then, $\forall j > i$, the element x_{n_j} belongs to $N(y)$ so that $x_{n_i} \to y$. □

5.1.2 Sequences and Functions

Let $f : X \to Y$ be a function between two topological spaces and x^* a limit point of the domain of f, $x^* \in D\mathbb{D}_f$. Consider a sequence $(x_n) \subseteq \mathbb{D}_f$, such that $x_n \to x^*$. One can ask what happens to the sequence of images $(f(x_n))$. If $\lim_{x \to x^*} f(x) = y^*$, then $f(x_n) \to y^*$. The limit of the sequence of images can be inferred from the limit of the function. To do the opposite, that is, to infer the limit of the function from the limit of the sequences of images, we need the existence of a countable base of neighbourhoods of x^*.

Theorem 5.2 *Let $f : X \to Y$ be a function between two topological spaces (X, T_X) and (Y, T_Y) and consider $x^* \in D\mathbb{D}_f$. Assume (X, T_X) is first-countable. Then $\lim_{x \to x^*} f(x) = y^*$ if and only if $\forall (x_n) \subseteq \mathbb{D}_f$ such that $x_n \to x^*$, it is $f(x_n) \to y^*$.*

Proof First, we will prove that the limit of the function is the limit of the sequence of images. This proof does not require any assumption on the topology. Suppose that $\lim_{x \to x^*} f(x) = y^*$. Therefore, $\forall N(y^*), \exists N(x^*)$ such that $f(N(x^*)) \subseteq N(y^*)$. Now since $x_n \to x^*, \exists n$ such that $\forall k > n, x_k \in N(x^*)$, then $\forall k > n, f(x_k) \in N(y^*)$.

To prove the statement in the other direction, we will show that if $\lim_{x \to x^*} f(x) \neq y^*$, then one can build a sequence (x_n) such that $x_n \to x^*$ but $f(x_n) \not\to y^*$. By assumption, $\exists N(y^*)$, such that $\forall N(x^*)$, $f(N(x^*)) \not\subseteq N(y^*)$. Now since the space is first-countable, consider the countable local base $\{B_k\}$ that generates all neighbourhoods of x^*, and consider $N_k(x^*) = B_1 \cap B_2 \cap ... \cap B_k$. Let $x_k \in N_k(x^*)$ and $f(x_k) \notin N(y^*)$. By the definition of local base, for any $N(x^*), \exists B_k$ such that $B_k \subseteq N(x^*)$, all x_h with $h \geq k$ belong to $N(x^*)$ and $x_k \to x^*$. But $f(x_k) \not\to y^*$ because none of the x_k belong to $N(y^*)$. □

Using this theorem, we can study the properties of the limit of a function at one point using the behaviour of the images of the sequences converging to it. In practise, it

means that any general property we prove about the limit of sequences also applies
to the limit of functions.

Sequences can be defined in a recursive fashion. Given a topological space X,
consider a function $f : X \to X$ and a starting element x_0. Then we can consider
the sequence $(x_n = f(x_{n-1}))$. When it is important to stress the initial point, this
sequence can be denoted as $(f^n x_0)$ where f^n represents the recursive composition
of the function f taken n times. Assume that $x_n \to x^*$ and that $\lim_{x \to x^*} f(x) =
f^*$. Because the sequence of images under f is a subsequence $(f(x_n) = x_{n+1})$,
Theorem 5.2 implies $f^* = x^*$. If the function f is continuous, then the limit x^*
solves $f(x^*) = x^*$. The point x^* is a *fixed point* of the function f. It is important
to remember that even if the equation $f(x) = x$ admits one or more solutions, that
is, even if the function f has one or more fixed points, we cannot conclude that any
sequence $(f^n x_0)$ is convergent.

5.1.3 Uniqueness of Limit

The notion of the limit of a sequence acquires a much stronger characterisation when
Hausdorff spaces are considered.

Theorem 5.3 *Consider the sequence (x_n) in a Hausdorff topological space. Then,
if the sequence has a limit, this limit is unique.*

Proof Assume that there are two distinct limits y and y'. Since the space is Hausdorff,
there are two neighbourhoods $N(y)$ and $N(y')$ such that $N(y) \cap N(y') = \emptyset$. This
leads to a contradiction, since both neighbourhoods should eventually contain all
elements of the sequence. $\square$

In a Hausdorff space, if $x_n \to y$ and the sequence is nonconstant, then the derivative
set of its image is the singlet $D\{x_n\} = \{y\}$. The opposite is not true, as the following
example shows.

▶ **Example 5.2** *(Limit points of a sequence with no limit)* Consider the Euclidean
topology on $\mathbb{R}$ and the sequence

$$x_n = \begin{cases} \frac{1}{n} & \text{if } n \text{ is even,} \\ n & \text{if } n \text{ is odd.} \end{cases}$$

$D\{x_n\} = \{0\}$, but the sequence has no limit.

5.2 Sequences in Metric Spaces

Any metric space (X, d) is equipped with a topology induced by the distance function. Since the induced topology is Hausdorff, if the limit of a sequence exists, it is unique. For simplicity, the definition of the limit of a sequence is restated using open balls, the natural base of the topology, instead of generic neighbourhoods.

Definition 5.5 Consider the metric space (X, d) and let $(x_n) \subseteq X$ be a sequence of elements of X, then $x_n \to x \in X$ if $\forall \epsilon > 0, \exists N_\epsilon \in \mathbb{N}$ such that $d(x, x_n) < \epsilon$ if $n > N_\epsilon$.

This definition does not add anything to Definition 5.2. In fact, if Definition 5.2 applies, then Definition 5.5 does so too, because open balls are just particular neighbourhoods. Analogously, if Definition 5.5 applies, since any neighbourhood of y contains an open ball that contains y (see the properties of a base in Sect. 2.2), Definition 5.2 also applies. From Definition 5.5 it immediately follows that a convergent sequence is bounded.

Theorem 5.4 *If* $x_n \to x$ *then* $\{x_n\}$ *is bounded.*

Proof Take $\epsilon > 0$. There exists a N such that $\forall n > N$, it is $d(x, x_n) < \epsilon$, that is, $x_n \in B(x, \epsilon)$. Let M be the maximum among the distances $d(x_h, x)$ with $h \leq n$ and ϵ. Clearly $\forall n, x_n \in B(x, M)$, so that the set $\{x_n\}$ is bounded. $\qquad \square$

In a metric space, given a nonconstant sequence (x_n), if $x \in D\{x_n\}$ then there exists a subsequence that converges to x. To build it, consider the sequence of open balls with radius $1/m$ and centre x. Let x_{n_1} be the first element of the sequence (x_n) belonging to $B(x, 1)$ and x_{n_m} be the first element of the sequence (x_n) with an index greater than n_{m-1} belonging to $B(x, 1/m)$. The sequence x_{n_m} is a subsequence of (x_n). Since open balls are a base of the topology, for any neighbourhood $N(x)$ of x, there exists a ball $B(x, \rho)$ such that $B(x, \rho) \subseteq N(x)$. By construction, $\forall m > 1/\rho$, $x_{n_m} \in B(x, \rho)$, so that $x_{n_m} \to x$.

5.2.1 Cauchy Sequences and Complete Spaces

In a metric space, we can evaluate the distance between any two elements of a sequence. We can then investigate what happens to this distance *asymptotically*, that is, when considering terms as the index increases. Of particular importance are those sequences whose elements are progressively closer together. They take their name from the French mathematician Augustin Louis Cauchy (1789–1857).

Definition 5.6 *(Cauchy sequence)* In a metric space (X, d), the sequence (x_n) is said to be a *Cauchy sequence* if $\forall \epsilon > 0, \exists n_\epsilon \in \mathbb{N}$ such that $\forall n, m > n_\epsilon, d(x_n, x_m) < \epsilon$.

The Cauchy property implies two things: the set of elements of the sequence is bounded, and it has at most one limit point.

Theorem 5.5 *If (x_n) is a Cauchy sequence, then the set $\{x_n\}$ is bounded.*

Proof Since the sequence is Cauchy, $\exists n$ such that $\forall m, l \geq n$, $d(x_m, x_l) < 1$. Set $L = \max\{1, d(x_j, x_n) \mid j = 1, \ldots, n - 1\}$, then, $\forall k$, $x_k \in B(x_n, L)$ and the statement follows.

Theorem 5.6 *If (x_n) is a Cauchy sequence, then the set $\{x_n\}$ cannot have two limit points.*

Proof Assume y and z are two limit points of $\{x_n\}$ and let $\delta = d(y, z)$. For any n, $\exists l, m > n$ such that $x_l \in B(y, \delta/3)$ and $x_m \in B(z, \delta/3)$. From the triangle inequality $d(y, x_l) + d(x_l, x_m) + d(x_m, z) \geq \delta$. Then $d(x_l, x_m) \geq \delta - d(y, x_l) - d(x_m, z) > \delta/3$ so that the sequence cannot be Cauchy. $\square$

It turns out that converging sequences are always Cauchy.

Theorem 5.7 *In any metric space (X, d), every convergent sequence is Cauchy.*

Proof Let $x_n \to y$. For any $\epsilon > 0$, $\exists N_\epsilon$ such that $\forall n > n_\epsilon$, $d(x_n, y) < \epsilon/2$. Then $\forall n, m > n_\epsilon$, using the triangle inequality, $d(x_n, x_m) \leq d(x_n, y) + d(x_m, y) < \epsilon$ and the assertion follows. $\square$

However, in general, the opposite is not true. That is, Cauchy sequences are not always converging sequences. A counterexample is provided below.

▶ **Example 5.3** *(Fibonacci numbers and induction)* Leonardo *filius* Bonacci or Fibonacci was an Italian mathematician of the thirteenth century. Born in Pisa, Fibonacci visited several Mediterranean cities to study Arab and Indian mathematics. He became famous in the West and managed to obtain a permanent salary from the emperor Frederick II just to pursue his study of algebra. In his 1202 book *Liber Abaci*, Fibonacci introduced the sequence (a_n) defined by the recursive relation $a_n = a_{n-1} + a_{n-2}$ with initial values $a_1 = a_2 = 1$. The first elements of the sequence read $(1, 1, 2, 3, 5, 8, 13, 21, \ldots)$. Consider the sequence formed by the quotient of subsequent "Fibonacci numbers" $(f_n = a_{n+1}/a_n)$. I want to prove that (f_n) is a Cauchy sequence. To see it notice that $a_n^2 - a_{n+1}a_{n-1} = \pm 1$. This can be proved by induction: first, notice that this is true for $n = 2$, indeed $a_2^2 - a_3 a_1 = 1 - 2 = -1$; then if we assume that it is true for n, substituting the definition of Fibonacci numbers, we have

$$a_{n+1}^2 - a_{n+2}a_n = a_{n+1}^2 - a_{n+1}a_n - a_n^2 =$$
$$a_{n+1}(a_{n+1} - a_n) - a_n^2 = -(a_n^2 - a_{n+1}a_{n-1}),$$

so that it is also true for $n + 1$. Furthermore, notice that if $n > 5$, then $a_n > n$. This can also be proved by induction. The proof is trivial and is left to the reader (see Exercise 5.4). On the basis of these considerations, one has

$$|f_{n+1} - f_n| = \left|\frac{a_{n+2}a_n - a_{n+1}^2}{a_n a_{n+1}}\right| = \frac{1}{|a_n a_{n+1}|} \leq \frac{1}{n(n+1)},$$

and for the triangle inequality, this implies that

$$|f_{n+h} - f_n| \leq \sum_{j=1}^{h} |f_{n+j} - f_{n+j-1}| \leq \sum_{j=1}^{h} \frac{1}{(n+j)(n+j-1)} =$$

$$\sum_{j=1}^{h} \frac{1}{n+j-1} - \frac{1}{n+j} = \frac{1}{n} - \frac{1}{n+h} \leq \frac{1}{n}$$

which, in turn, implies that if $h, k > n$, $|f_h - f_k| < 1/n$, so that the sequence is Cauchy: for any $\epsilon > 0$ in Definition 5.6, take $n_\epsilon = [1/\epsilon] + 1$, where $[x]$ denotes the integer part of x. The elements of the sequence are rational numbers but the sequence (f_n) does not have any limit in $\mathbb{Q}$. To see it, notice that by applying the recursive formula to the Fibonacci numbers one has $f_{n+1} = 1 + 1/f_n$, so that if $f_n \to \tilde{f}$, it should be $\tilde{f} = 1 + 1/\tilde{f}$. However, no rational number could possibly satisfy this equation. In fact, if $\tilde{f} = p/q$, with p and q mutually prime natural numbers, it should be $p^2 = q(p + q)$, which is absurd because if p has k as a factor, q cannot have k as a factor. In $\mathbb{R}$, the limit of the sequence (f_n) is the positive solution of the quadratic function $\tilde{f}^2 - \tilde{f} - 1$, namely $\tilde{f} = (1 + \sqrt{5})/2$ which is also known as the *golden ratio*. This is a fixed point of the function $g(x) = 1 + 1/x$.

If we consider sequences in a compact subset, Theorem 5.7 is valid in both directions. In fact, a Cauchy sequence in a compact subset has two possibilities: it is constant and then converging, or it has an infinite number of points. In the latter case, according to Theorem 2.15, it has at least one limit point. Since Theorem 5.6 excludes the possibility that it can have more than one limit point, it follows that the set has exactly one limit point, which is precisely the limit of the sequence.

Theorem 5.8 *A Cauchy sequence on a compact set $K \subseteq X$ in (X, d) is convergent.*

Proof Let $(x_n) \subseteq K$ be a Cauchy sequence and consider the sets obtained by dropping the first n elements, $E_n = \{x_{n+1}, x_{n+2}, ..., x_{n+k}, ...\}$. The closure of these sets $\bar{E}_n$ are compact sets because they are closed subsets of a compact set. Moreover, $\bar{E}_{n+1} \subseteq \bar{E}_n$. Since $\bar{E}_n \neq \emptyset \; \forall n$, according to Theorem 2.13 their intersection is not empty. Let $E^* = \cap_n \bar{E}_n \neq \emptyset$. Suppose $y, y' \in E^*$ and let $d(y, y') = \eta > 0$. This implies that $\forall n$, $\text{diam}(E_n) \geq \eta$. But this contradicts the hypothesis, as $\exists n_\eta$ such that $\forall n, m > n_\eta, d(x_n, x_m) < \eta/2$, which implies that $\text{diam}(E_{n_\eta}) \leq \eta/2 < \eta$. Thus η

must be zero and E^* contains only one point. Denote it y^*. Note that $\epsilon > 0$, $\exists n_\epsilon$ such that $\mathrm{diam}(\bar{E}_{n_\epsilon}) < \epsilon$. Thus, $\forall x_n$ such that $n > n_\epsilon$, $d(x_n, y^*) < \epsilon$, which proves the assertion. $\square$

Theorem 5.8 has the following important consequence.

Corollary 5.1 *Any Cauchy sequence in* $(\mathbb{R}, |\cdot|)$ *is convergent.*

Proof Let (x_n) be a Cauchy sequence in $\mathbb{R}$. Since it is bounded, by Theorem 5.5, there exists a bounded and closed interval that contains it. This interval is compact (see Theorem 2.18) and, from Theorem 5.8, this implies that (x_n) is convergent. $\square$

The previous corollary is not true in $\mathbb{Q}$, because the bounded closed intervals in $\mathbb{Q}$ are not compact. Indeed, in $\mathbb{Q}$ we can build a bounded sequence (an infinite set of points) that does not have any limit in $\mathbb{Q}$ because the limit is an irrational number. This was the situation that we encountered in Example 5.3. See Corollary 5.5 for an extension of the same result to $(\mathbb{R}^n, \|\cdot\|)$. Those metric spaces in which the Cauchy property implies convergence are generally more useful. They comply with the natural idea that considering smaller and smaller nested sets whose diameter progressively goes to zero, one is actually identifying a single point belonging to the intersection of all of them. Spaces that do not have this property are perceived as "lacking something".

Definition 5.7 *(Complete space)* A metric space (X, d) is *complete* if every Cauchy sequence is convergent.

By Corollary 5.1, the metric space $(\mathbb{R}, |\cdot|)$ is complete. According to Theorem 5.8, if $K \subseteq X$ is compact, then the subspace (K, d) is complete. In other words, compact subspaces of metric spaces are always complete.

▶ **Example 5.4** *(The metric space of bounded functions is complete)* In the metric space $(B(A), d_\infty)$ defined in Example 3.1, a sequence (f_n) of bounded functions is Cauchy if $\forall \epsilon > 0$, $\exists n_\epsilon$ such that $\forall l, m \geq n_\epsilon$, $d_\infty(f_l, f_m) < \epsilon$, that is,

$$\sup_{a \in A} \{|f_l(a) - f_m(a)|\} < \epsilon.$$

This, in turn, implies that $\forall l, m \geq n_\epsilon$, $|f_l(a) - f_m(a)| < \epsilon$ for any a. Thus, if we assume that (f_n) is Cauchy, $\forall a \in A$, the sequence $(f_n(a))$ is a Cauchy sequence of real numbers. Since $\mathbb{R}$ is complete (see Corollary 5.1), this sequence is convergent. Let $g(a) = \lim_{n \to \infty} f_n(a)$. We can prove that the function g is bounded. For the triangle inequality,

$$|g(a)| \leq d_\infty(g, 0) \leq d_\infty(g, f_{n_\epsilon}) + d_\infty(f_{n_\epsilon}, 0)$$

where 0 is the constant zero function and a a generic point. Since f_{n_ϵ} is bounded, $\exists M > 0$ such that $d_\infty(f_{n_\epsilon}, 0) < M$. Moreover, $\forall a, |f_l(a) - f_{n_\epsilon}(a)| < \epsilon$ when $l \geq n_\epsilon$, so that $d_\infty(f_l, f_{n_\epsilon}) \leq \epsilon$. Since the distance is a continuous function, it is $d_\infty(g, f_{n_\epsilon}) \leq \epsilon$. Thus, $\forall a, |g(a)| \leq \epsilon + M$ and $g \in (B(A), d_\infty)$. Finally, note that $\forall \epsilon > 0$, if $n \geq n_\epsilon$,

$$d_\infty(g, f_n) \leq d_\infty(g, f_{n_\epsilon}) + d(f_n, f_{n_\epsilon}) \leq 2\epsilon,$$

which proves that $f_n \to g$ (in the metric d_∞). Thus $(B(A), d_\infty)$ is a complete metric space. The use of $\mathbb{R}$ as the image space of the functions is not important. The same analysis can be applied to functions with images in a generic complete metric space. For an extension of this result to continuous functions when the image space is $\mathbb{R}$ see Example 5.19. We will return on the problem of sequences of functions in Sect. 5.6.1.

However, it is important to stress that completeness is a specific property of the metric considered and is not associated with the induced topology. In other terms, two metrics can define the same topology, but one is complete and the other is not.

▶ **Example 5.5** *(Completeness and topology)* Let $X = \{1/n \mid n \in \mathbb{N}\}$. The topology induced by the Euclidean distance in X is the discrete topology. Indeed notice that any singlet $\{x\} \subseteq X$ is an open set, because there always exists a Euclidean neighbourhood of x whose intersection with X is just x. However, the set X is not complete with respect to the Euclidean metric, because the sequence $(1/n)$ is a Cauchy sequence in X, but does not have any limit in X.

On the other hand, consider the function $d : X \times X \to \mathbb{R}_{\geq 0}$,

$$d(x, y) = \begin{cases} 0 & \text{if } x = y, \\ 1 & \text{otherwise.} \end{cases}$$

It is immediate to see that this function defines a metric on X and it induces on it the discrete topology. This time, however, the space X is complete. In this metric, any Cauchy sequence is a sequence that is eventually constant, and thus convergent.

Since the topology induced by the metric is first-countable, see Lemma 4.1, the equivalence between the limit of a function in a point and the limit of the image of a sequence converging to that point discussed in Theorem 5.2 is always valid in a metric space. Moreover, it turns out that, in a complete metric space, contractions, as defined in Sect. 3.4, have a single fixed point. This result was attributed to the Polish mathematician Stefan Banach (1892–1945), who first stated it in 1922. The Italian mathematician Renato Caccioppoli (1904–1959) arrived independently at the same conclusion a few years later.

Theorem 5.9 (Banach fixed point theorem) *Let (X, d) be a complete metric space and ϕ a contraction defined over it. Then there exists one and only one* fixed point

$x^* \in X$ such that $\phi(x^*) = x^*$ and any recursive sequence $(\phi^n x_0)$ converge to x^* for any initial $x_0 \in X$.

Proof For any $x_0 \in X$, consider the recursively defined sequence

$$\begin{cases} s_0 = x_0, \\ s_{n+1} = \phi(s_n). \end{cases}$$

Because ϕ is a contraction, $\forall m \geq 1$,

$$d(s_{m+1}, s_m) = d(\phi(s_m), \phi(s_{m-1})) \leq k d(s_m, s_{m-1}) \leq k^m d(s_1, s_0).$$

For any two integers m and n such that $m > n$,

$$d(s_m, s_n) \leq d(s_m, s_{m-1}) + d(s_{m-1}, s_n) \leq \sum_{h=n}^{m-1} d(s_{h+1}, s_h).$$

Then, using the previous result and the expression of the geometric progression in Example 1.5,

$$d(s_m, s_n) \leq \sum_{h=n}^{m-1} k^h d(s_1, s_0) \leq \frac{k^n(1 - k^{m-n})}{1 - k} d(s_1, s_0) < \frac{k^n}{1 - k} d(s_1, s_0).$$

Thus, $\forall \epsilon > 0$, if l is such that $d(s_1, s_0)k^l/(1 - k) < \epsilon$, that is, if

$$l > (\log(1 - k) + \log \epsilon - \log d(\phi(x_0), x_0))/\log(k),$$

then $\forall n, m > l$, $d(s_m, s_n) < \epsilon$. We have proved that the sequence (s_n) is Cauchy. Since X is complete, $\exists x^*$ such that $\lim_{n \to \infty} s_n = x^*$. Moreover, since x^* is the limit of the sequence, $\forall \epsilon > 0$ there exists a n_ϵ such that if $n \geq n_\epsilon$, $d(s_n, x^*) < \epsilon/2$. Then

$$d(\phi(x^*), x^*) \leq d(\phi(x^*), s_{n_\epsilon+1}) + d(s_{n_\epsilon+1}, x^*) =$$
$$d(\phi(x^*), \phi(s_{n_\epsilon})) + d(s_{n_\epsilon+1}, x^*) < d(x^*, s_{n_\epsilon}) + d(s_{n_\epsilon+1}, x^*) < \epsilon.$$

Since this is true $\forall \epsilon > 0$, then $d(\phi(x^*), x^*) = 0$, that is, $\phi(x^*) = x^*$. Now assume that there exists an x^{**} such that $x^{**} = \phi(x^{**})$. We have $d(x^{**}, x^*) = d(\phi(x^{**}), \phi(x^*)) \leq k d(x^{**}, x^*)$ which is possible only if $d(x^{**}, x^*) = 0$, that is $x^{**} = x^*$ so that the fixed point is unique.   □

▶ **Example 5.6** *(Golden ratio)* In the Euclidean metric space $(\mathbb{R}, |.|)$, consider the sequence (x_n) defined by recursion as

$$\begin{cases} x_0 = a, \\ x_{n+1} = \sqrt{1 + x_n}, \end{cases}$$

with $a > -1$. Let us investigate whether $f(x) = \sqrt{1+x}$ is a contraction. For any two real numbers x and y, notice that

$$|f(x) - f(y)| = \frac{|x - y|}{\sqrt{1+x} + \sqrt{1+y}}$$

so that if $x, y \in \mathbb{R}_{\geq 0}$, $|f(x) - f(y)| \leq |x - y|/2$ and f is a contraction in the metric space $(\mathbb{R}_{\geq 0}, |.|)$. If $a > -1$, then $x_n > 0$ for any $n > 0$, so that the sequence is eventually in $\mathbb{R}_{\geq 0}$. Using Theorem 5.9 we can conclude that the sequence converges and its limit is the unique solution of the equation $x = \sqrt{1+x}$ in $\mathbb{R}_{\geq 0}$. It is immediate to verify that this limit is the golden ratio defined in Example 5.3.

5.3 Sequences in $\mathbb{R}$

The real numbers are characterised by a field structure, which defines the two operations of addition and multiplication, and an order relation (see Sect. 1.5). The Euclidean topology of Definition 2.7 is Hausdorff and second-countable, and the metric space $(\mathbb{R}, |.|)$ (see also Example 3.1) is complete (see Corollary 5.1). The continuity of the distance function (see Sect. 3.2) guarantees that if $x_n \to x$, then $|x_n| \to |x|$. The interaction of the limit of sequences with the usual addition and multiplication operations can be easily addressed.

Theorem 5.10 *Let $(x_n), (y_n) \subseteq \mathbb{R}$ be two sequences of real numbers. If $x_n \to x$ and $y_n \to y$, then $x_n + y_n \to x + y$ and $x_n y_n \to xy$.*

Proof Consider the addition. For any $\epsilon > 0$ there is an n such that if $k > n$ then $|x - x_k| < \epsilon/2$ and $|y - y_k| < \epsilon/2$. Therefore, by the triangle inequality, $|x + y - x_k - y_k| \leq |x - x_k| + |y - y_k| \leq \epsilon$ and the statement is proved. The proof for the multiplication develops along similar lines and is left to the reader. $\square$

Concerning the ratios, we have the following unsurprising result.

Theorem 5.11 *Let $(x_n), (y_n) \subseteq \mathbb{R}$, if $x_n \to x$ and $y_n \to y \neq 0$, then $x_n/y_n \to x/y$.*

Proof To prove this theorem, it suffices to prove that if $y_n \to y \neq 0$ then $1/y_n \to 1/y$. If this is the case, the result follows from the direct application of Theorem 5.10 to the sequences (x_n) and $(1/y_n)$.

To prove the statement, note first that if $y_n \to y$, then there exists a n' such that if $n > n'$, then $|y_n| > |y|/2$. For any $\epsilon > 0$, let $n_\epsilon > n'$ be such that if $n > n_\epsilon$ then $|y - y_n| < y^2 \epsilon/2$. Thus for $n > n_\epsilon$,

$$\left| \frac{1}{y} - \frac{1}{y_n} \right| = \frac{|y - y_n|}{|y||y_n|} < 2 \frac{|y - y_n|}{y^2} < \epsilon$$

and the assertion is proved. □

Similarly to what happens with the arithmetic operations, the ordering relation is preserved by the limit.

Lemma 5.1 *Consider the sequences (x_n) and (y_n) and assume that it is eventually $x_n \geq y_n$. Then if $x_n \to x$ and $y_n \to y$, it is $x \geq y$.*

Proof We will proceed by contradiction. Let us assume that $x < y$ and set $\epsilon = (y - x)/2 > 0$. Then $\exists n_\epsilon$ such that $\forall n > n_\epsilon$, $x_n < x + \epsilon = (x + y)/2$ and $y_n > y - \epsilon = (x + y)/2$. Thus, for $n > n_\epsilon$, $x_n < y_n$, which contradicts the hypothesis. □

The previous theorem can be used to compare a sequence with a constant sequence. For example, if the sequence (a_n) converges and eventually $a_n \geq 00$, then its limit is nonnegative. In the same way, the nonpositivity of the limit of a sequence follows from the nonpositivity of its elements, and it is sufficient that they are eventually nonpositive. Conversely, the strong version of the order relation is generally not preserved in the limit. If eventually it is $a_n > b_n$, it could well be $\lim_{n \to \infty} a_n = \lim_{n \to \infty} b_n$.

▶ **Example 5.7** *(Limits violate strict order)* Consider the sequences $(a_n = 1/n)$ and $(b_n = 1/n^2)$. For $n > 1$ it is $a_n > b_n$. However, their limits are equal: $\lim_{n \to \infty} a_n = \lim_{n \to \infty} b_n = 0$.

The term-by-term comparison of the elements of sequences leads to a simple criterion to establish their convergence.

Theorem 5.12 (Comparison theorem) *Let (a_n), (b_n), and (c_n) be three sequences in $\mathbb{R}$. If it is eventually $a_n \leq c_n \leq b_n$, $a_n \to l$, and $b_n \to l$, then $c_n \to l$.*

Proof For any $\epsilon > 0$, $\exists n_\epsilon$ such that if $n > n_\epsilon$, $l - \epsilon < a_n$ and $b_n < l + \epsilon$. Thus, for $n > n_\epsilon$, $l - \epsilon < c_n < l + \epsilon$, and the assertion is proved. □

Useful results are available for sequences with ordered elements.

Definition 5.8 The sequence $(x_n) \subseteq \mathbb{R}$ is *monotonically increasing* if, $\forall n$, $x_{n+1} \geq x_n$, and *monotonically decreasing* if, $\forall n$, $x_{n+1} \leq x_n$.

The sequence is said to be *strictly monotonically increasing* or *strictly monotonically decreasing* if in the inequalities above, $\geq$ and $\leq$ are replaced with $>$ and $<$, respectively.

A *monotonic sequence* is a sequence that increases or decreases monotonically. Bounded monotonic sequences are converging sequences.

Theorem 5.13 (Monotone convergence) *Suppose the sequence (x_n) is monotonic, increasing or decreasing, then it converges if it is, respectively, bounded above or below.*

Proof Consider an increasing sequence (x_n). Since it is bounded above, let $x^* = \sup\{x_n\}$. By the definition of supremum, $\forall \epsilon > 0$, $\exists k$ such that $0 \leq x^* - x_k < \epsilon$, but since the sequence is increasing $\forall n > k$ it is $0 \leq x^* - x_n < \epsilon$, whence the convergence. For a decreasing sequence bounded below the proof is analogous. $\square$

The next example is the first of a series of examples aimed at deriving well-known functions and discussing their properties using the theory of converging sequences. Readers are suggested to study these examples as they illustrate several standard methods used to deal with sequences and their limits.

▶ **Example 5.8** *(The natural logarithm as a limit of a sequence)* Take any $x > 0$ and consider the sequence $(f_n(x))$ with

$$f_n(x) = n \left(\sqrt[n]{x} - 1 \right).$$

We will prove that $(f_n(x))$ converges. First of all, notice that $\forall z > 0$,

$$n(z^{n+1} - 1) - (n+1)(z^n - 1) = (1 - z)^2 \sum_{h=1}^{n} h z^{h-1}.$$

This can be easily verified by expanding the square on the right hand side and simplifying the common terms of the resulting summations. The right-hand side is the sum of nonnegative terms, so the left-hand side is also nonnegative. Thus, we can conclude that

$$n(z^{n+1} - 1) \geq (n+1)(z^n - 1).$$

Set $z = x^{1/(n(n+1))}$ and observe that $f_{n+1}(x) \leq f_n(x)$, that is, the sequence $(f_n(x))$ is decreasing. Moreover, from the previous inequality, by dividing both sides by z^n and rearranging the terms, we get the following inequality:

$$n(z - 1) \geq \frac{z^n - 1}{z^n}.$$

Substituting $z = \sqrt[n]{x}$ we realise that $f_n(x) \geq (x - 1)/x$ so that the sequence $(f_n(x))$ is bounded below. Thus, according to Theorem 5.13, the sequence has a limit. The *natural logarithm* of x is defined as $\log x = \lim_{n \to \infty} f_n(x)$ and satisfies the *logarithm inequality*

$$(x - 1)/x \leq \log x \leq f_1(x) = x - 1, \quad \forall x > 0.$$

The function $\log x$ has several notable properties. First, $\log 1 = 0$, $\log x > 0$ if $x > 1$ and $\log x < 0$ if $x < 1$.

Second, note that $f_n(1/x) = -f_n(x)/\sqrt[n]{x}$. Thus, taking the limit and for the continuity of the power function, $\log 1/x = -\log x$.

Third, given two positive numbers x and y, via a simple rearrangement of terms, one gets

$$f_n(xy) - f_n(x) - f_n(y) = n(\sqrt[n]{x} - 1)(\sqrt[n]{y} - 1).$$

The right-hand side corresponds to $f_n(x) f_n(y)/n$ so that

$$\lim_{n \to \infty} f_n(xy) - f_n(x) - f_n(y) = \lim_{n \to \infty} \frac{f_n(x) f_n(y)}{n} = 0.$$

We can conclude that $\log xy = \log x + \log y$.

Fourth, $\forall \delta > 0$, $\log(x + \delta) - \log x = \log(1 + \delta/x) > 0$, so that the function f is strictly increasing.

Fifth, using the properties and inequalities above, it is easy to see that for any $\epsilon > 0$, if $\delta < x\epsilon/(1 + \epsilon)$, $|\log(x \pm \delta) - \log x| < \epsilon$. Thus, $\log x$ is a continuous function at any point $x > 0$.

Sixth, given any couple of natural numbers $p, q \in \mathbb{N}$, for the fourth property it is $\log x^p = p \log x$, and because $f_n(x^{1/q}) = f_{qn}(x)/q$, $\log x^{1/q} = 1/q \log x$. Given any $z > 0$ and a sequence of positive rational numbers (p_n/q_n) such that $p_n/q_n \to z$, $\log x^{p_n/q_n} = p_n/q_n \log x$, which, taking the limit, for the continuity of the involved functions, gives $\log x^z = z \log x$.

Seventh, the range of the function is the whole $\mathbb{R}$. The function being continuous, its range is connected. Thus, it suffices to prove that it is unbounded. To prove it, we will show that $\forall z \in \mathbb{R}$, $\exists y > 0$ such that $\log y < z$. For the second property, this implies that $\log 1/y > -z$. Fix an integer m and a real quantity $\delta > 0$ such that $1 + z/m - \delta/m > 0$. This is possible for any real number z. Then define $y = (1 + z/m - \delta/m)^m$. Since the sequence $(f_n(x))$ is decreasing, $\forall n > m$,

$$f_n(y) \leq f_m(y) = m\left(\sqrt[m]{y} - 1\right) = z - \delta.$$

Taking the limit, $\lim_{n \to \infty} f_n(y) = \log y < z$.

The inequality $\log x \leq x - 1$ was used by the Hungarian mathematician George Pólya (1887–1985) to prove the arithmetic mean–geometric mean (AM-GM) inequality.[1]

▶ **Example 5.9** *(Concavity of the logarithm and AM–GM inequality)* For any x, $y \in \mathbb{R}_{>0}$ and $\lambda \in [0, 1]$, using the logarithm inequality and the properties of $\log x$ derived in Example 5.8, one has

$$\lambda \left(\frac{x}{y}\right)^{1-\lambda} + (1-\lambda)\left(\frac{y}{x}\right)^{\lambda} \geq \lambda \left(1 + (1-\lambda)\log\frac{x}{y}\right) + (1-\lambda)\left(1 + \lambda\log\frac{y}{x}\right) = 1.$$

Multiplying both sides by $x^{\lambda} y^{1-\lambda}$ we obtain the *generalised AM-GM inequality*

$$\lambda x + (1-\lambda)y \geq x^{\lambda} y^{1-\lambda}.$$

Applying the natural logarithm function to both sides of the inequality and using the property of the logarithm, one gets

$$\log(\lambda x + (1-\lambda)y) \geq \lambda \log x + (1-\lambda)\log y,$$

that is, the natural logarithm is a concave function (see Definition 1.23). This result can be used to easily prove the AM–GM inequality discussed in Theorem 1.14.

▶ **Example 5.10** *(Inverse of exponential and Euler's e)* Given two positive real numbers $a \neq 1$ and y, we are interested in the real number x such that $a^x = y$. This is the inverse of the exponential function defined in Theorem 1.13. To find the solution, we apply the logarithm to both sides of the equation to obtain $x \log a = \log y$. Thus, the solution of the problem is the unique positive real number $x = \log y / \log a$. The function $\log_a x = \log x / \log a$ is known as the *logarithm in base a*. It is proportional to the natural logarithm defined in Example 5.8 and has the same properties, with the only difference that if $a < 1$, then $\log a < 0$ and $\log_a(x)$ is a strictly decreasing and convex function. In particular, if $e > 1$ is the unique real number such that $\log e = 1$, then e^x is the inverse function of $\log x$. The number e takes its name from the Swiss mathematician Leonhard Euler 1707–1783. Its first computation was performed by the Swiss mathematician Jacob Bernoulli in 1683. He obtained it while looking for the formula of continuous compound interest. Since $a^x = e^{x \log a}$, for the continuity of the composition of continuous functions, the function a^x is continuous not only with respect to x, but also with respect to a.

---

[1] We reported another famous proof due to Cauchy, based on forward-backward induction, in Theorem 1.14.

The following result summarises the behaviour of the power, exponential, and logarithm of sequences. It follows directly from Theorem 5.2 and the continuity of the functions involved.

Theorem 5.14 *Let (x_n) and (y_n) be such that $x_n \to x$ and $y_n \to y$. If x is positive, then $x_n^{y_n} \to x^y$. If x and y are both positive, then $\log_{y_n} x_n \to \log_y x$.*

Proof If x is positive, then (x_n) is eventually positive. Since $\log x$ and its inverse e^x are continuous functions, Theorem 5.2 implies that

$$\lim_{n\to\infty} x_n^{y_n} = \lim_{n\to\infty} e^{y_n \log x_n} = e^{\lim_{n\to\infty} y_n \log x_n} = e^{x \log y} = x^y,$$

where we have used Theorem 5.10.

Analogously, if x and y are both positive, then (x_n) and (y_n) are eventually positive,

$$\lim_{n\to\infty} \log_{y_n} x_n = \lim_{n\to\infty} \frac{\log x_n}{\log y_n} = \frac{\log x}{\log y} = \log_y x,$$

where we have used Theorem 5.11 for quotients.   □

▶ **Example 5.11** *(Notable limit: $\log(n)/n$)* For any $n \geq 1$, $\log(n)/n > 0$. From the inequality derived in Example 5.8, it follows that $\log x < x$, thus $\log x = \log(\sqrt{x})^2 = 2\log\sqrt{x} < 2\sqrt{x}$. Then $\log(n)/n \leq 2/\sqrt{n}$, and since $\lim_{n\to\infty} 2/\sqrt{n} = 0$, for Theorem 5.12, $\lim_{n\to\infty} \log(n)/n = 0$.

▶ **Example 5.12** *(Notable limit: Euler's e)* Consider the sequence $(a_n = (1+1/n)^n)$. Using the binomial expansion,

$$a_n = \sum_{h=0}^{n} \binom{n}{h} \frac{1}{n^h} = \sum_{h=0}^{n} \frac{1}{h!} \left(1 - \frac{1}{n}\right) \cdots \left(1 - \frac{h-1}{n}\right).$$

The same expression for a_{n+1} reads

$$a_{n+1} = \sum_{h=0}^{n+1} \frac{1}{h!} \left(1 - \frac{1}{n+1}\right) \left(1 - \frac{2}{n+1}\right) \cdots \left(1 - \frac{h-1}{n+1}\right).$$

The summation that defines a_{n+1} contains more terms than the summation that defines a_n and theirs are larger, as $1/n > 1/(n+1)$. Thus we can conclude that $a_{n+1} > a_n$. At the same time, $1 - h/n < 1$ and

$$a_n < \sum_{h=0}^{n} \frac{1}{h!} = 1 + 1 + \frac{1}{1 \cdot 2} + \frac{1}{1 \cdot 2 \cdot 3} + \frac{1}{1 \cdot 2 \cdot 3 \cdot 4} + \dots.$$

Notice that if $h > 2$, $h! > 2^{h-1}$. The inequality can be simply obtained by replacing each number greater than 2 in the factorial with 2. Then

$$\sum_{h=0}^{n} \frac{1}{h!} < 1 + \sum_{h=1}^{n} \frac{1}{2^{h-1}} = 1 + \frac{1-(1/2)^n}{1/2} = 3 - \frac{1}{2^{n-1}} < 3,$$

and one has $a_n < 3$, $\forall n$. Thus, the sequence (a_n) is strictly increasing and is bounded above. According to Theorem 5.13 it has a limit, which lays between 2 (in fact $a_1 = 2$) and 3. We denote this limit with e. Now consider the sequence $(b_n = (1 - 1/n)^n)$. It is immediate to verify that $b_n = (n-1)/n1/a_{n-1}$. Since $(n-1)/n \to 1$ and $1/a_{n-1} \to 1/e$, $b_n \to 1/e$. It remains to prove that e is the base of the inverse of the $\log x$ function defined in Example 5.10. That is, that $\log e = 1$. This proof is postponed to Sect. 5.6.1. More general versions of the limits discussed here are studied in Example 5.14.

▶ **Example 5.13** *(Notable limit: $n^{1/n}$)* Using the continuity of the exponential (inverse logarithm) function and the result in Example 5.11 one has

$$\lim_{n \to \infty} \sqrt[n]{n} = e^{\lim_{n \to \infty} \log \sqrt[n]{n}} = e^{\lim_{n \to \infty} \log(n)/n} = e^0 = 1.$$

Using Theorem 1.14, one can derive a definition of the exponential function as the limit of a sequence.

▶ **Example 5.14** *(Exponential function as a limit of sequences)* For any $x \in \mathbb{R}$, consider the sequences $(g_n(x) = (1+x/n)^n)$ and $(h_n(x) = (1-x/n)^{-n})$. Whatever the value of x, for n sufficiently large, $1+x/n$ and $1-x/n$ are both positive. Consider such an n and apply the AM–GM inequality to a set of $n+1$ points, n equal to $1+x/n$ and one equal to 1, to obtain

$$\frac{1+n(1+x/n)}{n+1} = 1 + \frac{x}{n+1} \geq \sqrt[n+1]{\left(1+\frac{x}{n}\right)^n}.$$

Taking the $n+1$th power of both sides, the previous expression reduces to $g_{n+1}(x) \geq g_n(x)$. Analogously, applying the AM–GM inequality to a set of $n+1$ points of which n are equal to $1 - x/n$ and one is equal to 1 one gets

$$\left(1 - \frac{x}{n+1}\right)^{n+1} \geq \left(1 - \frac{x}{n}\right)^n,$$

which, taking the inverse of both sides, implies $h_{n+1}(x) \leq h_n(x)$. Also, for sufficiently large n, $g_n(x)/h_n(x) = (1 - x^2/n^2)^n < 1$.

Summarising, we have found that, after possibly discarding a fixed number of initial terms, $(g_n(x))$ is an increasing sequence, $(h_n(x))$ is a decreasing sequence, and

$g_n(x) \leq h_n(x)$. Hence both sequences converge. Define $\exp x = \lim_{n\to\infty} h_n(x)$. Notice that $(h_n(x)/g_n(x))^n$ is a subsequence of $h_n(x)$: the subsequence obtained by considering only integers that are the square power of other integers. Thus, $(h_n(x)/g_n(x))^n \to \exp x$ and, by Theorem 5.14,

$$\lim_{n\to\infty} h_n(x)/g_n(x) = \lim_{n\to\infty} h_n(x)^{1/n} = 1,$$

so that $g_n(x) \to \exp x$. Next, we prove a peculiar multiplicative property of $\exp x$. Note that

$$\frac{g_n(x)g_n(y)}{g_n(x+y)} = \left(1 + \frac{xy}{n(x+y+n)}\right)^n.$$

If $|x| < 1$ the order relation between $g_n(x)$ and $h_n(x)$, derived above, is valid for any n so that

$$1 + x = g_1(x) \leq g_n(x) = \left(1 + \frac{x}{n}\right)^n \leq h_n(x) \leq h_1(x) = \frac{1}{1-x}$$

and we obtain the *exponential inequality* $1 + x \leq \exp x \leq 1/(1-x)$. When n is sufficiently large, $|xy/(x+y+n)| < 1$. Hence, eventually, using the previous inequality,

$$1 + \frac{xy}{x+y+n} \leq \frac{g_n(x)g_n(y)}{g_n(x+y)} \leq \left(1 + \frac{xy}{x+y+n}\right)^{-1}.$$

But since both the left and right sides tend to one, according to Theorem 5.12, $g_n(x)g_n(y)/g_n(x+y) \to 1$, that is $\exp x + y = (\exp x)(\exp y)$. This property is the same of the exponential function defined in Definition 1.19. In fact, the function defined in this example is precisely the inverse of the $\log x$ function defined in Example 5.8. That is, $\exp 1 = e$. We do not prove it now. Notice that by using the expression of f_n in Example 5.8 one gets $f_n(g_n(x)) = g_n(f_n(x)) = x$. However, this is insufficient to conclude that the limits of the sequences define functions that are the inverse of each other. To prove it, one has to deal with a double limit, $\lim_{n\to\infty} \lim_{m\to\infty} f_n(g_m(x))$, which is in general problematic, see Example 5.27. The proof is postponed to Sect. 5.6.1, when the notion of uniform convergence will be available.

A sequence that does not have any limit is generally named *divergent*. The order relation on $\mathbb{R}$ suggests extending the usual notion of a limit to include special cases of divergent sequences. For this purpose, the extended real number system $\bar{\mathbb{R}}$ of Definition 2.11 is adopted.

Definition 5.9 We will write $\lim_{n\to+\infty} x_n = +\infty$ if $\forall z \in \mathbb{R}, \exists N_z$ such that $\forall n > N_z, x_n > z$. Analogously we write $\lim_{n\to+\infty} x_n = -\infty$ if $\forall z \in \mathbb{R}, \exists N_z$ s.t. $\forall n > N_z, x_n < z$.

Sometimes, with a slight abuse of language, a sequence (x_n) such that $\lim_{n\to+\infty} x_n = \pm\infty$ is said to *diverge* to plus or minus infinity. Saying that $\lim_{n\to\infty} x_n = +\infty$ is different from saying that the set $\{x_n\}$ is unbounded above, as the following example clarifies.

▶ **Example 5.15** *(Divergence of unbounded sequences)* Consider the sequence $(x_n = n^{[n]_2})$, where $[\cdot]_2$ is the modulus two introduced in Example 1.4: $[n]_2 = 1$ if n is odd and $[n]_2 = 0$ if n is even. The set $\{x_n\}$ has no (finite) limit points and is unbounded but (x_n) does not have any finite or infinite limit.

5.3.1 Upper and Lower Limits

Given a sequence of real numbers (x_n), one can consider the set of points that are the limit of at least one of its subsequences. If the sequence is convergent, this set contains a single point (see Theorem 5.1). Alternatively, this set can contain multiple points or be empty. Based on Definition 5.9 we can consider the "enlarged set" $E(x_n) \subseteq \bar{\mathbb{R}}$ of the limits of all subsequences of (x_n) in $\bar{\mathbb{R}}$. This set is never empty. If the set $\{x_n\}$ is not bounded above, then $+\infty \in E(x_n)$. In fact, one can easily build a subsequence that has $+\infty$ as a limit by considering progressively larger elements of $\{x_n\}$. If the set $\{x_n\}$ is unbounded below, then $-\infty \in E(x_n)$. In this case, one can easily build a subsequence having $-\infty$ as a limit by considering smaller and smaller elements of $\{x_n\}$. If the set $\{x_n\}$ is bounded, then it is contained in a closed interval. If it is made by a finite number of isolated points, at least one of these points is the limit of a constant subsequence. Otherwise, it is an infinite subset of a compact set, and according to Theorem 2.15 it has at least one limit point. Thus, one can build a subsequence converging to this point (see the discussion after Theorem 5.4).

In general, $D\{x_n\} \subseteq E(x_n)$. The supremum and infimum of the set $E(x_n)$ are used to build two special types of limits for sequences in $\mathbb{R}$.

Definition 5.10 *(Upper and lower limits)* Consider a sequence of real numbers $(x_n) \subseteq \mathbb{R}$ and let $E(x_n) \subseteq \bar{\mathbb{R}}$ be the set of limit points of subsequences of (x_n). Then its *upper limit*, or *limit superior*, x^* and its *lower limit*, or or *limit inferior*, x_* are defined as

$$x^* = \limsup_{n\to\infty} x_n = \sup E(x_n), \quad x_* = \liminf_{n\to\infty} x_n = \inf E(x_n).$$

The upper and lower limits of a sequence can be $+\infty$ or $-\infty$ if the set $\{x_n\}$ is unbounded above or below, respectively. By definition, $x_* \leq x^*$. Moreover, for any sequence (x_n) and any $\epsilon > 0$, there must be only a finite number of elements above $x^* + \epsilon$ or below $x_* - \epsilon$, otherwise there would be subsequences that have a limit larger than x^* or lower than x_*, which is absurd. At the same time, if $x_* < x^*$, for any sufficiently small $\epsilon > 0$, there must be an infinite number of elements outside $(x_* + \epsilon, x^* - \epsilon)$. This consideration immediately leads to the following conclusion.

Corollary 5.2 *Given a sequence (x_n), let x^* and x_* be its upper and lower limits. Then $x_n \to l$ if and only if $x^* = x_* = l$.*

An alternative way to define the upper and lower limits is to consider the sequence obtained by successively dropping the initial elements of the sequence and computing the supremum and infimum of the set of remaining points. When the number of dropped points goes to infinity, the supremum and infimum of the set define the upper and lower limits. The following theorem presents an alternative definition of these limits.

Theorem 5.15 *Given a sequence (x_n), consider $M_n = \sup\{x_h | h \geq n\}$ and $m_n = \inf\{x_h | h \geq n\}$, with $M_n, m_n \in \mathbb{R}$. Then*

$$\lim_{n \to \infty} M_n = \limsup_{n \to \infty} x_n, \quad \lim_{n \to \infty} m_n = \liminf_{n \to \infty} x_n.$$

Proof Consider the upper limit. If the set $\{x_n\}$ is unbounded from above, then $\forall n$, $M_n = +\infty = x^*$, and the statement is proved. If instead $\{x_n\}$ is bounded above, then x^* is finite and for any $\epsilon > 0$, eventually $x_n < x^* + \epsilon$, that is, $M_n \leq x^* + \epsilon$. Moreover, there must be an infinite set of points greater than $x^* - \epsilon$, or the upper limit would be lower than x^*. Thus, $\forall n$, $M_n > x^* - \epsilon$. In summary, eventually $x^* - \epsilon \leq M_n \leq x^* + \epsilon$ and the statement follows. The proof of the other case is identical. $\square$

Lemma 5.1 is still valid for the upper and lower limits, while, in general, they cannot be added or multiplied as in Lemma 5.10. They interact with arithmetic operations similarly to the supremum and infimum, so that, for instance, given two sequences (a_n) and (b_n) (see Exercise 5.30),

$$\liminf_{n \to \infty} a_n + \liminf_{n \to \infty} b_n \leq \liminf_{n \to \infty}(a_n + b_n) \leq$$

$$\leq \limsup_{n \to \infty}(a_n + b_n) \leq \limsup_{n \to \infty} a_n + \limsup_{n \to \infty} b_n.$$

In some cases, the upper and lower limits of the ratio of two sequences can be bounded by the respective limits of the ratio of their increments. This is a result due to the Italian mathematician Ernesto Cesàro (1859–1906) and is similar to L'Hôpital rule for the limit of functions that we will encounter in Chap. 6.

Theorem 5.16 (Stolz-Cesàro theorem) *Consider (a_n) and (b_n) in $\mathbb{R}$ and let (b_n) be positive, strictly increasing and unbounded. Then*

$$\limsup_{n \to \infty} \frac{a_n}{b_n} \leq \limsup_{n \to \infty} \frac{a_{n+1} - a_n}{b_{n+1} - b_n}, \quad \liminf_{n \to \infty} \frac{a_n}{b_n} \geq \liminf_{n \to \infty} \frac{a_{n+1} - a_n}{b_{n+1} - b_n},$$

and if $\lim_{n \to \infty}(a_{n+1} - a_n)/(b_{n+1} - b_n) = l$, then $\lim_{n \to \infty} a_n/b_n = l$.

Proof Only the first statement needs to be proved, as the second directly follows by considering the sequence $(-a_n)$ instead of (a_n). If $\lim \sup_{n \to \infty} (a_{n+1} - a_n)/(b_{n+1} - b_n) = +\infty$ the statement is trivial. Assume $\lim \sup_{n \to \infty} (a_{n+1} - a_n)/(b_{n+1} - b_n) = L \in \mathbb{R}$. Then $\forall \epsilon > 0$ eventually $a_{h+1} - a_h \leq (b_{h+1} - b_h)(L + \epsilon)$. Adding k subsequent inequalities and simplifying terms one gets $a_{h+k} - a_h \leq (b_{h+k} - b_h)(L + \epsilon)$ which can be rewritten as

$$\frac{a_{h+k}}{b_{h+k}} - \frac{a_h}{b_{h+k}} \leq (L + \epsilon)(1 - \frac{b_h}{b_{h+k}}).$$

Because b_{h+k} is increasing and unbounded, for sufficiently large k, $a_h/b_{h+k} < \epsilon$ so that $a_{h+k}/b_{h+k} \leq L + 2\epsilon$. The previous inequality applies eventually, thus $\lim \sup_{n \to \infty} a_n/b_n < L + 2\epsilon$. Since this is true $\forall \epsilon > 0$, the assertion follows. □

For an application of this result, see Example 5.18 below. This result can be used to prove a theorem about the convergence of the averages of the elements of a sequence.

Definition 5.11 *(Cesàro mean)* Consider the sequence (x_n), and let $s_n = \sum_{k=1}^{n} x_k$ be its $n$tn *partial sum*, that is the sum of its first n elements. The *Cesàro mean* of (x_n) is the sequence (c_n) with $c_n = s_n/n$.

From Theorem 5.16 we have the following.

Corollary 5.3 *If the sequence (x_n) converges to l, then its Cesàro mean (c_n) converges to l as well.*

Proof Apply Theorem 5.16 to the sequences (s_n) and (n). □

In other words, if the sequence is convergent, the arithmetic mean of its elements converges to its limit. Note that the Cesàro mean of a divergent sequence can converge; see Exercise 5.32.

5.3.2   Infinity and Infinitesimals

Consider two sequences with the same limit. If this limit is finite and different from zero, we know that, according to Theorem 5.11, the ratio of their terms converges to 1.[2] If the limit of the two sequences is zero or $\pm\infty$, this conclusion is generally not warranted. In this case, the ratio of the terms of the two sequences can have quite

---

[2] Notice that, depending on how these sequences are defined, it could be that the ratio of their terms is not defined for a finite number of terms. As already discussed, this issue is totally irrelevant regarding the limit of the sequence and it does not deserve any treatment. We will always rule out this possibility by considering an appropriate starting index for the elements of the sequence.

Fig. 5.1 The relation
between an arc and its sinus

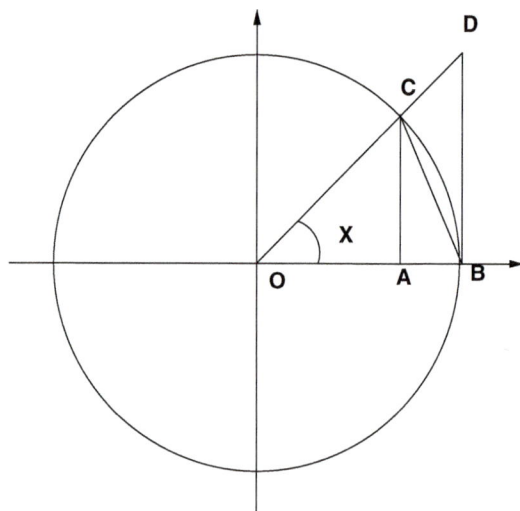

peculiar behaviours. We want a comparison criterion between sequences that can capture not only the fact that they have the same limit but also that they tend to this limit in a similar way.

Definition 5.12 Consider two sequences (a_n) and (b_n) with the same limit, finite or infinite. The two sequences are *asymptotically equivalent* if $\lim_{n\to\infty} a_n/b_n = 1$ and we write $a_n \sim b_n$.

The asymptotic equivalence is an equivalence relation among all sequences that have the same limit. In particular, if $a_n \sim b_n$ and $b_n \sim c_n$, then $a_n \sim c_n$. This relation is of some interest only when the sequences are infinite or infinitesimal. In all other cases, Theorem 5.11 guarantees that there is only one equivalence class.

▶ **Example 5.16** *(Notable limit: $n\sin(1/n)$)* We shall prove that the sequences $(a_n = 1/n)$ and $(b_n = \sin(1/n))$ are asymptotically equivalent. Consider an arc of length x with $0 < x < \pi/2$ on the unit circle as displayed in Fig. 5.1. The triangle OBC has an area equal to $(\sin x)/2$ and is contained in the circular sector OCB, whose area is $x/2$. In turn, the latter is contained in the triangle OBD whose area is $(\tan x)/2$. Thus,

$$0 < \frac{1}{2}\sin x < \frac{x}{2} < \frac{1}{2}\tan x$$

which, by Theorem 5.12, implies that $\lim_{n\to\infty} \sin(1/n) = 0$ and, due to the well-known Pythagorean formula $\sin^2 x + \cos^2 x = 1$, that $\lim_{n\to\infty} \cos(1/n) = 1$. For large n, $1/n < \pi/2$ so that, dividing the previous inequalities by $\sin x$, eventually

$$1 < \frac{1/n}{\sin(1/n)} < \frac{1}{\cos(1/n)}.$$

The sequences on the left and on the right converge to 1. Thus, according to Theorem 5.12, $\lim_{n\to\infty} a_n/b_n = 1$ and the statement is proved.

Often it is useful to compare the convergence to zero of two infinitesimal sequences. The idea is to identify the sequences that converge "faster" to zero, that is, the sequence whose elements are eventually closer to zero than the elements of the other sequence.

Definition 5.13 *(Order of infinitesimal)* Consider two sequences (a_n) and (b_n) that converge to 0. The first sequence is an *infinitesimal of higher order* than the second sequence if $\lim_{n\to\infty} a_n/b_n = 0$. The two sequences are *infinitesimal of the same order* if there exist two constants $A, B > 0$ such that $A \leq \lim \inf_{n\to\infty} |a_n/b_n|$ and $\lim \sup_{n\to\infty} |a_n/b_n| \leq B$.

The sequence (b_n) is an infinitesimal of lower order than (a_n) if the latter is an infinitesimal of higher order than the former. Note that if the ratio of the two infinitesimals absolutely converges to a finite positive number, that is, $\lim_{n\to\infty} |a_n/b_n| = l > 0$, then the two infinitesimals are of the same order. In fact, this is a special case of Definition 5.13, with $A = B = l$. In particular, two asymptotically equivalent sequences converging to zero are infinitesimals of the same order. The asymptotic relationship between sequences is sometimes described using *Landau symbols*, named after the German mathematician Edmund Landau (1877–1938).

Definition 5.14 *(Landau symbols)* Given two infinitesimal sequences (a_n) and (b_n), we will write $a_n = o(b_n)$ (small "o") if (a_n) is an infinitesimal of higher order than (b_n). We will write $a_n = O(b_n)$ (big "O") if there exists a constant $k \geq 0$ such that eventually $|a_n| \leq k|b_n|$.

For example, we write $1/n^2 = o(1/n)$ and $1/n^2 = O((-1)^n/n^2)$. Definition 5.14 introduces an order relation on infinitesimals. In fact, if $a_n = O(b_n)$ and $b_n = O(c_n)$, then $a_n = O(c_n)$. This can be easily proved by noticing that the constant k that must exist for the last relation to be valid is just the product of the constants of the first two relations. The big "O" is also reflexive and is antisymmetric with respect to the equivalence classes constituted by infinitesimals of the same order according to Definitions 5.13. However, this relation is not complete. That is, given two sequences (a_n) and (b_n), it could be that neither $a_n = O(b_n)$ nor $b_n = O(a_n)$.

▶ **Example 5.17** (*Big O incomplete ordering*) Consider the sequence $(a_n = 1/n^2)$ and the sequence (b_n) with

$$b_n = \begin{cases} 1/n & \text{if } n \text{ is even;} \\ 1/n^3 & \text{if } n \text{ is odd.} \end{cases}$$

Clearly, neither $a_n = O(b_n)$ nor $b_n = O(a_n)$ is true.

It is straightforward to obtain a relation for infinite sequences similar to the one defined for infinitesimals. If the sequence (a_n) diverges to $\pm\infty$, then the sequence $(1/a_n)$ converges to zero. The opposite is not true, of course, as one can immediately see when considering the sequence $((-1)^n/n)$, which converges to zero, but whose inverse has no limit.

Definition 5.15 (*Order of infinite*) Consider two sequences (a_n) and (b_n) both converging to $+\infty$ or to $-\infty$. The first sequence is said an *infinite of higher order* of the second sequence if $(1/a_n)$ is an infinitesimal of higher order than $(1/b_n)$. The two sequences are *infinite of the same order* if $(1/a_n)$ and $(1/b_n)$ are infinitesimals of the same order.

Landau symbols can be used to denote the asymptotic relation between infinite sequences using sequences of reciprocal terms. Thus, if (a_n) and (b_n) are both converging to $+\infty$, we will write $1/a_n = o(1/b_n)$ to mean that the first sequence is an infinite of higher order than the second.

▶ **Example 5.18** (*Sum of the power of first integers*) It can be easily proved by induction that the sum of the first n integers is equal to $n(n+1)/2$. This implies that the sum grows asymptotically proportionally to n^2. In other terms, if we denote the sum of the kth power of the first n integers with $s_n(k)$, that is,

$$s_n(k) = \sum_{h=1}^{n} h^k,$$

we know that $\lim_{n\to\infty} s_n(1)/n^2 = 1/2$. It is possible to generalise this result. Fix a constant k and consider the sequences $(s_n(k))$ and (n^{k+1}). Then

$$\frac{s_n(k) - s_{n-1}(k)}{n^{k+1} - (n-1)^{k+1}} = \frac{n^k}{n^{k+1} - (n-1)^{k+1}} = \frac{n^k}{n^k(k+1) + n^{k-1}(k+1)k/2 + \dots},$$

where we used the binomial expansion of $(n-1)^{k+1}$ to rewrite the denominator. The missing terms contain a power of n lower than $k-1$. Dividing numerator and denominator by n^k and using Theorem 5.16,

$$\lim_{n \to \infty} \frac{s_n(k)}{n^{k+1}} = \lim_{n \to \infty} \frac{s_n(k) - s_{n-1}(k)}{n^{k+1} - (n-1)^{k+1}} = \frac{1}{k+1}.$$

Therefore, we conclude that for large n, $s_n(k) \sim n^{k+1}/(k+1)$.

5.4 Sequences in Normed Spaces

For sequences defined over normed spaces, it is natural to ask how the notion of limit interacts with the operations on vectors. The following result is a direct consequence of Theorem 4.11.

Corollary 5.4 *(Equivalence of limits) If ρ and ρ' are two norms on $\mathbb{R}^n$, then for any sequence $(\mathbf{x}_h)$, $\lim_{h \to +\infty} \rho'(\mathbf{x}_h) = 0$ if and only if $\lim_{h \to +\infty} \rho(\mathbf{x}_h) = 0$.*

Proof According to Definition 4.6 and due to Theorem 4.11, there are two $c, C > 0$ such that $0 \leq \rho'(\mathbf{x}) < C\rho(\mathbf{x})$ and $0 \leq \rho(\mathbf{x}) < c\rho'(\mathbf{x})$. The statement follows immediately from Theorem 5.12 by comparing the two sequences $(\rho'(\mathbf{x}_h))$ and $(\rho(\mathbf{x}_h))$ in $\mathbb{R}$. □

Thus, the notion of limit in a finite-dimensional normed space does not depend on the specific definition of the adopted norm.[3] The following discussion mainly focuses on Euclidean spaces $(\mathbb{R}^n, \| \cdot \|)$. In this section, the expression x^j represents the jth component of the vector $\mathbf{x}$. An important consequence of Theorem 5.8 is that finite-dimensional normed spaces on $\mathbb{R}$ are complete metric spaces.

Corollary 5.5 *Any Cauchy sequence in $(\mathbb{R}^n, \| \cdot \|)$ is convergent.*

Proof Let $(\mathbf{x}_h)$ be a Cauchy sequence in $\mathbb{R}^n$. Fix $\epsilon > 0$ and consider h^* so that for $h > h^*$, $\|\mathbf{x}_{h^*} - \mathbf{x}_h\| < \epsilon$. Then consider

$$L = \max\{\|\mathbf{x}_1 - \mathbf{x}_{h^*}\|, \|\mathbf{x}_2 - \mathbf{x}_{h^*}\|, \ldots, \|\mathbf{x}_{h^*-1} - \mathbf{x}_{h^*}\|, \epsilon\}.$$

Clearly, $\{\mathbf{x}_h\} \subset B(\mathbf{x}_{h^*}, L)$ so that the sequence is bounded. Therefore, it is a subset of an n-cell which, according to Theorem 4.9, is compact. For Theorem 5.8 this implies that the sequence is convergent. □

---

[3] This should not surprise the reader, as the "limit" is a topological notion and equivalent norms induce the same topology.

The previous corollary is not true, for example, for linear spaces defined on $\mathbb{Q}$, that is, for the n-tuple with rational components. This is because the n-cell in $\mathbb{Q}^n$ is not compact. Some intuitive properties can be derived on the basis of the general idea that the limit of a sequence of vectors is the limit of the sequence of their components.

Theorem 5.17 *Consider a sequence* $(\mathbf{x}_h) \subset \mathbb{R}^n$. *Then* $\lim_{h \to +\infty} \mathbf{x}_h = \mathbf{x}$ *if and only if* $\forall i = 1, \ldots, n$, $\lim_{h \to +\infty} x_h^i = x^i$.

Proof Let us start by proving the theorem when the norm is the Euclidean one (see Sect. 4.2.1). First of all notice that for the triangle inequality

$$0 \le \|\mathbf{x}_h - \mathbf{x}\| \le \sum_{i=1}^{n} |x_h^i - x^i|.$$

Thus, if $\forall i$, $x_h^i - x^i \to 0$, then $\|\mathbf{x}_h - \mathbf{x}\| \to 0$. Conversely, $\forall i = 1, \ldots, n$,

$$0 \le |x_h^i - x^i| \le \sqrt{\sum_{j=1}^{n} (x_h^j - x^j)^2} = \|\mathbf{x}_h - \mathbf{x}\|$$

so that if $\|\mathbf{x}_h - \mathbf{x}\| \to 0$, $x_h^i - x^i \to 0$. For Corollary 5.4, the result applies to any norm.   □

Limits are preserved by sum of vectors, multiplication by scalars, and inner product.

Theorem 5.18 *Consider two sequences* $(\mathbf{x}_h)$, $(\mathbf{y}_h) \subset \mathbb{R}^n$ *such that* $\lim_{h \to +\infty} \mathbf{x}_h = \mathbf{x}$ *and* $\lim_{h \to +\infty} \mathbf{y}_h = \mathbf{y}$, *then:*

1. *(convergence of vector sum and product by scalar)* $\lim_{h \to +\infty} \mathbf{x}_h + \mathbf{y}_h = \mathbf{x} + \mathbf{y}$; $\lim_{h \to +\infty} c\mathbf{x}_h = c\mathbf{x}$, *with* $c \in \mathbb{R}$;
2. *(convergence of inner product)*, $\lim_{h \to +\infty} \mathbf{x}_h \cdot \mathbf{y}_h = \mathbf{x} \cdot \mathbf{y}$.

Proof These properties can be easily proved starting from Definition 5.5 for the Euclidean norm (see Sect. 4.2.1). For instance consider the last point. If $\mathbf{x} = 0$ or $\mathbf{y} = 0$, the result is trivial. Assume that both limits are different from the zero vector. Then, by the triangle inequality,

$$\|\mathbf{x} \cdot \mathbf{y} - \mathbf{x}_h \cdot \mathbf{y}_h\| = \|\mathbf{x} \cdot \mathbf{y} - \mathbf{x} \cdot \mathbf{y}_h + \mathbf{x} \cdot \mathbf{y}_h - \mathbf{x}_h \cdot \mathbf{y}_h\| \le \|\mathbf{x}\|\|\mathbf{y} - \mathbf{y}_h\| + \|\mathbf{y}_h\|\|\mathbf{x} - \mathbf{x}_h\|.$$

Consider $\forall \epsilon > 0$ and let h_ϵ be such that $\forall h > h_\epsilon$, $\|\mathbf{y} - \mathbf{y}_h\| < \epsilon/(2\|\mathbf{x}\|)$. Moreover let $M = \sup\{\|\mathbf{y}_h\|\}$ and h'_ϵ be such that $\forall h > h'_\epsilon$, $\|\mathbf{x} - \mathbf{x}_h\| < \epsilon/(2M)$. Then for

$h > \max\{h_\epsilon, h'_\epsilon\}$,

$$\|\mathbf{x}\mathbf{y} - \mathbf{x}_h\mathbf{y}_h\| < \|\mathbf{x}\| \frac{\epsilon}{2\|\mathbf{x}\|} + \|\mathbf{y}_h\| \frac{\epsilon}{2M} < \epsilon.$$

The other points can be proved similarly. For Corollary 5.4, the results apply to any norm. $\square$

The next result concerns the convex cones defined in Sect. 1.6. It will be used in Theorem 7.17.

Theorem 5.19 *The convex cone is a topologically closed set.*

Proof We prove that the convex cone $C \subseteq \mathbb{R}^n$ is closed by showing that all its limit points belong to C itself. Specifically, we will show that the limit of any converging sequence of elements of C belongs to C. We will proceed by induction. If $n = 1$, then the convex cone is a half-line, which is closed. Assume that any convex cone is closed when the number of generating vectors is $n - 1$, and consider the cone C generated by $\{\mathbf{a}_1, \ldots, \mathbf{a}_n\}$. If $-\mathbf{a}_i \in C$ for all $i = 1, \ldots, n$, then C is a subspace and is closed. Without loss of generality, assume that $-\mathbf{a}_n \notin C$ and let $\bar{C} \subseteq C$ be the cone generated by $\{\mathbf{a}_1, \ldots, \mathbf{a}_{n-1}\}$. Then any element $\mathbf{z} \in C$ can be written as $\mathbf{z} = \bar{\mathbf{z}} + \alpha \mathbf{a}_n$ with $\bar{\mathbf{z}} \in \bar{C}$ and $\alpha \geq 0$.

Consider a sequence $(\mathbf{z}_h)$ of elements in C, converging to $\mathbf{z}$. We have to prove that $\mathbf{z} \in C$. Let $\mathbf{z}_h = \bar{\mathbf{z}}_h + \alpha_h \mathbf{a}_n$, with $\bar{\mathbf{z}}_h \in \bar{C}$ and $\alpha_h \geq 0$. Assume that the sequence (α_h) is unbounded above. Then there exists a subsequence (α_i) which converges to plus infinity. The subscript i is used instead of h to denote the indexes of the subsequence. Consider the sequence $(\mathbf{z}_i / \alpha_i)$. Since $\{\mathbf{z}_i\}$ is convergent, it is bounded, thus $\lim_{i \to \infty} \mathbf{z}_i / \alpha_i = \bar{\mathbf{z}}_i / \alpha_i + \mathbf{a}_k = \mathbf{0}$, which implies that $\lim_{i \to \infty} \bar{\mathbf{z}}_i / \alpha_i = -\mathbf{a}_n$. Because $\bar{C}$ is closed, this would imply that $-\mathbf{a}_n \in \bar{C}$, which is ruled out by hypothesis. Thus, the set $\{\alpha_h\}$ is bounded and, consequently, there exists a subsequence (α_i) convergent to some α_0. This implies that $(\mathbf{z}_i - \alpha_i \mathbf{a}_n)$ is a convergent sequence. Let $\bar{\mathbf{z}}$ be its limit. Because by assumption $\bar{C}$ is closed, $\bar{\mathbf{z}} \in \bar{C}$, so that $\mathbf{z} = \alpha_0 \mathbf{a}_n + \bar{\mathbf{z}} \in C$. Since the original sequence $(\mathbf{z}_h)$ is convergent, and because the limit of a subsequence of a convergent sequence is equal to the limit of the sequence, we can conclude that the original sequence converges to the same limit and the statement is proved. $\square$

We conclude this section with an example of infinite-dimensional complete normed spaces. These spaces have a specific name.

Definition 5.16 *(Banach space)* A *Banach space* is a normed space (V, ρ) that is complete with respect to the metric d_ρ derived from the norm.

In Example 5.4 we have seen that the space of bounded functions with the image in $\mathbb{R}$ is a complete metric space and thus is a Banach space. We can extend the same result to the case of continuous functions.

▶ **Example 5.19** *(The metric space of bounded continuous functions is complete)*
Let (X, T) be a topological space, and let $C(X)$ be the set of continuous bounded
functions $f : X \to \mathbb{R}$. We will prove that $(C(X), d_\infty)$ is complete. First of all,
notice that, with reference to Example 3.1 and from the discussion in Example 5.4,
since $C(X) \subseteq B(X)$, we know that for any Cauchy sequence of functions (f_n) in
$C(X)$ and $\forall a \in X$ there exists $g(a) = \lim_{n\to\infty} f_n(a)$. The function g is bounded
and $f_n \to g$. It remains to prove that g is continuous. Fix $\epsilon > 0$ and let $N_n(a)$ be
the neighbourhood of a such that $\forall x \in N_n(a)$, $|f_n(x) - f_n(a)| < \epsilon/3$. For any n,
by the triangular inequality,

$$|g(a) - g(x)| \leq |g(a) - f_n(a)| + |f_n(a) - f_n(x)| + |f_n(x) - g(x)|.$$

Choose n so that $d_\infty(g, f_n) < \epsilon/3$, then if $x \in N_n(a)$, $|g(a) - g(x)| < \epsilon$, which
proves the continuity of g. Thus, $(C(A), d_\infty)$ is a Banach space.

5.5 Series in $\mathbb{R}$

Consider the sequence $(a_n) \in \mathbb{R}$ and call $s_n = \sum_{i=1}^{n} a_i$, the sum of the first n
elements of the sequence, the partial sum of order n. Partial sums define a new
sequence (s_n). We are interested in its limit.

Definition 5.17 If the sequence of partial sums (s_n) has a limit s, we say that the
series defined by the terms (a_n) and denoted by $\sum_{n=1}^{\infty} a_n$ has a limit or converges,
and we write $\sum_{n=1}^{\infty} a_n = s$.

Sometimes, for brevity, the limit of the series is denoted by $\sum_n a_n$, omitting the
summation range. If a series converges, the sequence of its elements is said to be
summable. If the sequence of partial sums is not convergent, we will say that the
series is *divergent*. It could diverge to $\pm\infty$ according to Definition 5.9.

▶ **Example 5.20** *(Geometric Series)* Consider $x \in \mathbb{R}$ and define the *geometric series*
of *ratio* x as the sum of its successive powers,

$$\sum_{n=0}^{\infty} x^n = 1 + x + x^2 + x^3 + \dots.$$

Is this series convergent? If $x \neq 0, 1$, its partial sum is obtained from the expression
of the geometric progression in Example 1.5,

$$s_n = \frac{x^{n+1} - 1}{x - 1}.$$

It is immediate to see that it diverges to $+\infty$ if $x > 1$. The same is true when $x = 1$, in which case $s_n = n$. If $x = -1$ the sequence of partial sums oscillates between $+1$ and 0, without converging. If $x < -1$, the sequence still oscillates, but this time the range of the oscillations is increasing and $\lim \inf_{n \to \infty} s_n = -\infty$ while $\lim \sup_{n \to \infty} s_n = +\infty$. Finally, for $-1 < x < 1$ the series converge and

$$\sum_{n=0}^{\infty} x^n = \frac{1}{1-x}.$$

The next result follows from the completeness of $\mathbb{R}$.

Theorem 5.20 $\sum_{n=1}^{\infty} a_n$ *converges if and only if* $\forall \epsilon > 0$, $\exists n_\epsilon \in \mathbb{N}$ *such that* $\forall n, m > n_\epsilon$, $|\sum_{i=n}^{m} a_i| < \epsilon$.

Proof The statement follows by noticing that $|\sum_{i=n}^{m} a_i|$ is simply $|s_n - s_m|$ so that the Theorem is a re-statement of the fact that a sequence in $\mathbb{R}$ converges if and only if it is a Cauchy sequence. □

The previous theorem implies that if the series $\sum_{n=1}^{\infty} a_n$ is convergent, then $\lim_{n \to \infty} |a_n| = 0$. However, the latter is not sufficient to imply the convergence of the series. Conversely, the convergence of the series of absolute values is a sufficient condition for the convergence of the original series. If the absolute values of the terms of the series form a converging series, the series is *absolute convergent*.

Theorem 5.21 (Absolute convergence) *If* $\sum_{n=1}^{\infty} |a_n|$ *converges, then* $\sum_{n=1}^{\infty} a_n$ *converges.*

Proof The statement follows directly from Theorem 5.20 by noticing that for any n and m, the triangle inequality, Theorem 4.3, implies that $\forall \epsilon > 0$, if $\sum_{i=n}^{m} |a_i| < \epsilon$, then $|\sum_{i=n}^{m} a_i| < \epsilon$. □

In other words, an absolutely convergent series is convergent. If the terms of the sequence are positive, $a_n > 0$, then the sequence of partial sums increases monotonically.

Theorem 5.22 *If eventually* $a_n \geq 0$ *or* $a_n \leq 0$, *then* $\sum_{n=1}^{\infty} a_n$ *converges if and only if the set of partial sums* $\{s_n\}$ *is bounded.*

Proof The statement follows directly from Theorem 5.13. □

There exist several results in the literature that provide sufficient conditions for the convergence or divergence of series. In common parlance, these conditions are called "test" or "criteria". A general convergence criterion can be obtained by comparing the series term by term.

Theorem 5.23 (Comparison test) *Assume that eventually $a_n \geq b_n \geq 0$. Then, if $\sum_n a_n$ converges, so does $\sum_n b_n$. While if $\sum_n b_n$ diverges, so does $\sum_n a_n$.*

Proof For the first statement, let $\sum_n a_n = a$, and note that $\sum_{n=1}^m b_n \leq \sum_{n=1}^m a_n \leq a$, so that, from Theorem 5.22, the result follows. The second statement is proved analogously. □

The fact that the terms of the series are positive is essential to prove the previous theorem. In practise, this theorem is often used in conjunction with Theorem 5.21 to prove convergence via absolute convergence.

5.5.1 Series with Decreasing Terms

When the terms of the series decrease in magnitude, there are several results that can be used to prove its convergence. The next theorem shows that the convergence of the series with positive decreasing terms can be deduced by analysing the behaviour of a subset of terms. The idea is to pick appropriately spaced terms and sum them with appropriate weights. This result is due to the German mathematician Oscar Schlömilch (1823–1901).

Theorem 5.24 (Schlömilch's condensation test) *Let (a_n) with $a_n > 0$ be monotonically decreasing. Consider a strictly increasing sequence of natural numbers (u_n), with $u_0 = 1$. Let $\Delta u_n = u_{n+1} - u_n$ and assume that there exists a constant $\exists c > 0$ such that $\forall n$, $\Delta u_n / \Delta u_{n-1} < c$. Then $\sum_{n=1}^\infty a_n$ converges if and only if $\sum_{n=0}^\infty \Delta u_n a_{u_n}$ converges.*

Proof Since the series (a_n) has positive decreasing terms, $\forall n$,

$$\Delta u_{n+1} a_{u_{n+1}} / c < \Delta u_n a_{u_{n+1}} \leq a_{u_n} + a_{u_n+1} + a_{u_n+2} + \ldots + a_{u_{n+1}-1} \leq \Delta u_n a_{u_n},$$

where, in the first inequality, we have used the assumed relation between the successive increments of the sequence (u_n). By summing the inequality above over n, the terms in the middle cover all the elements of the series, so that

$$\frac{1}{c} \sum_{n=1}^\infty \Delta u_n a_{u_n} < \sum_{n=1}^\infty a_n \leq \sum_{n=0}^\infty \Delta u_n a_{u_n}.$$

The statement follows from the comparisons test, Theorem 5.23. □

The previous theorem has a famous special case.

Corollary 5.6 *(Cauchy's condensation test) Let (a_n) with $a_n > 0$ be monotonically decreasing. Then $\sum_n a_n$ converges if and only if $\sum_k 2^k a_{2^k}$ converges.*

Proof Take $u_n = 2^n$ in Theorem 5.24.   □

▶ **Example 5.21** *(Harmonic series)* Probably the simplest series that one can imagine is the *harmonic series* $\sum_{n=1}^{+\infty} 1/n$. We can generalise the analysis of the convergence of this series to the so called *p-series* or *hyperharmonic series*

$$\sum_{n=1}^{+\infty} \frac{1}{n^p}, \quad \text{with} \quad p > 0.$$

To investigate its convergence, apply Corollary 5.6 and consider the "condensed" series

$$\sum_{k=0}^{\infty} 2^k \frac{1}{2^{kp}} = \sum_{k=0}^{\infty} (2^k)^{1-p} = \sum_{k=0}^{\infty} (2^{1-p})^k.$$

This is a geometric series that converges if $2^{1-p} < 1$, i.e. $p > 1$. We can conclude that the *p*-series converge if and only if the parameter p is greater than one. In particular, the harmonic series with $p = 1$ does not converge.

If the terms of a series are the product of the terms of a bounded series and the elements of a monotonic positive sequence that converges to zero, the convergence of the first series can be proved by the following test originally derived by the German mathematician Johann Peter Gustav Lejeune Dirichlet (1805–1859).

Theorem 5.25 (Dirichlet's test) *Let (a_n) be a positive decreasing sequence that converges to zero, that is $a_{n-1} \geq a_n \geq 0$ and $a_n \to 0$. Let (b_n) denotes the term of a bounded sequence, that is, $\exists M > 0$ such that $\forall n$, $|\sum_{h=1}^{n} b_h| < M$. Then the series $\sum_{n=1}^{+\infty} a_n b_n$ is convergent.*

Proof Define the partial sums $S_n = \sum_{h=1}^{n} a_h b_h$ and $B_n = \sum_{h=1}^{n} b_h$. Using summation by parts, Lemma 1.1,

$$S_n = a_n B_n + \sum_{h=1}^{n-1} B_h (a_h - a_{h+1}).$$

By hypothesis $\lim_{n \to +\infty} a_n B_n = 0$. Therefore, the limit of the composed series $\sum_{n=1}^{+\infty} a_n b_n$ is equal to the limit of $\sum_{n=1}^{+\infty} B_n (a_n - a_{n+1})$. Note that $|B_n (a_n - a_{n+1})| < M(a_n - a_{n+1})$, and the series $\sum_{n=1}^{+\infty} M(a_n - a_{n+1})$ converges to Ma_1. Then, by Theorem 5.23, the series $\sum_{n=1}^{+\infty} B_n (a_n - a_{n+1})$ is absolutely convergent and thus, by Theorem 5.21, convergent.   □

The previous theorem is often applied to series with terms of decreasing magnitude and of alternating sign. If (a_n) is a positive decreasing sequence that converges to zero, the series $\sum_{n=1}^{+\infty}(-1)^n a_n$ converges. In this case, the role of the bounded series in Theorem 5.25 is played by the series $\sum_{n=1}^{+\infty}(-1)^n$ which takes alternating values of -1 and 0. This special case is named *Leibnitz criterion* after the German mathematician Gottfried Wilhelm (von) Leibniz (1646–1716).

▶ **Example 5.22** *(Alternating harmonic series)* Consider the alternating harmonic sequence $\sum_{n=1}^{+\infty}(-1)^{n+1}/n^p$. According to Theorem 5.25, and differently from the harmonic series, this series converges $\forall p > 0$. In particular, as we will see in Example 6.4, when $p = 1$ its limit is $\log 2$.

If (a_n) is a positive and decreasing sequence and $s = \sum_{n=1}^{+\infty}(-1)^n a_n$, then for the partial sum we have $s - s_n = \sum_{h=n+1}(-1)^h a_h$ so that $s - s_n \geq -a_{n+1}$ if n is even and $s - s_n \leq a_{n+1}$ if n is odd. Thus, in general, $|s - s_n| < a_{n+1}$. This represents a simple way to measure the approximation to s provided by the truncated series. Another common test for sequences with decreasing terms requires the notion of integral and is presented in Theorem 8.8.

5.5.2 Tests Based on the Asymptotic Behaviour of Terms

In this section, two theorems are presented that provide sufficient conditions for the convergence or divergence of series based on the asymptotic behaviour of their terms, that is, on the speed with which the term a_n goes to zero when n increases.

Theorem 5.26 (Root test) *Given the series $\sum_n a_n$, consider*

$$\alpha = \limsup_{n \to \infty} \sqrt[n]{|a_n|}.$$

Then if $\alpha < 1$ the series converges and if $\alpha > 1$ the series diverges.

Proof Assume that $\alpha < 1$. Then there exist $\beta \in (0, 1)$ and $n_0 \in \mathbb{N}$ such that $\forall n > n_0$, $\sqrt[n]{|a_n|} < \beta$, which implies $|a_n| < \beta^n$. Since the series $\sum_n \beta^n$ is convergent (see Example 5.20), for the comparison test (Theorem 5.23), the original series is absolutely convergent and, thus, convergent.

Conversely, assume that $\alpha > 1$. Then there exist $\beta > 1$ and an unbounded sequence of integers (m) such that $\sqrt[m]{|a_m|} > \beta$. This implies $|a_m| > \beta^m > 1$ so that the sequence $(|a_n|)$ does not converge to zero. Since this is a necessary condition for the convergence of the series (see Theorem 5.20), the assertion is proved.   □

The root test is powerful, as it allows one to decide the convergence or divergence of $\sum_n a_n$ exclusively based on the upper limit of the sequence $(\sqrt[n]{|a_n|})$. The only case where the test does not provide an answer is when $\limsup_{n\to\infty} \sqrt[n]{|a_n|} = 1$. A less general, but sometimes easier to apply, criterion is described below.

Theorem 5.27 (Ratio test) *Consider the series $\sum_n a_n$. If*

$$\limsup_{n\to\infty} |a_{n+1}/a_n| < 1,$$

then the series converges. Instead, if

$$\liminf_{n\to\infty} |a_{n+1}/a_n| > 1,$$

then the series diverges.

Proof In the first case, there exist $\beta \in (0, 1)$ and $n_0 \in \mathbb{N}$ such that $\forall n > n_0$, $|a_{n+1}/a_n| < \beta$, that is $|a_{n+1}| < \beta|a_n|$ and, by recursion, $|a_{n_0+k}| < \beta^k|a_{n_0}|$. The series $\sum_n \beta^n|a_{n_0}|$ is a converging geometric series, so that the original series converges for the comparison test (Theorem 5.23).

In the second case, $\exists n^*$ such that $\forall n > n^*$, $|a_{n+1}| > |a_n|$. Thus, the sequence of absolute values $(|a_n|)$ increases and (a_n) cannot converge to zero. □

The ratio test is not conclusive in all cases in which the upper limit of the sequence $(|a_{n+1}/a_n|)$ is greater than 1 and its lower limit is lower than 1.

▶ **Example 5.23** *(Failure of the ratio test)* The ratio test can fail spectacularly in providing an answer. Consider the series $\sum_n a_n$ with

$$a_n = \begin{cases} 1/2^n & \text{if } n \text{ is odd;} \\ 1/3^n & \text{if } n \text{ is even.} \end{cases}$$

This series is clearly convergent, as its terms are bounded above by those of the geometric series of ratio $1/2$ (see Example 5.20). However, a_{n+1}/a_n is $1/3(2/3)^n$ when n is even and $1/2(3/2)^n$ when n is odd. Thus $\liminf_{n\to\infty} \left|\frac{a_{n+1}}{a_n}\right| = 0$ but $\limsup_{n\to\infty} \left|\frac{a_{n+1}}{a_n}\right| = +\infty$ and the test is not conclusive. On the contrary, $\forall n$, $\sqrt[n]{|a_n|} < 1/2$, and using the root test we can confirm the convergence of the series.

The fact that the root test is stricter than the ratio test is general. Suppose that the condition for convergence of the ratio test is satisfied: $\limsup_{n\to\infty} |a_{n+1}/a_n| = \delta < 1$. This means that for sufficiently large n and for all j, $|a_{n+j}| < \delta^j|a_n|$, which implies $\sqrt[n+j]{|a_{n+j}|} < \delta^{j/(n+j)}|a_n|^{1/(n+j)}$. The right-hand side converges to δ when j goes to infinity, so the condition for convergence of the root test is satisfied.

We conclude this section with a further characterisation of the exponential function as a limit of a series.

▶ **Example 5.24** *(Exponential function as a series)* Let x be a real number and consider

$$S(x) = \sum_{n=0}^{\infty} \frac{x^n}{n!}.$$

Applying the ratio test to $a_n = x^n/n!$, we see that

$$\limsup_{n \to \infty} \left| \frac{a_{n+1}}{a_n} \right| = \lim_{n \to \infty} \frac{|x|}{n+1} = 0.$$

Therefore, we can conclude that the series is convergent for any x. Consider the function $S(x)$ that assigns to each point x the value of the associated series. First of all, notice that $S(0) = 1$. Using the expression of the power of the binomial,

$$S(x+y) = \sum_{n=0}^{\infty} \frac{(x+y)^n}{n!} = \sum_{n=0}^{\infty} \sum_{k=0}^{n} \frac{x^k}{k!} \frac{y^{n-k}}{(n-k)!}.$$

By rearranging the terms of the double summation,

$$S(x+y) = \sum_{k=0}^{\infty} \sum_{h=0}^{\infty} \frac{x^k}{k!} \frac{y^h}{h!} = S(x)S(y).$$

The relation $S(x+y) = S(x)S(y)$ is the defining property of the exponential function. We can prove that $S(x) = e^x$, where e is Euler's number e defined in Example 5.12. Applying the binomial expansion note that

$$\left(1 + \frac{1}{n}\right)^n = \sum_{h=0}^{n} \binom{n}{h} \frac{1}{n^h} = \sum_{h=0}^{n} \frac{1}{h!} \left(1 - \frac{1}{n}\right)\left(1 - \frac{2}{n}\right) \cdots \left(1 - \frac{h-1}{n}\right) < \sum_{h=0}^{n} \frac{1}{h!}.$$

So, taking the limit, $e \le S(1)$. At the same time, if $m < n$

$$\left(1 + \frac{1}{n}\right)^n \ge \sum_{h=0}^{m} \frac{1}{h!} \left(1 - \frac{1}{n}\right)\left(1 - \frac{2}{n}\right) \cdots \left(1 - \frac{h-1}{n}\right).$$

Indeed, the expansion of the power on the left-hand side contains $n + 1$ terms while the summation on the right-hand side contains only $m + 1$ terms. Taking the limit $n \to \infty$ of both sides, we get

$$e \ge \sum_{h=0}^{m} \frac{1}{h!}.$$

Since this is valid $\forall m$, it should be $e \ge S(1)$, and the assertion is proved.

There are several methods to sum the elements of a sequence that can be found in applications. The most common is discussed in the next example.

▶ **Example 5.25** *(Cesàro summation)* Given a sequence (a_n), consider the sequence of partial sums (s_n), $s_n = \sum_{k=1}^{n} a_k$. If

$$\lim_{n \to \infty} \frac{1}{n} \sum_{k=1}^{n} s_k = l \in \mathbb{R},$$

the sequence (a_n) is said to be *Cesàro summable* and l is its *Cesàro summation*. In other words, a sequence is Cesàro summable if the Cesàro mean of its partial sums converges. Using Corollary 5.3, it is immediate to see that a summable sequence is also Cesàro summable. The opposite is not true. A sequence can be Cesàro summable, but not summable, see Exercise 5.36.

The next example discusses a useful property characterising converging series.

▶ **Example 5.26** *(Kronecker's lemma)* Let $\sum_n a_n$ be a converging series and consider an increasing unbounded sequence of positive terms $0 < b_1 < b_2 < \dots$. Then

$$\lim_{n \to \infty} \frac{1}{b_n} \sum_{k=1}^{n} b_k a_k = 0.$$

To see it, define the partial sum $s_n = \sum_{k=1}^{n} a_k$, setting $s_0 = 0$. Then using summation by parts, Lemma 1.1,

$$\sum_{k=1}^{n} b_k (s_k - s_{k-1}) + \sum_{k=1}^{n} s_k (b_{k+1} - b_k) = b_n s_n,$$

where we used the fact that $s_0 = 0$. Note that $\sum_{k=1}^{n} b_k (s_k - s_{k-1}) = \sum_{k=1}^{n} b_k a_k$. Thus, reorganising terms and dividing by b_n,

$$\frac{1}{b_n} \sum_{k=1}^{n} b_k a_k = s_n - \frac{1}{b_n} \sum_{k=1}^{n} s_k (b_{k+1} - b_k).$$

Define $c_n = \sum_{k=1}^{n} s_k (b_{k+1} - b_k)$. Note that because the series is convergent,

$$\lim_{n \to \infty} \frac{c_n - c_{n-1}}{b_n - b_{n-1}} = \lim_{n \to \infty} s_n = l.$$

Therefore, by Theorem 5.16, $\lim_{n \to \infty} s_n - c_n/b_n = 0$, and the statement is proved. This result, attributed to the German mathematician Leopold Kronecker (1823–

1891), is also useful in the study of sequences. In fact, it implies that for any sequence (a_n), if the series $\sum_n a_n/b_n$ converges, then $\lim_{n\to\infty} \sum_{k=1}^{n} a_k/b_n = 0$.

5.6 Sequences and Series of Functions

Consider the set of functions $\mathcal{F}$ from a set A to a topological space (X, T) and let (f_n) be a sequence of functions, that is, a map from $\mathbb{N}$ to $\mathcal{F}$. For any $a \in A$, we can consider the sequence $(f_n(a))$ of elements of X.

Definition 5.18 *(Pointwise convergence)* If $\forall a \in A$ the sequence of images $(f_n(a))$ is convergent, we say that the sequence of functions (f_n) is *pointwise convergent*, and we write $f = \lim_{n\to\infty} f_n$, where the function $f : A \to X$ is defined using the limit of the sequences of images $f(a) = \lim_{n\to\infty} f_n(a)$.

For example, the sequence of functions (x^n) pointwise converges to the zero function in $[0, 1)$. When dealing with sequences of functions, an interesting question is whether some property of the functions composing the sequence is preserved when taking the limit. The answer is generally negative. Consider a sequence of functions (f_n) that map a topological space X in a topological space Y, and assume that the sequence is pointwise convergent $f_n \to f$. Now, take a converging sequence (x_m) in X, $x_m \to x$. Assuming that the limit exists, define $y = \lim_{m\to\infty} f(x_m)$ and, for each n, $y_n = \lim_{m\to\infty} f_n(x_m)$. Can we conclude that $y_n \to y$? To prove this, we should be able to swap the order of the limits, that is, it should be $\lim_{n\to\infty} \lim_{m\to\infty} f_n(x_m) = \lim_{m\to\infty} \lim_{n\to\infty} f_n(x_m)$. The following example shows that this inversion is not always possible.

▶ **Example 5.27** *(Limits cannot be swapped)* For any $n \in \mathbb{N}$ consider the sequence $(n/(n+m))$ with $m \in \mathbb{N}$. This is a sequence of sequences. Now notice that

$$\lim_{n\to\infty}\left(\lim_{m\to\infty}\frac{n}{n+m}\right) = 0$$

while

$$\lim_{m\to\infty}\left(\lim_{n\to\infty}\frac{n}{n+m}\right) = 1.$$

The fact that, in general, changing the order of limits changes the final result implies that even if all functions that form the pointwise converging sequence $f_n \to f$ are continuous, the continuity of the limit function f is not guaranteed.

▶ **Example 5.28** *(Continuous functions having a discontinuous limit)* Consider the sequence of real functions (f_n) defined as

$$f_n(x) = \begin{cases} -1 & \text{if } x \leq -\frac{1}{n}, \\ nx & \text{if } -\frac{1}{n} < x < \frac{1}{n}, \\ 1 & \text{if } x \geq \frac{1}{n}. \end{cases}$$

All functions f_n are continuous on $\mathbb{R}$, but the limit

$$f(x) = \lim_{n \to \infty} f_n(x) = \begin{cases} -1 & \text{if } x < 0, \\ 0 & \text{if } x = 0, \\ 1 & \text{if } x > 0, \end{cases}$$

is not continuous in $x = 0$.

5.6.1 Uniform Convergence

If the image space of the functions is a metric space, we can impose a stricter requirement for the convergence of a sequence of functions.

Definition 5.19 *(Uniform convergence)* The sequence of functions $(f_n(x))$ with images in a metric space (Y, d) *uniformly converges* to the function $f(x)$ if, $\forall \epsilon > 0$, $\exists n \in \mathbb{N}$ such that if $m > n$, $d(f_m(x), f(x)) < \epsilon$, $\forall x$.

A sequence of functions that is uniformly convergent is clearly also pointwise convergent. However, in the case of uniform convergence, we can actually swap the order of the limits discussed above.

Theorem 5.28 *Let (f_n) be a uniformly convergent sequence of functions from a topological space (X, T) to a metric space (Y, d). Then if f_n is continuous $\forall n$, the limit function $f = \lim_{n \to \infty} f_n$ is continuous.*

Proof Take $\epsilon > 0$. For the triangle inequality,

$$d(f(x), f(y)) \leq d(f(x), f_n(x)) + d(f_n(x), f_n(y)) + d(f(y), f_n(y)).$$

Since the sequence of functions is uniformly convergent, $\forall x \in X$, there exists and n such that $d(f(x), f_n(x)) < \epsilon/3$ and $d(f(y), f_n(y)) < \epsilon/3$. Furthermore, since f_n is continuous, there is a neighbourhood $N(x)$ of x, such that $\forall y \in N(x)$, $d(f_n(x), f_n(y)) < \epsilon/3$. Thus, substituting in the previous inequality, we see that $\forall y \in N(x)$, $d(f(x), f(y)) < \epsilon$, and the theorem is proved.   □

A useful criterion for the uniform convergence of a sequence of real-valued functions was first discovered by the Italian mathematician Ulisse Dini (1845–1918).

Theorem 5.29 (Dini theorem) *Consider a sequence of continuous functions* (f_n), $f_n : K \subseteq X \to \mathbb{R}$ *with* K *compact, pointwise converging to a continuous function* f. *Assume that the sequence* $(f_n(x))$ *is monotonic, increasing or decreasing, in the same direction, for all* $x \in K$. *Then the sequence* (f_n) *is uniformly convergent.*

Proof For definiteness, let us assume that the sequence $(f_n(x))$ is increasing $\forall x \in K$. Let $g_n(x) = f(x) - f_n(x) \geq 0$. For any $\epsilon > 0$ define $E_n(\epsilon) = \{x \in K, g_n(x) < \epsilon\}$. By hypothesis, $\forall x$, the sequence $(g_n(x))$ decreases in n. Thus $E_n(\epsilon) \subseteq E_{n+1}(\epsilon)$. Again, by hypothesis, $g_n(x)$ are continuous functions, so $E_n(\epsilon)$ is an open set (the preimage of $(-\infty, \epsilon)$). The sequence $(g_n(x))$ is pointwise convergent to 0, therefore, $\forall x \in K, \exists n_x$ such that $x \in E_{n_x}(\epsilon)$ and $\cup_x E_{n_x}(\epsilon)$ is a cover of K. Since K is compact, there exists a finite subcover and because they are nested, the subcover is made by a single open set $E_{n'}(\epsilon)$. Thus, $\forall n \geq n'$ and $\forall x$, $f(x) - f_n(x) < \epsilon$, and the statement is proved. If the sequence is decreasing you can repeat the same proof using $g_n(x) = f_n(x) - f(x)$.   □

▶ **Example 5.29** *(The base of the natural logarithm)* Using Theorem 5.29, it is immediate to realise that the sequence (f_n) defined in Example 5.8 and the sequences (g_n) in Example 5.14 are uniformly convergent in any closed interval $[a, b]$ of the real line. Now we want to prove that the exp function in Example 5.14 is the inverse of the log function in Example 5.8. That is, we have to prove that $\forall x \in \mathbb{R}$, $f(\lim_{n\to\infty} g_n(x)) = x$. Notice that $\forall n \in \mathbb{N}$, $f_n(g_n(x)) = x$. Therefore, given the continuity of f, we have to prove that $\lim_{n\to\infty} f(g_n(x)) - f_n(g_n(x)) = 0$. Consider the closed interval $[g_1(x), \exp x]$. In Example 5.14 we proved that $g_n(x)$ belongs to this interval for any n. In this interval, the sequence (f_n) is uniformly convergent, thus given any $\epsilon > 0$, there is a n_ϵ such that if $n > n_\epsilon$, $f_n(y) - f(y) < \epsilon$ for any y in the interval. In particular, for $y = g_n(x)$. This proves the statement. Using similar arguments, you can easily prove that $\log(e) = 1$ (see Example 5.12) so that the log function defined in Example 5.8 is actually the logarithm in base e.

The following theorem offers an alternative definition of uniform convergence.

Theorem 5.30 *The sequence of functions* (f_n), *with* $f_n : X \to Y$ *and* (Y, d) *metric space, is uniformly convergent to* f *if and only if*

$$\lim_{n\to\infty} \sup_{x \in X} d(f_n(x), f(x)) = 0.$$

Proof Consider $\epsilon > 0$. If (f_n) is uniformly convergent, for n sufficiently large, $d(f_n(x), f(x)) < \epsilon/2$ for any x. This implies $\sup_{x \in X} d(f_n(x), f(x)) < \epsilon$. At the same time, if for n sufficiently large, $\sup_{x \in X} d(f_n(x), f(x)) < \epsilon$, then it must be $d(f_n(x), f(x)) < \epsilon$ for any x. $\qquad\square$

▶ **Example 5.30** *(Banach spaces of bounded and continuous functions)* Theorem 5.30 connects the notion of uniform converge with the convergence under the metric d_∞ introduced in Examples 3.1. However, the completeness of the space of bounded functions rests on the completeness of the image space, as discussed in Examples 5.4. Moreover, Theorem 5.28 guarantees that when the image space is complete, the space of bounded continuous functions is complete too, see Example 5.19.

Similarly to what was done for numerical series, we can investigate the convergence of the series of functions $\sum_{n=1}^{\infty} f_n$. The series is pointwise or uniformly convergent if the sequence of partial sums of functions is so. A series of functions is *normally convergent* if $\sum_{n=1}^{\infty} \sup_x\{|f_n(x)|\}$ is convergent. Normal convergence implies uniform convergence.

Theorem 5.31 (Weiersrass M-test) *Consider the series of functions $\sum_{n=1}^{\infty} f_n$ and assume there exists a sequence (M_n) of positive numbers such that $\forall x$, $|f_n(x)| \leq M_n$. Then, if $\sum_{n=1}^{\infty} M_n$ is convergent, the series of functions is uniformly convergent.*

Proof By hypothesis, the series $\sum_{n=1}^{\infty} |f_n(x)|$ is convergent for any x. Thus, by Theorem 5.21, the original series is pointwise convergent. Let $f(x) = \sum_{n=1}^{\infty} f_n(x)$. Then notice that

$$\sup_x \left| \sum_{i=1}^{n} f_i(x) - f(x) \right| = \sup_x \left| \sum_{i=n+1}^{\infty} f_i(x) \right| \leq \sup_x \sum_{i=n+1}^{\infty} |f_i(x)| \leq \sum_{i=n+1}^{\infty} M_n.$$

Since the series $\sum_{n=1}^{\infty} M_n$ is convergent, the latter expression is lower than any $\epsilon > 0$ for n sufficiently large (see Theorem 5.20) thus, from Theorem 5.30, the statement follows. $\qquad\square$

Exercises

Exercise 5.1 Let $N_1, N_2 \subset \mathbb{N}$ be two infinite disjoint subsets such that $N_1 \cup N_2 = \mathbb{N}$. Consider a sequence (x_n) in a topological space (X, T) and the two subsequences (x_{n_1}) and $(x)_{n_2}$ with $n_i \in N_i$, $i = 1, 2$. Prove that if $x_{n_1} \to l$ and $x_{n_2} \to l$, then $x_n \to l$.

Exercise 5.2 In the Euclidean topology on $\mathbb{R}$, use the definition of limit to prove that $\lim_{n\to\infty} 1/n^k = 0$ for any positive integer k and that $\lim_{n\to\infty} (n^h + n^k)/n^k = 1$ for two positive integers k, h such that $k > h$. *Hint: You can use a base of the topology.*

Exercise 5.3 In the Euclidean topology on $\mathbb{R}$, use the definition of limit to prove that $\lim_{n\to\infty} 1/\sqrt[k]{n} = 0$ for any positive integer k and that $\lim_{n\to\infty}(\sqrt[h]{n} + \sqrt[k]{n})/\sqrt[k]{n} = 1$ for two positive integers k, h such that $k < h$. *Hint: Remember that $\sqrt[k]{n} = n^{1/k}$.*

Exercise 5.4 Considering Example 5.3, develop a recursive proof of the fact that $a_n > n$ if $n > 5$.

Exercise 5.5 Consider the metric space $\mathbb{L}$ of bounded functions $f : [-1, 1] \to \mathbb{R}$ with the sup distance (see Example 3.1). In this space, build a sequence of continuous functions (f_n) and a sequence of non-continuous functions (g_n) converging to $f(x) = x^2$.

Exercise 5.6 Let A be a closed bounded subset of $\mathbb{R}$ and $I = [a, +\infty)$ an unbounded interval. Prove that they are complete subspaces of $(\mathbb{R}, |\cdot|)$. *Hint: Use the theorems Luke.*

Exercise 5.7 Let F be the set of all binary sequences $f : \mathbb{N} \to \{0, 1\}$. Consider the metric space (F, d) where d is defined as

$$d(f, g) = \lim_{h\to\infty} \sum_{n=1}^{h} \frac{|f(n) - g(n)|}{2^n}, \; f, g \in F.$$

Is the metric space (F, d) complete? If so, prove it. Otherwise, provide a counterexample. *Hint: This corresponds to a metric space you should already know*

Exercise 5.8 Prove that the following functions are contractions in the indicated intervals:

$$f_1(x) = \sqrt{x}, x \in [1, +\infty); \quad f_2(x) = x^2, x \in [0, 1/3]; \quad f_3(x) = x^3, x \in [0, 1/\sqrt{6}].$$

Exercise 5.9 Consider the recursively defined sequence $a_1 = 2, a_{n+1} = 2 + \sqrt{a_n}$. Use the Banach fixed point theorem to prove that this sequence converges. Compute its limit.

Exercise 5.10 Prove that, in any metric space, if f and g are contractions, then their composition is a contraction.

Exercise 5.11 Consider the recursively defined sequence $a_1 = \sqrt{2}, a_{n+1} = \sqrt{2 + \sqrt{a_n}}$ Use the Banach fixed point theorem to prove that this sequence converges to one root of the equation $x^4 - 4x^2 - x + 4 = 0$ laying between $\sqrt{3}$ and 2.

Exercise 5.12 Consider the map $f(x) = x + x^{-1}$ from $[1, \infty)$ to itself. Prove $|f(x) - f(y)| < |x - y|$ for any couple $x, y \in [1, \infty)$. Show that, nevertheless, the map has no fixed points. Why the Banach fixed point theorem does not apply?

Exercise 5.13 Consider a sequence of real numbers (x_n). Prove that if $a_n \to a$, then $|a_n| \to |a|$.

Exercise 5.14 Use the definition of the limit of a sequence in $\mathbb{R}$ to prove that

- if $-1 < x < 1$, then $\lim_{h \to \infty} x^h = 0$;
- if $|x| > 1$, the sequence (x^h) is unbounded;
- if $x > 0$, then $\lim_{h \to \infty} x^{1/h} = 1$.

Hint: Remember the property of the log *function in Example 5.8*

Exercise 5.15 (AM–GM inequality) Using the concavity of the $\log x$ function derived in Example 5.9 and the Jensen inequality described in Corollary 1.1, proves Theorem 1.14. *Hint: Use rescaled weights that sum to one.*

Exercise 5.16 Consider the *arithmetic progression* $a_n = a_{n-1} + q$ with generic a_1 and $q \neq 0$. Proves that the sequence (a_{n+1}/a_n) converges to 1.

Exercise 5.17 Consider the *geometric progression* $a_n = q a_{n-1}$ with a generic a_1 and discuss its asymptotic behaviour as a function of a_1 and q.

Exercise 5.18 (Babilonian algorithm) For any $x > 0$, and $a \neq 0$, consider the recursively defined sequence

$$\begin{cases} a_1 = a, \\ a_{n+1} = \frac{1}{2}\left(a_n + \frac{x}{a_n}\right). \end{cases}$$

Prove that if $a > 0$, then $a_n \to \sqrt{x}$ and if $a < 0$, then $a_n \to -\sqrt{x}$. *Hint: Prove that the sequence is monotonic and bounded.*

Exercise 5.19 Using the result in Example 5.12, prove that the sequence (a_n) with

$$a_n = \left(1 + \frac{1}{n^k}\right)^n$$

converges to 1 if $k > 1$ and diverges if $k < 1$. *Hint: Use the inequality in Example 5.8.*

Exercise 5.20 Show that $\lim_{n \to \infty} p^{1/n} = 1$ for any $p > 0$.

Exercise 5.21 Let $0 \le a < 1$. Consider the sequence (x_n) defined by

$$x_{n+1} = ax_n + \frac{1}{n+1} \quad \text{and} \quad x_0 = 1.$$

Prove $\lim_{n\to\infty} x_n = 0$ and compute $\lim_{n\to\infty} x_n/a^n$. *Hint: For the first part, compare the nth element of the sequence with the corresponding fixed point of the map. For the second part, write the expression of the nth element of the sequence.*

Exercise 5.22 Compute the following limits:

$$\lim_{n\to\infty} \frac{n\sqrt{n} - n^2}{n+1}, \ \lim_{n\to\infty} \frac{4n + 2/n}{1/n^2 + 5n}, \ \lim_{n\to\infty} n - \sqrt{n + n^2}, \ \lim_{n\to\infty} \left(3^n + 4^n\right)^{\frac{1}{n}}.$$

Exercise 5.23 Consider the recursively defined sequence $a_1 = 2$, $a_{n+1} = (a_n + 6)/2$. Determine whether (a_n) converges and, in this case, compute its limit.

Exercise 5.24 Consider the recursively defined sequence $a_1 = 1$, $a_{n+1} = 3 - 1/a_n$. Establish if (a_n) converges and, in this case, compute its limit.

Exercise 5.25 Consider the recursively defined sequence $a_1 = \sqrt{2}$, $a_{n+1} = \sqrt{2 + a_n}$. Prove that the sequence converges and compute its limit.

Exercise 5.26 Establish if $\sum_{n=1}^{\infty} 2(\sqrt{n} - \sqrt{n-1}) - 1/\sqrt{n}$ converges or not. *Hint: Remember that $\sum_n 1/n^p$ converges when $p > 1$.*

Exercise 5.27 Let $P_n(x)$ and $Q_m(x)$ be two polynomials in $\mathbb{R}$ of order n and m respectively. Consider the sequence $(P_n(h)/Q_m(h))$ for $h \in \mathbb{N}$ sufficiently large such that $Q_m(h) \neq 0$ (does this h exist?). Prove that if $m > n$, then $\lim_{h\to\infty} P_n(h)/Q_m(h) = 0$. What happens if $m = n$? And if $m < n$? *Hint: See what happens for low degree polynomials and try to generalise.*

Exercise 5.28 Discuss for which values of the real parameters a, b, c, d, f the following limit exists:

$$\lim_{n\to\infty} \left(\frac{an + b}{cn + d}\right)^{fn}.$$

Hint: Analyse what happens in a few cases giving simple values to the parameters. In general use Example 5.14.

Exercise 5.29 Consider a sort of Fibonacci sequence built using recursive convex combination. The sequence starts with two numbers x_1 and x_2 in $\mathbb{R}$ and, for any $n > 2$, $x_n = \lambda x_{n-1} + (1 - \lambda)x_{n-2}$ with $\lambda \in (0, 1)$. Does the sequence have a limit? If so, can you compute it?

Exercise 5.30 Given two sequences (a_n) and (b_n), prove that

$$\liminf_{n\to\infty} a_n + \liminf_{n\to\infty} b_n \leq \liminf_{n\to\infty} a_n + b_n \leq$$

$$\leq \limsup_{n\to\infty} a_n + b_n \leq \limsup_{n\to\infty} a_n + \limsup_{n\to\infty} b_n.$$

Exercise 5.31 Find the upper and lower limits of the sequence defined by the recursion

$$a_n = \begin{cases} a_{n-1} & \text{if } n \text{ is even,} \\ 1/2 + a_{n-1} & \text{if } n \text{ is odd,} \end{cases}$$

and with $a_1 = 0$.

Exercise 5.32 Prove that the Cesàro mean of the sequence (a_n) with $a_n = 1$ if n is odd and $a_n = 0$ if n is even, converges to $1/2$.

Exercise 5.33 With reference to Definitions 5.13 and 5.14, prove that if $a_n = O(b_n)$ and $b_n = O(a_n)$, then the two sequences are infinitesimal of the same order.

Exercise 5.34 If possible, establish the asymptotic order relation between these infinitesimal sequences: $(1/n)$, $(1/(2+n))$, $1 - \cos(1/n)$, $\log(n^2)/n^2$. *Hint: Use what you know about notable limits.*

Exercise 5.35 Given $k \in \mathbb{R}$ and $\alpha, \beta \in [\frac{1}{2}, 1)$, consider a sequence (x_n, y_n) in $\mathbb{R}^2$ recursively defined by the relation

$$(x_{n+1}, y_{n+1}) = \begin{cases} (y_n, (1 - 2\alpha)x_n) & \text{if } n \text{ is even,} \\ ((1 - 2\beta)y_n, x_n) & \text{if } n \text{ is odd,} \end{cases}$$

with $(x_0, y_0) = (k, k)$. Determine whether (x_n, y_n) converges and, in that case, compute its limit.

Exercise 5.36 Prove that the Cesàro summation of the sequence $(a_n = (-1)^n)$ is $1/2$. Note that this sequence is not summable.

Exercise 5.37 Considering Examples 5.12 and 5.24, find an upper bound for the difference $e - \sum_{k=0}^{n} 1/k!$ as a function of n.

Exercise 5.38 For any $a > 1$ proves that

$$\frac{1}{a} \leq \sum_{n=1}^{\infty} \left(\frac{n}{an+1}\right)^n \leq \frac{1}{a-1}.$$

Exercise 5.39 Investigate the convergence of the following series:

1. $\displaystyle\sum_k \frac{1}{k + \sqrt{k}}$ 6. $\displaystyle\sum_k \frac{k^k}{k!}$ 11. $\displaystyle\sum_k \left(\frac{3k}{k+3}\right)^k$

2. $\displaystyle\sum_k \frac{1}{k + k^2}$ 7. $\displaystyle\sum_k (-1)^{k-1}\frac{k}{3^k}$ 12. $\displaystyle\sum_k \frac{1}{k(k+1)}$

3. $\displaystyle\sum_k \frac{k^2}{k!}$ 8. $\displaystyle\sum_k \frac{(k+1)!}{2^k k!}$ 13. $\displaystyle\sum_k \left(\frac{1}{2k+1} - \frac{1}{2k+3}\right)$

4. $\displaystyle\sum_k (-1)^{k-1}\frac{k^2}{k!}$ 9. $\displaystyle\sum_k \frac{4k}{1+k^2}$ 14. $\displaystyle\sum_k \frac{a^k}{k}, a \in (0, 1)$

5. $\displaystyle\sum_k \frac{k}{(k+1)!}$ 10. $\displaystyle\sum_k \frac{2^k}{k3^k}$ 15. $\displaystyle\sum_k \frac{1}{\log k}$

Exercise 5.40 Using the definition of e in Example 5.12 and the definition of the log function in Example 5.8, prove that $\log e = 1$.

Exercise 5.41 Let (a_n) with $a_n > 0$ be monotonic decreasing. Use Theorem 5.24 to prove that $\sum_n a_n$ converges if and only if $\sum_k 3^k a_{3^k}$ converges.

Differential Calculus of Functions of One Variable

6

6.1 Limit of Real Functions

The Euclidean topology on $\mathbb{R}$ has been introduced using the order relation in Definition 2.7 and derived in Example 3.1 as the induced topology of the complete Euclidean metric (see Corollary 5.1). It is Hausdorff (that is, different points have disjoint neighbourhoods, see Theorem 2.17) and second-countable (open intervals with rational boundaries are a base of the topology, see Theorem 2.8). Thus, the limit of a function at a point $x \in \mathbb{R}$ can be defined using the limit of the images, under the function, of the sequence converging to x (see Theorem 5.2). Hence, we can easily derive how the limits of real functions interact with the arithmetic operations and the order relation.

Theorem 6.1 *Let $\lim_{x \to x_0} f(x) = y_0$ and $\lim_{x \to x_0} g(x) = y_0'$. Then*

- $\lim_{x \to x_0} \alpha f(x) + \beta g(x) = \alpha y_0 + \beta y_0'$;
- $\lim_{x \to x_0} f(x) g(x) = y_0 y_0'$;
- *if $y_0' \neq 0$, $\lim_{x \to x_0} f(x)/g(x) = y_0/y_0'$;*
- *if $f(x) \geq g(x)$ in a neighbourhood of x_0, then $y_0 \geq y_0'$ (see Lemma 5.1);*
- *if a function $h(x)$ is such that $f(x) \geq h(x) \geq g(x)$ in a neighbourhood of x_0, and $y_0 = y_0'$, then $\lim_{x \to x_0} h(x) = y_0$ (see Theorem 5.12).*

Proof Consider the case of a linear combination of the two functions f and g in the statement. For any sequence $(z_n) \to x_0$, $(f(z_n)) \to y_0$ and $(g(z_n)) \to y_0'$. From the property of the limit of sequences, $(\alpha f(z_n) + \beta g(z_n)) \to \alpha y_0 + \beta y_0'$. This is the sequence of images of (z_n) under the function $\alpha f(x) + \beta g(x)$ and since this is true for any sequence $(z_n) \to x_0$, because of Theorem 5.2, we have proved the first statement. The other statements are proved in similar ways. □

© The Author(s), under exclusive license to Springer Nature Switzerland AG 2023
G. Bottazzi, *Advanced Calculus for Economics and Finance*, Classroom Companion: Economics, https://doi.org/10.1007/978-3-031-30316-6_6

In the previous chapters, two important results on compact sets in $\mathbb{R}$ were derived. The first result is that a set $K \subseteq \mathbb{R}$ is compact if and only if any infinite subset of K has at least one limit point and all its limit points belong to K. This is usually called the **Bolzano–Weierstrass** theorem and was proved in Sect. 2.3 for a Hausdorff, second-countable topological space. The second result is that a set $K \subseteq \mathbb{R}$ is compact if and only if it is closed and bounded. This is usually called the **Heine–Borel** theorem and was specifically proved for $\mathbb{R}$, in Theorem 2.18.

The set of real numbers is often extended by adding the special points $\pm\infty$ to obtain the set $\bar{\mathbb{R}}$; see Definition 2.11. The notion of the limit of a function is extended to include these points using the appropriate neighbourhoods.

Definition 6.1 *(Infinite limits)* We write $\lim_{x \to x_0} f(x) = +\infty$ if $\forall M > 0$, there exists a neighbourhood of x_0, $N(x_0)$, such that $f(N(x_0)) > M$. Analogously, we write $\lim_{x \to x_0} f(x) = -\infty$ if $\forall M > 0$, there exists a neighbourhood of x_0, $N(x_0)$, such that $f(N(x_0)) < -M$.

Definition 6.2 *(Limits at infinity)* We write $\lim_{x \to +\infty} f(x) = y$ if for any neighbourhood of y, $N(y)$, $\exists M > 0$ such that $f((M, +\infty)) \subseteq N(y)$ and $\lim_{x \to -\infty} f(x) = y$ if for any neighbourhood of y, $N(y)$, $\exists M > 0$ such that $f((-\infty, -M)) \subseteq N(y)$.

In the definition of limit, instead of considering an open neighbourhood of the point, one can consider *left and right neighbourhoods*, defined as open intervals of which the point is the supremum and infimum, respectively.

Definition 6.3 *(Left and right limit)* Let f be a real function and x_0 be an interior or boundary point of its domain. We will say that y is the *left limit* of f in x_0 and denote it with $\lim_{x \to x_0^-} f(x) = y$ if $\forall N(y)$, neighbourhood of y, $\exists \delta > 0$ such that $f((x_0 - \delta, x_0)) \subseteq N(y)$. Analogously, we will say that y is the *right limit* of f in x_0 and denote it with $\lim_{x \to x_0^+} f(x) = y$ if $\forall N(y)$ neighbourhood of y, $\exists \delta > 0$ such that $f((x_0, x_0 + \delta)) \subseteq N(y)$.

Clearly, if $\lim_{x \to x_0^-} f(x) = \lim_{x \to x_0^+} f(x) = y$, then $\lim_{x \to x_0} f(x) = y$. The notion of left and right limits can be extended to $\bar{\mathbb{R}}$ without effort, for example, to define a left limit equal to $+\infty$. The specific definition of these limits should be understood at this point.

▶ **Example 6.1** *(Left and right infinite limits)* Consider the function $f(x) = 1/(x - x_0)$, with $x_0 \in \mathbb{R}$. For any $M > 0$, if $x \in (x_0, x_0 + 1/M))$ then $f(x) > M$ and if $x \in (x_0 - 1/M, x_0)$ then $f(x) < -M$. We can conclude that $\lim_{x \to x_0^+} f(x) = +\infty$ and $\lim_{x \to x_0^-} f(x) = -\infty$. For any $\epsilon > 0$, if $x > x_0 + 1/\epsilon$ or $x < x_0 - 1/\epsilon$, then $|f(x)| < \epsilon$. We can conclude that $\lim_{x \to \pm\infty} f(x) = 0$.

Theorem 5.13 on the limit of monotonic sequences can be applied to derive an intuitive characterisation of the limits of functions. For example, if $f(x)$ increases monotonically in a left neighbourhood (a, x_0) of x_0, then $\lim_{x \to x_0^+} = \sup_{(a,x_0)} f$. Similarly, if there exists an $M > 0$ such that $f(x)$ increases monotonically for any $x > M$, then $\lim_{x \to +\infty} = \sup_{\{x > M\}} f$.

▶ **Example 6.2** *(Notable limits of functions)* In Example 5.3 it was shown that $(x - 1)/x \leq \log x \leq x - 1$, that is $1/(x + 1) \leq \log(1 + 1/x) \leq 1/x$. Therefore, $\lim_{x \to +\infty} x \log(1 + 1/x) = 1$ and $\lim_{x \to +\infty} (1 + 1/x)^x = e$.

The argument in Example 5.3 proves that $\lim_{x \to +\infty} \log(x)/x = 0$, so that $\lim_{x \to +\infty} x^{1/x} = 1$.

Example 5.13 makes clear that for sufficiently small and positive x, $0 \leq \sin x \leq x \leq \tan x$. By symmetry, this means that $\forall x$ in a neighbourhood of 0, $|\sin x| \leq |x| \leq |\tan x|$. Dividing by $\sin x$, for the comparison theorem, this implies that $\lim_{x \to 0} \sin(x)/x = 1$ and for the properties of sine and cosine,

$$\lim_{x \to 0} \frac{\cos(x) - 1}{x^2} = -\lim_{x \to 0} \frac{\sin^2 x}{x^2 (\cos(x) + 1)} = -\frac{1}{2}.$$

6.2   Continuity of Real Functions

The definition of continuity for functions defined over topological spaces is provided in Sect. 2.5. In essence, a function f is continuous at a point x if its limit in x is equal to its value in x. It is continuous in a subset A if it is continuous at all points of A. The property of the limit of functions in Theorem 6.1 guarantees that the sum, product, and reciprocal (when different from zero) of continuous functions are continuous functions. The same is true for the composition of continuous functions (see Theorem 2.24). For continuous functions, the inverse image of an open set is open (see Theorem 2.21), the inverse image of a closed set is closed (see Theorem 2.23), the image of a connected set is connected (see Theorem 2.25), and the image of a compact set is compact (see Theorem 2.26). A continuous function in a compact set has both a maximum and a minimum. This is usually named the **Weierstrass extreme value** theorem and was proved in Theorem 2.30. It is a simple consequence of the fact that a continuous function maps compact sets to compact sets and that, in $\mathbb{R}$, a compact set is closed and bounded. The boundedness guarantees that the supremum and infimum of the image exist. Closure guarantees that they are part of the set, so that they are the image of some point. Then, if we consider a real function f defined over a closed interval $[a, b]$, $f([a, b]) = [\min_{[a,b]} f, \max_{[a,b]} f]$, where $\max_{[a,b]} f$ and $\min_{[a,b]} f$ are the maximum and minimum of the function f in $[a, b]$. However, this fact alone is not sufficient to guarantee continuity. The next example shows that it is easy to define a function that maps a closed interval to a closed interval but it is not continuous.

▶ **Example 6.3** *(Not continuous closed function)* Consider the function

$$f(x) = \begin{cases} \sin \frac{1}{x} & x \neq 0, \\ 0 & x = 0. \end{cases}$$

Note that $f([-2/\pi, 2/\pi]) = [-1, 1]$, but the function is clearly not continuous in 0. In fact, in any neighbourhood of 0, it takes all the values in the interval $[-1, 1]$.

However, according to Theorem 2.28, if the function is monotonic, the fact that it maps an interval into an interval is sufficient to imply continuity. Moreover, in Theorem 2.31, we prove that concave and convex functions are continuous. Thus, many functions we encounter in practice are continuous in a large part of their domain. Polynomials, rational functions, power, exponential, logarithmic, and trigonometric functions are piecewise monotonic and their domains can be partitioned into intervals in which the behaviour of the function is monotonic and in which the function takes all values between the extrema.

▶ **Example 6.4** *(Brouwer's fixed point theorem for real functions)* Let $I = [0, 1]$ and consider $f : I \to I$. We can show that there exists at least one *fixed point* $x^* \in I$ such that $f(x^*) = x^*$. Consider $g(x) = f(x) - x$. If $g(0) = 0$ or $g(1) = 0$, then $x = 0$ or $x = 1$ is the point we are looking for, respectively. Otherwise, $g(0) > 0$ and $g(1) < 0$ but since g takes all the values in $[\min_I g, \max_I g]$, there is at least one point x^* for which $g(x^*) = 0$, that is, $f(x^*) = x^*$. A generalised version of this result is discussed in Appendix B.

Although the continuous behaviour of the functions we are used to consider seems to be the norm rather than the exception, it is possible that there exist isolated points in which the continuity condition is violated. For practical purposes, we classify these points into three groups (Fig. 6.1).

Definition 6.4 Let $f : [a, b] \to \mathbb{R}$ and let $x_0 \in (a, b)$ be a point in which f is not continuous.

- If $\lim_{x \to x_0} f(x)$ exists but is different from $f(x_0)$, then x_0 is a *discontinuity of the first kind* or *trivial*;
- if $\lim_{x \to x_0} f(x)$ does not exist but $\lim_{x \to x_0^-} f(x)$ and $\lim_{x \to x_0^+} f(x)$ exist with different values, then x_0 is a *discontinuity of the second kind* or *jump*;
- if one of the two limits, right and left, does not exist, then x_0 is a *discontinuity of the third kind* or *essential*.

Trivial discontinuities can be eliminated so that the function becomes continuous simply by setting $f(x) = \lim_{y \to x} f(y)$. On the contrary, the second kind of dis-

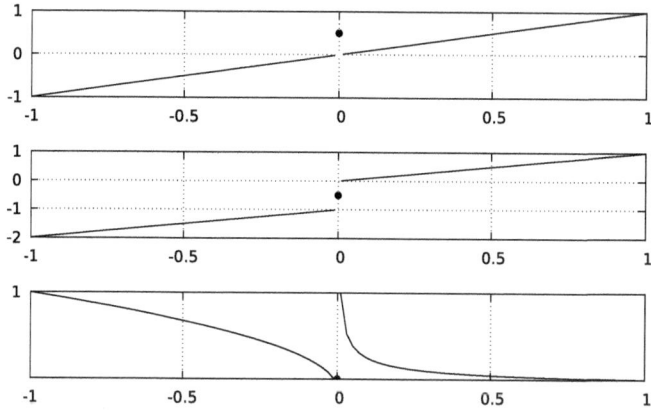

Fig. 6.1 Examples of trivial (top), jump (centre), and essential (bottom) discontinuities. The black dot represents the value of the function in 0. In the case of jump or essential discontinuities, the position of the dot is not relevant in the characterisation of the discontinuity

continuity cannot be eliminated. At this point, the function has a jump, and its values to the left and to the right of the point are separated by a finite amount. If $\lim_{y \to x+} f(y) = f(x)$ the function is said to be *right continuous* in x, while if $\lim_{y \to x-} f(y) = f(x)$ the function is said to be *left continuous*. A nice result exists about the discontinuity of monotonic functions.

Theorem 6.2 *Let $f : E \subseteq \mathbb{R} \to \mathbb{R}$ be monotonic. Then it can only have discontinuities of the second kind, and their number is at most countable.*

Proof Take a point $x \in \mathrm{int}E$ and let E_x^- and E_x^+ be the set of elements of E which are, respectively, lower and greater than x, formally $E_x^- = \{y \in E | y < x\}$ and $E_x^+ = \{y \in E | y > x\}$. Without loss of generality, assume that the function f is increasing. Then the set $f(E_x^-)$ is bounded above, for instance by $f(x')$ with $x' > x$. In the same way, the set $f(E_x^+)$ is bounded below. Furthermore, since the function is increasing, $\lim_{y \to x-} f(y) = \sup f(E_x^-)$ and $\lim_{y \to x+} f(y) = \inf f(E_x^+)$. According to Definition 1.16, both the left and right limits exist. If they are equal, the function is continuous in x. Otherwise, the function has a discontinuity of the second kind. For a decreasing function, simply exchange the sup and the inf in the last equation.

Let $D \subseteq E$ be the set of points in which the function f is discontinuous. For any $x \in D$, define the open interval $I_x = (f^-(x), f^+(x))$ where $f^\pm(x) = \lim_{y \to x\pm} f(y)$. If $x_1, x_2 \in D$ and $x_2 > x_1$, then $f^-(x_2) \geq f^+(x_1)$ so that the intervals do not overlap. Now, for any $x \in D$, let y_x be a rational element of I_x, $y_x \in I_x \cap \mathbb{Q}$. The intervals do not overlap, so that all the y_x are different. Since the set D is in one-to-one relationship with a subset of $\mathbb{Q}$, it is countable. □

With respect to Theorem 2.28, here we have dropped the assumption that the image of the function is an interval, allowing for the presence of jumps. The third kind of discontinuity is often encountered when one of the two limits, left or right, is infinite.

Another useful notion when dealing with real functions is uniform continuity. The definition of uniformly continuous functions is provided in Sect. 3.3. A function is uniformly continuous if the function values can be made as close as we want by taking sufficiently closed points of its domain. The **Heine–Cantor** theorem, proved in Sect. 3.3 for any metric space, guarantees that continuous functions on a compact set are uniformly continuous.

6.3 Differential Analysis

The next definition introduces the notion of derivative, which will be extensively studied in the remainder of this chapter.

Definition 6.5 *(Derivative)* Let $f : (a, b) \subseteq \mathbb{R} \to \mathbb{R}$. For $x \in (a, b)$, consider the function $\Phi(y) = (f(y) - f(x))/(y - x)$ with $y \in (a, b) \setminus \{x\}$. If $\lim_{y \to x} \Phi(y)$ exists we call it the *derivative* of f in x and denote it by $df(x)/dx$ (Leibniz notation), $d/dx f(x)$, or $f'(x)$ (notation initially proposed by the Italian mathematician Giuseppe Luigi Lagrange, 1736–1813). If the derivative is defined in x, the function is *differentiable* in x. If it is defined $\forall x \in (a, b)$, the function is *differentiable* in (a, b).

The derivative is defined at an interior point of an open interval that belongs to the domain of the function. Since any open set is the union of open intervals (see Theorem 2.9), we can say that the derivative of a function can be defined at any interior point in its domain. It is immediate to see that the derivative of the constant function is zero and the derivative of the linear function $f(x) = cx$ is c. The *derivative function* f' assigns to each point x, in which the derivative of f is defined, the value of the derivative.

▶ **Example 6.5** *(Derivatives of sine, cosine, natural logarithm, and exponential functions)* From the trigonometric formula for the sum of two angles,

$$\frac{\sin(x + h) - \sin x}{h} = \sin x \frac{\cos h - 1}{h} + \cos x \frac{\sin h}{h}.$$

Then, using the limits in Example 6.2,

$$\sin' x = \lim_{h \to 0} (\sin(x + h) - \sin x)/h = \cos x.$$

A similar argument shows that $\cos' x = -\sin x$. From the logarithm inequality $(x-1)/x \leq \log x \leq x-1$,

$$\frac{1}{x+h} \leq \frac{\log(x+h) - \log x}{h} = \frac{1}{h} \log\left(1 + \frac{h}{x}\right) \leq \frac{1}{x},$$

so that, by comparison theorem, $\log' x = 1/x$.

Note that $(\exp(x+h) - \exp x)/h = \exp x (\exp(h) - 1)/h$ and from the exponential inequality, $h \leq \exp(h) - 1 \leq h(1-h)$. By the comparison theorem, $\exp' x = \exp x$.

For real functions of one real variable, being differentiable implies being continuous.[1]

Theorem 6.3 *If $f : (a, b) \to \mathbb{R}$ is differentiable in $x_0 \in (a, b)$, then f is continuous in x_0.*

Proof Since the derivative exists, by hypothesis $\lim_{x \to x_0}(f(x) - f(x_0))/(x - x_0) = f'(x) < +\infty$ and obviously $\lim_{x \to x_0}(x - x_0) = 0$. Therefore, by Theorem 6.1,

$$\lim_{x \to x_0}(f(x) - f(x_0)) = \lim_{x \to x_0}\frac{f(x) - f(x_0)}{x - x_0}(x - x_0) = 0.$$

□

A definition similar to 6.5 exists for the boundary of an open interval, based on the notion of left and right limits of Definition 6.3.

Definition 6.6 *(Right and left derivatives)* Let $f : (a, b) \to \mathbb{R}$. The *right derivative* in a and the *left derivative* in b are, respectively,

$$f'^{-}(b) = \lim_{x \to b^-} \Phi(x) \quad \text{and} \quad f'^{+}(a) = \lim_{x \to a^+} \Phi(x),$$

with $\Phi(x)$ as in Definition 6.5.

The function f is said to be *left or right differentiable* in x if the left or right derivatives of f at x exist, respectively.

If the right and left derivatives of a function f exist at a point x and are equal, then the derivative $f'(x)$ is defined and equal to the left and right derivatives. If the right and left derivatives at x are different, the derivative of f in x is not defined. With a little abuse of language, sometimes we say that a function is differentiable in

---

[1] As we will see in Chap. 7, for a function of many variables, the relationship between its continuity and the existence of its derivatives is more complex.

a closed interval $[a, b]$ if it is differentiable in (a, b) and, in addition, it is left and right differentiable in the upper and lower endpoints, respectively.

Theorem 6.4 *(Right and left derivatives of concave and convex functions) A concave or convex function f in an open interval $I \in \mathbb{R}$ always possesses right and left derivatives. If f is concave, $f'^{+} \leq f'^{-}$, while if f is convex, $f'^{+} \geq f'^{-}$.*

Proof Let f be a concave function in an open interval $I \in \mathbb{R}$. According to Theorem 1.16, given any three points $x < y < z$ in I, $f(y) \geq (z - y)/(z - x)f(x) + (y - x)/(z - x)f(z)$. So, $\forall x \in I$ and sufficiently small $h' > h > 0$,
$$\frac{f(x + h') - f(x)}{h'} \leq \frac{f(x + h) - f(x)}{h} \leq \frac{f(x) - f(x - h)}{h} \leq \frac{f(x) - f(x - h')}{h'}.$$

Therefore, the function $g^{+}(h) = (f(x + h) - f(x))/h$ is decreasing and bounded above and the function $g^{-}(h) = (f(x) - f(x - h))/h$ is increasing and bounded below. This implies that $\lim_{h \to 0^{+}} g^{\pm}(h) = f'^{\pm}(x)$ exist and are ordered as in the statement. If the function f is convex, consider the concave function $-f$. □

The proof of the previous theorem also clarifies that if the function f is concave, $\forall y \in I$, $f(y) \leq f(x) + f'^{\pm}(x)(y - x)$. If the function f is convex, $f(y) \geq f(x) + f'^{\pm}(x)(y - x)$.

The derivative introduced in Definitions 6.5 and 6.6 is based on the notion of limit, so all theorems valid for the latter apply to the former. In particular, if the derivative of f and g exists in x_0, then $\forall \alpha, \beta$ constant, the derivative of $\alpha f(x) + \beta g(x)$ exists in x_0 and is equal to $\alpha f'(x_0) + \beta g'(x_0)$. The product of functions requires a specific formula.

Theorem 6.5 *(Product rule) Consider two functions f and g differentiable in x_0. Then their product $h(x) = f(x)g(x)$ is differentiable in x_0 and $h'(x_0) = f'(x_0)g(x_0) + f(x_0)g'(x_0)$.*

Proof The derivative of h in x is the limit $x \to x_0$ of $(f(x)g(x) - f(x_0)g(x_0))/(x - x_0)$. By adding and subtracting the same quantity, this expression can be rewritten as
$$f(x)\frac{g(x) - g(x_0)}{x - x_0} + \frac{f(x) - f(x_0)}{x - x_0}g(x_0).$$

Because f and g are differentiable, and hence continuous, the limit $x \to x_0$ of the first term is $f(x_0)g'(x_0)$ and that of the second term is $f'(x_0)g(x_0)$. □

If a function $f(x)$ is different from zero, then the quotient $f(x)/f(x)$ is constant and its derivative is zero. Based on this simple consideration and Theorem 6.5, it is immediate to derive the following.

Corollary 6.1 *(Reciprocal and quotient rules) If f is a differentiable function in x_0 and $f(x_0) \neq 0$, then its reciprocal $1/f$ is differentiable in x_0 and*

$$\frac{d}{dx}\frac{1}{f(x_0)} = -\frac{f'(x_0)}{f(x_0)^2}.$$

If f and g are differentiable functions in x_0 and $g(x_0) \neq 0$, then their quotient f/g is differentiable in x_0 and

$$\frac{d}{dx}\frac{f(x_0)}{g(x_0)} = \frac{f'(x_0)g(x_0) - f(x_0)g'(x_0)}{g(x_0)^2}.$$

▶ **Example 6.6** *(Derivative of tangent and hyperbolic functions)* Consider the *hyperbolic sine and cosine* functions

$$\sinh x = \frac{e^x - e^{-x}}{2} \quad \text{and} \quad \cosh x = \frac{e^x + e^{-x}}{2}.$$

Using the derivative of e^x from Example 6.5, it is immediate to see that $\sinh' x = \cosh x$ and $\cosh' x = \sinh x$. Their behaviour is similar to the trigonometric functions of the same name, as $\cosh^2 x - \sinh^2 x = 1$. While trigonometric functions represent points in the unit circle whose coordinates solve the equation $x^2 + y^2 = 1$, hyperbolic functions represent the points on a unit hyperbola whose coordinates solve the equation $x^2 - y^2 = 1$.

Using Theorem 6.5, the results in Example 6.5, and the quotient rule, and remembering that $\sin^2 x + \cos^2 x = 1$, it is easy to derive the expression of the derivative of the trigonometric tangent, $\tan x = \sin x / \cos x$,

$$\tan' x = \frac{\sin' x}{\cos x} - \frac{\sin x \cos' x}{\cos^2 x} = \frac{1}{\cos^2 x},$$

and for the hyperbolic tangent, $\tanh x = \sinh x / \cosh x$,

$$\tanh' x = \frac{\sinh' x}{\cosh x} - \frac{\sinh x \cosh' x}{\cosh^2 x} = \frac{1}{\cosh^2 x}.$$

The following theorem states the so-called *chain rule*, that is, the way in which the derivative of composed functions is computed.

Theorem 6.6 *(Chain rule) Let $f(a, b) \to \mathbb{R}$ be continuous in (a, b) and differentiable in $x_0 \in (a, b)$ and let $g : \mathbb{I}_f \to \mathbb{R}$ be differentiable in $f(x_0)$. Then the composed function $h(x) = g(f(x))$ is differentiable in x_0 and $h'(x_0) = g'(f(x_0))f'(x_0)$.*

Proof From the definition of derivative, multiplying and dividing by the same quantity, one gets the following.

$$\lim_{x \to x_0} \frac{g(f(x)) - g(f(x_0))}{x - x_0} = \lim_{x \to x_0} \frac{g(f(x)) - g(f(x_0))}{f(x) - f(x_0)} \frac{f(x) - f(x_0)}{x - x_0}.$$

For continuity, when $x \to x_0$, $f(x) \to f(x_0)$ so that the limits of the two fractions exist separately.   □

▶ **Example 6.7** *(Derivative of power and exponential functions)* Consider the function $f(x) = x^\alpha$. By the property of the natural logarithm, $f(x) = e^{\alpha \log x}$. Using the chain rule of the derivative, $f'(x) = \alpha e^{\alpha \log x}/x = \alpha x^{\alpha - 1}$. Analogously, consider $f(x) = \alpha^x$. Then $f(x) = e^{x \log \alpha}$, whence $f'(x) = e^{x \log \alpha} \log \alpha = \alpha^x \log \alpha$. Note that if $\alpha = e$, the result in Example 6.5 is recovered.

After observing that $f^{-1}(f(x)) = x$, the chain rule can be used to obtain the derivative of the inverse function.

Corollary 6.2 *If the function f is invertible and differentiable in a neighbourhood of a point x_0 and $f'(x_0) \neq 0$, then $(f^{-1})'(f(x_0)) = 1/f'(x_0)$.*

▶ **Example 6.8** *(Derivative of inverse trigonometric and hyperbolic functions)* Let $y(x) = \arcsin x$ be the inverse sine function defined from $[-1, 1]$ to $[-\pi/2, \pi/2]$. It exists because, in the interval $[-\pi/2, \pi/2]$, the sine function is continuous and strictly monotonic. Using the results in Example 6.5 $x'(y) = \cos y$, we have

$$\frac{d \arcsin x}{dx} = \frac{1}{\cos y} = \frac{1}{\sqrt{1 - \sin^2 y}} = \frac{1}{\sqrt{1 - x^2}}.$$

Analogously, let $y(x) = \arccos x$ be the inverse cosine function defined from $[-1, 1]$ to $[0, \pi]$. In this case,

$$\frac{d \arccos x}{dx} = -\frac{1}{\sin y} = -\frac{1}{\sqrt{1 - \cos^2 y}} = -\frac{1}{\sqrt{1 - x^2}}.$$

For the inverse tangent function $y(x) = \arctan x$ defined from $\mathbb{R}$ to $[-\pi/2, \pi/2]$, we have

$$\frac{d \arctan x}{dx} = \cos^2 y = \frac{1}{1 + \tan^2 y} = \frac{1}{1 + x^2}.$$

The expression of the derivative of the hyperbolic inverse functions, arcsinh$x$, defined from $\mathbb{R}$ to $\mathbb{R}$, arccosh$x$, defined from $[1, +\infty)$ to $\mathbb{R}_{>0}$, and arctanh$x$ defined from $(-1, 1)$ to $\mathbb{R}$ is obtained with similar procedures. The reader can convince himself that arcsinh$'x = 1/\sqrt{1 + x^2}$, arccosh$'x = 1/\sqrt{x^2 - 1}$, and arctanh$'x = 1/(1 - x^2)$.

▶ **Example 6.9** *(Geometric interpretation of the derivative)* Although in this book I do not indulge in geometric interpretations of mathematical results, it is probably useful to discuss the geometric analogue of the notion of derivative. In the Cartesian reference system, a straight line has equation $y = \alpha x + \beta$, where α is said to be the *angular coefficient* or slope and β is the *intercept*. The points (x, y) of the line passing through two points (x_1, y_1) and (x_2, y_2) solve $(x-x_1)/(x_2-x_1) = (y-y_1)/(y_2-y_1)$. After rearranging terms, assuming $x_1 \neq x_2$, the equation becomes

$$y = y_1 + \frac{y_2 - y_1}{x_2 - x_1}(x - x_1).$$

Consider two points that belong to the curve $y = f(x)$, having coordinates $(x_1, f(x_1))$ and $(x_2, f(x_2))$. By substitution, the line passing through them is defined by

$$y - f(x_1) = \frac{f(x_2) - f(x_1)}{x_2 - x_1}(x - x_1).$$

When we take x_1 and x_2 close to each other (see Fig. 6.2), the straight line approximates the tangent to the curve at these points. In the limit, the angular coefficient of the tangent is the derivative of the function f.

▶ **Example 6.10** *(Differentiable and Lipschitz continuous functions)* Consider a Lipschitz continuous real-valued function f in $I = (a, b)$. Then $\exists k > 0$, such that for any two points $x, x' \in I$, $|f(x) - f(x')| < k|x - x'|$ (see Definition 3.7). This implies that if the function f is differentiable, its derivative belongs to $(-k, k)$. Thus, a function whose derivative is not bounded cannot be Lipschitz continuous. Consider the function $f(x) = |x|^{3/2} \sin(1/x)$ if $x \in (-1, 1) \setminus \{0\}$ and $f(0) = 0$. The function

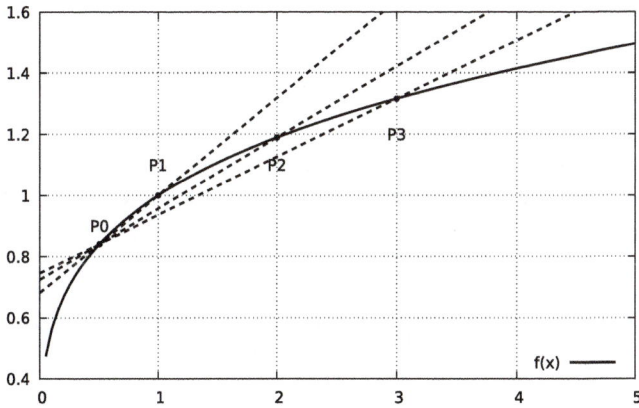

Fig. 6.2 Successive approximations of the tangent to $f(x)$ in P_0. The line passing through P_0 and $P_i, i = 1, 2, 3$, becomes similar to the tangent of $f(x)$ in P_0 as the point P_i moves closer to P_0

is differentiable in $(-1, 1)$. In fact, in $x = 0$, $\lim_{h \to 0} f(h)/h = 0$, and for $x \neq 0$, the function is the product and composition of differentiable functions. However, its derivative $f'(x) = (3/2)x/|x|^{1/2} \sin(1/x) + 1/|x|^{1/2} \cos(1/x)$, due to the second term, is not bounded in $(-1, 1)$. Thus, $f(x)$ cannot be Lipschitz continuous in $(-1, 1)$.

6.3.1 Higher-Order Derivatives

The procedure to obtain the derivative of a function can be applied in a recursive manner. If the derivative function $f'(x)$ exists in a neighbourhood of x, we can consider

$$f''(x) = f^{(2)}(x) = \lim_{y \to x} \frac{f'(y) - f'(x)}{y - x}.$$

This limit, if it exists, is the *second(-order) derivative* or *derivative of order two* of the function f in x. The same procedure can be repeated as long as the function is defined and the limit exists. Often, the derivative of order n is denoted by $f^{(n)}(x)$. In this notation, $f^{(0)}(x)$ stands for the original function. Given an open set $A \subseteq R$, the set of functions that can be derived n times in A and whose n^{th} derivative is a continuous function is denoted by $C^n(A)$. In particular, functions belonging to $C^1(A)$ are said to be *continuously differentiable* in A. Functions that can be derived any number of times belong to $C^\infty(A)$.

6.3.2 Derivatives and Function Behaviour

The derivative can provide useful information on the local behaviour of the function.

Theorem 6.7 *Let $f : (a, b) \to \mathbb{R}$ be differentiable in $x \in (a, b)$. If $f'(x) > 0$, then f is strictly increasing in x. If $f'(x) < 0$, then f is strictly decreasing in x. If x is a local maximum or minimum, then $f'(x) = 0$.*

Proof The proof of the first two points is based on Definition 1.17. If $\lim_{y \to x}(f(x) - f(y))/(x - y) = f'(x) > 0$, there exists $\delta > 0$ such that if $|y - x| < \delta$ and $y \neq x$, then $(f(y) - f(x))/(y - x) > 0$, that is, $(f(y) - f(x))(y - x) > 0$. Analogously, if $f'(x) = l < 0$, there exists $\delta > 0$ such that if $|y - x| < \delta$ then $(f(y) - f(x))/(y - x) < 0$.

Regarding the last point, if a differentiable function is neither strictly increasing nor strictly decreasing in x, its derivative in x cannot be positive or negative and must be zero.   □

Positive and negative values of the derivative are sufficient to imply local monotonic behaviour. If in an open interval $f'(x) \neq 0$, the function f is bijective and can be inverted in that interval. If $f'(x) = 0$, then the point x is said to be a *stationary point* of the function f. Note that the fact that x is a stationary point of f is generally not sufficient to imply that x is an extremal point of f (see Corollary 6.4). The following results show how the behaviour of a differentiable function in an interval can be related to the value of its derivative at some specific points in the interval.

Theorem 6.8 *(Rolle) If f is continuous in $[a, b]$ and differentiable in (a, b) and $f(a) = f(b)$, then there exists $x \in (a, b)$ such that $f'(x) = 0$.*

Proof If $\forall x$, $f(x) = f(a)$, then the function is constant and $f'(x) = 0$ $\forall x$. Otherwise, since the function is continuous, it has a global maximum point x_M and a global minimum point x_m in $[a, b]$. One of the two points must be internal, otherwise the function would be constant. For definiteness, let x_m be internal. Then $f'(x_m) = 0$ by Theorem 6.7.   □

Theorem 6.9 *(Cauchy) Let f, g be continuous in $[a, b]$ and differentiable in (a, b); then there is a point $x \in (a, b)$ such that*

$$(f(a) - f(b)) g'(x) = (g(a) - g(b)) f'(x).$$

Proof Define $h(x) = (f(a) - f(b))g(x) - (g(a) - g(b))f(x)$ and apply Theorem 6.8 to h.   □

Theorem 6.10 *(Mean value or Lagrange) Let f be continuous in $[a, b]$ and differentiable in (a, b). Then, $\exists x \in (a, b)$ such that*

$$f(b) - f(a) = f'(x)(b - a).$$

Proof Apply Theorem 6.9 setting $g(x) = x$.   □

The previous theorem provides a simple criterion for identifying contractions.

Corollary 6.3 *(Derivative and contraction) Let f be differentiable in (a, b) and assume $\sup_{x \in (a,b)} |f'(x)| < 1$. Then f is a contraction on (a, b).*

Proof Let $k = \sup_{x \in (a,b)} |f'(x)| < 1$. According to Theorem 6.10, for any two points $x, y \in (a, b)$ there exists a point $z \in (a, b)$ such that $|f(x) - f(y)| = |f'(z)||x - y|$ which implies $|f(x) - f(y)| \le k|x - y|$, whence the assertion.   □

Theorem 6.11 *Let f be differentiable in (a, b) and consider a sub-interval $(x_1, x_2) \subset (a, b)$. Let λ be such that $f'(x_1) < \lambda < f'(x_2)$. Then $\exists x_0 \in (x_1, x_2)$ such that $f'(x_0) = \lambda$.*

Proof Consider the function $g(x) = f(x) - \lambda x$. Since g is continuous it has a maximum in $[x_1, x_2]$. Moreover $g'(x_1) = f'(x_1) - \lambda < 0$ and so the function is decreasing at x_1, while $g'(x_2) = f'(x_2) - \lambda > 0$ and so the function is increasing at x_2. As a consequence, the function reaches its minimum in an interior point $x_0 \in (x_1, x_2)$. Then $g'(x_0) = f'(x_0) - \lambda = 0$, which proves the assertion.   □

The previous theorem states that if f is differentiable in (a, b), then f' cannot have a discontinuity of the second kind (a jump) in (a, b). However, it does not assume or imply that the derivative is a continuous function (Fig. 6.3).

▶ **Example 6.11** *(Discontinuous derivative)* Consider the function

$$f(x) = \begin{cases} x^2 \sin(1/x) & x \neq 0, \\ 0 & x = 0. \end{cases}$$

Since f is the composition of continuous functions, it is continuous for $x \neq 0$. Moreover, $\lim_{x \to 0} x^2 \sin(1/x) = 0$, so that the function is continuous in 0. Let us consider its derivative. For $x = 0$, $\lim_{h \to 0}(f(h) - f(0))/h = \lim_{h \to 0} h \sin(1/h) = 0$. Using the chain and product rules at all other points, finally

$$f'(x) = \begin{cases} 2x \sin(1/x) - \cos(1/x) & x \neq 0, \\ 0 & x = 0, \end{cases}$$

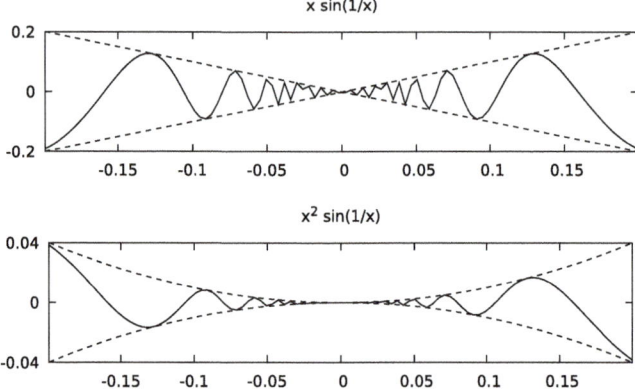

Fig. 6.3 **Top:** The function $x \sin(1/x)$ is continuous on the entire real axis but not differentiable in 0. **Bottom:** The function $x^2 \sin(1/x)$ is continuous and differentiable on the entire real axis, but its derivative is not continuous in 0

so that the function is differentiable in $\mathbb{R}$. However, because of the oscillatory behaviour of the cosine function, $\lim_{x \to 0} f'(x)$ does not exist and the derivative is not continuous in $x = 0$.

The local monotonic behaviour of the derivative provides information on the local concavity or convexity of a function.

Theorem 6.12 *Let the function f be differentiable in the interval (a, b). Then f is concave if and only if f' is decreasing and convex if and only if f' is increasing.*

The same statements hold with the addition of the "strict" qualifier to both the monotonic behaviour of the derivative and the concavity/convexity property of the function.

Proof We start by proving that an increasing derivative implies convexity. Consider $x_1, x_3 \in (a, b)$ and set $x_2 = (1 - \lambda)x_1 + \lambda x_3$ for any $\lambda \in (0, 1)$. By Theorem 6.10 there exist two points $y_1 \in (x_1, x_2)$ and $y_2 \in (x_2, x_3)$ such that $f(x_2) - f(x_1) = f'(y_1)(x_2 - x_1)$ and $f(x_3) - f(x_2) = f'(y_2)(x_3 - x_2)$. As f' is increasing, $f'(y_2) \geq f'(y_1)$, that is,

$$\frac{f(x_3) - f(x_2)}{x_3 - x_2} \geq \frac{f(x_2) - f(x_1)}{x_2 - x_1}.$$

Substituting the expression for x_2 in terms of λ proves the assertion. An analogous construction can be used to prove the concavity of a function starting from a decreasing derivative. The strict version of the statement is promptly derived along the same lines.   □

If the function admits a second derivative, then the monotonic behaviour of the first derivative is related to its sign. In this case, the previous theorem can be restated in terms of the sign of the second derivative. Because a positive derivative implies a strictly increasing function and a negative derivative a strictly decreasing one, we have the following.

Corollary 6.4 *Assume $f \in C^2((a, b))$, that is, the function f has first- and second-order continuous derivatives in an interval (a, b), and let $x \in (a, b)$. Then if $f'(x) = 0$ and $f''(x) > 0$, the function has a strict local minimum at x, while if $f'(x) = 0$ and $f''(x) < 0$, the function has a strict local maximum at x.*

Proof Assume $f'(x) = 0$. If $f''(x) > 0$, then the first derivative is a strictly increasing function in a neighbourhood of x, so it is negative for values lower than x and positive for values higher than x. It follows that, in a neighbourhood of x, the function f is strictly decreasing for values lower than x and strictly increasing for values higher than x. Thus, x is a strict local maximum. The other case is proved analogously.   □

Thus, for a function $f \in C^2$, if x is a local maximum, it must be $f''(x) \leq 0$, and if x is a local minimum, it must be $f''(x) \geq 0$.

Definition 6.7 *(Flex)* If a function f is convex (concave) in a right neighbourhood of x and concave (convex) in the left neighbourhood of x, then x is a *flex* or *inflection point*.

A flex is a point at which the function changes its convexity (or concavity). If the function f admits a second derivative at the flex x, it must be $f''(x) = 0$. However, the opposite is not true. Think of the function $f(x) = x^2$ in $x = 0$. Corollary 6.4 cannot be applied in a flex. In general, we can obtain more information about the local behaviour of the function by inspecting higher-order derivatives, if they exist (see Sect. 6.4). The problem of identifying extremal points will be treated in more general terms for functions of many variables in Chap. 7.

6.3.3 Derivatives and Limits

The next result can be useful for computing limits involving indeterminate expressions such as $0/0$ or ∞/∞. It takes advantage of the possibility, introduced by Theorem 6.9, of expressing the relative increments of two functions in a given interval with the ratio of their derivatives, computed at a suitable point. The theorem is stated in general terms for limits in the extended real number system. In fact, its statement is probably more complicated than its proof. The reader is advised to read it carefully.

Theorem 6.13 *(L'Hopital's rule) Consider a bounded or unbounded interval I and let $a = \inf I$, $b = \sup I$, $a, b \in \bar{\mathbb{R}}$. Let f and g be real differentiable functions in I, with $g'(x) \neq 0$, $\forall x \in (a, b)$.*
Assume $\lim_{x \to b^-} |f(x)| = \lim_{x \to b^-} |g(x)| \in \{0, +\infty\}$. If $\lim_{x \to b^-} f'(x)/g'(x) = l \in \bar{\mathbb{R}}$, then $\lim_{x \to b^-} f(x)/g(x) = l$.
Assume $\lim_{x \to a^+} |f(x)| = \lim_{x \to a^+} |g(x)| \in \{0, +\infty\}$. If $\lim_{x \to a^+} f'(x)/g'(x) = l \in \bar{\mathbb{R}}$, then $\lim_{x \to a^+} f(x)/g(x) = l$.

Proof We prove only the first statement, as the second is proved in the same way. According to Theorem 6.9, for any two points $x_1 < x_2 \in I$, $\exists y \in (x_1, x_2)$ such that

$$\frac{f(x_1) - f(x_2)}{g(x_1) - g(x_2)} = \frac{f'(y)}{g'(y)}.$$

Note that $g(x_1) - g(x_2) \neq 0$ because otherwise, according to Theorem 6.8, $\exists x_3 \in [x_1, x_2]$ such that $g'(x_3) = 0$, which is ruled out by hypothesis.

Suppose $\lim_{x \to b^-} |f(x)| = \lim_{x \to b^-} |g(x)| = +\infty$. Divide the numerator and denominator on the left-hand side by $g(x_2)$,

$$\frac{\frac{f(x_1)}{g(x_2)} - \frac{f(x_2)}{g(x_2)}}{\frac{g(x_1)}{g(x_2)} - 1} = \frac{f'(y)}{g'(y)}.$$

When $x_2 \to b^-$, both $f(x_1)/g(x_2)$ and $g(x_1)/g(x_2)$ converge to zero by hypothesis, so that

$$\lim_{x_2 \to b^-} \frac{f(x_2)}{g(x_2)} = \frac{f'(y)}{g'(y)},$$

for some $y \in (x_1, b)$. The left-hand side does not depend on x_1. Taking $x_1 \to b^-$, the right-hand side converges to l, proving the assertion.

Suppose instead that $\lim_{x \to b^-} f(x) = \lim_{x \to b^-} g(x) = 0$. Taking the limit $x_2 \to b^-$ in the first equation,

$$\frac{f(x_1)}{g(x_1)} = \frac{f'(y)}{g'(y)}$$

for some $y \in (x_1, b)$. When $x_1 \to b^-$, the right-hand side converges to l, proving the assertion. $\qquad \square$

In the previous theorem, we can have $b = +\infty$ and/or $a = -\infty$. Thus, the theorem can be applied to the computation of indeterminate expressions that arise when limits to $\pm\infty$ are considered. Moreover, it can be easily extended to proper limits.

Corollary 6.5 *Let f and g be real differentiable functions in the open bounded interval (a, b), with $g'(x) \neq 0$, $\forall x \in (a, b)$. If for some $x_0 \in (a, b)$, $\lim_{x \to x_0} |f(x)| = \lim_{x \to x_0} |g(x)|$ and they are both 0 or $+\infty$, and $\lim_{x \to x_0} f'(x)/g'(x) = l \in \bar{\mathbb{R}}$, then $\lim_{x \to x_0} f(x)/g(x) = l \in \bar{\mathbb{R}}$.*

Proof Apply Theorem 6.13 separately to the intervals (a, x_0) and (b, x_0). Since the left and right limits exist and are equal, the limit exists, too. $\qquad \square$

▶ **Example 6.12** *(Recursive application of L'Hopital's rule)* Sometimes, L'Hopital's rule must be applied multiple times. Consider the limit

$$\lim_{x \to 0} \frac{e^{3x} - 3e^x + 2}{x^3 + 2x^2}.$$

The limit of the derivatives of the numerator and denominator is still zero, so that

$$\lim_{x \to 0} \frac{3e^{3x} - 3e^x}{3x^2 + 4x} \sim \frac{0}{0}.$$

Taking again the derivative of the numerator and denominator,

$$\lim_{x \to 0} \frac{9e^{3x} - 3e^x}{4} = \frac{3}{2}.$$

By recursion, we can conclude that the original limit is also $3/2$. In all the passages above, one should check that the derivative of the denominator is different from zero in right and left neighbourhoods of the limit point.

▶ **Example 6.13** *(Change of variable with L'Hopital's rule)* To effectively apply L'Hopital's rule, sometimes a change of variable is necessary. We want to compute $\lim_{x \to 0} e^{-1/x^2}/x^n$, with $n \in \mathbb{N}$. A direct application of L'Hopital's rule is not useful. Assume n is even, set $n = 2k$, and consider the new variable $y = 1/x^2$. It is clear that when $x \to 0$, then $y \to +\infty$ and the above limit becomes

$$\lim_{y \to +\infty} \frac{e^{-y}}{1/y^k} = \lim_{y \to +\infty} \frac{y^k}{e^y}.$$

Applying the L'Hopital rule k times, the numerator reduces to the constant $k!$, while the denominator remains e^y. Thus, the limit is zero. When n is odd, set $n = 2k - 1$, then

$$\frac{e^{-1/x^2}}{x^{2k-1}} = x \frac{e^{-1/x^2}}{x^{2k}}.$$

On the basis of the previous result, both factors in the right-hand side go to zero when $x \to 0$, so the limit of their product is zero.

L'Hopital's rule in Theorem 6.13 exposes an important property of the derivative function. Assume that a function $f(x)$ is differentiable in the interval (a, b), apart possibly from a point $x_0 \in (a, b)$. Consider the interval (a, x_0). Then $\forall x \in (a, x_0)$ there exists a $y \in (x, x_0)$ such that $(f(x) - f(x_0))/(x - x_0) = f'(y)$. Thus,

$$\lim_{x \to x_0^-} \frac{f(x) - f(x_0)}{x - x_0} = \lim_{y \to x_0^-} f'(y).$$

This implies that if the left limit of the derivative function $f'(x)$ in x_0 exists, then it is equal to the left derivative of the function f in x_0. The same applies to the right derivative. Therefore, if the left and right limits of the derivative function f' at a point x_0 are different, the original function f cannot be derived in x_0. In other terms, the derivative function cannot be defined at a point where it has a jump discontinuity. In fact, this is exactly the conclusion that we reached in Theorem 6.11. For example, the so-called *Heaviside theta function* defined as

$$\theta(x) = \begin{cases} 0 & \text{if } x < 0, \\ 1 & \text{if } x \geq 0 \end{cases}$$

cannot be the derivative of any function.[2]

6.4   Taylor Polynomial and Power Series Expansion

The higher-order derivatives of a function at a point x internal to its domain can be used to build a polynomial that approximates the behaviour of the function in the neighbourhood of x. The approximation procedure and the estimation of the error involved are provided by the following theorem, based on the work of the English mathematician Brook Taylor (1685–1731).

Theorem 6.14 *(Taylor polynomial) Suppose $f : [a, b] \to \mathbb{R}$ and $f^{(n-1)}$ are continuous in $[a, b]$ and $f^{(n)}$ exist in (a, b). Let $x_0 \in (a, b)$ and define*

$$P_{n-1,x_0}(x) = \sum_{k=0}^{n-1} \frac{f^{(k)}(x_0)}{k!} (x - x_0)^k.$$

Then $\forall x \in (a, b)$ there exists a $\lambda \in (0, 1)$ such that

$$f(x) = P_{n-1,x_0}(x) + \frac{1}{n!} f^{(n)} (\lambda x_0 + (1 - \lambda)x) (x - x_0)^n.$$

Proof Consider the functions $\Phi(x) = f(x) - P^f_{n-1,x_0}(x)$ and $\Gamma(x) = (x - x_0)^n$. Both functions are differentiable in (a, b) and $\Phi(x_0) = \Gamma(x_0) = 0$. Now consider any $x \in (x_0, b)$. According to the Cauchy theorem (Theorem 6.9), there exists a $x_1 \in (x_0, x)$ such that

$$\frac{\Phi(x) - \Phi(x_0)}{\Gamma(x) - \Gamma(x_0)} = \frac{\Phi(x)}{\Gamma(x)} = \frac{\Phi'(x_1)}{\Gamma'(x_1)},$$

where we used the fact that $\Phi(x_0) = \Gamma(x_0) = 0$. The same theorem implies that there exists a $x_2 \in (x_0, x_1)$ such that

$$\frac{\Phi'(x_1) - \Phi'(x_0)}{\Gamma'(x_1) - \Gamma'(x_0)} = \frac{\Phi'(x_1)}{\Gamma'(x_1)} = \frac{\Phi''(x_2)}{\Gamma''(x_2)},$$

---

[2] Sometimes the Heaviside function is defined by assigning a different value at $x = 0$, most commonly $1/2$. Whatever the value of the function in zero, the impossibility of being the derivative of a function still remains.

where we used the fact that $\Phi'(x_0) = \Gamma'(x_0) = 0$. Thus,

$$\frac{\Phi(x)}{\Gamma(x)} = \frac{\Phi''(x_2)}{\Gamma''(x_2)}.$$

The procedure can be iterated to show that there exists a $x_n \in (x_0, x)$ such that

$$\frac{\Phi(x)}{\Gamma(x)} = \frac{\Phi^{(n)}(x_n)}{\Gamma^{(n)}(x_n)}.$$

Note that $\Phi^{(n)}(x) = f^{(n)}(x)$ and $\Gamma^{(n)}(x) = n!$ so that direct substitution proves the assertion by taking $\lambda = (x - x_n)/(x - x_0)$. The same proof can be repeated for $x \in (a, x_0)$.   □

It is important to stress that the value of λ in the previous theorem is not, in general, a constant but depends on the value of x. When $x_0 = 0$, the Taylor polynomial is named the *Maclaurin polynomial* after the Scottish mathematician Colin Maclaurin (1698–1746). The error of the polynomial approximation

$$R_{x_0}(x) = \frac{1}{n!} f^{(n)} (\lambda x_0 + (1 - \lambda)x) (x - x_0)^n$$

is often called *remainder*. Specifically, this is the remainder in the *Lagrange form*.[3] From the definition of R_{x_0}, $\lim_{x \to x_0} R_{x_0}(x)/(x - x_0)^{n-1} = 0$. In other terms, the remainder of the Taylor polynomial of order $n - 1$ is an infinitesimal of order greater than $n - 1$, $R_{x_0}(x) = o(|x - x_0|^{n-1})$. If the nth derivative is continuous, $\lim_{x \to x_0} R_{x_0}(x)/(x - x_0)^n = f^{(n)}(x_0)/n!$. In this case, $R_{x_0}(x) = O(|x - x_0|^n)$ and the remainder is an infinitesimal of order at least n. Sometimes, it is useful to have a global bound on the error. If $M = \sup_{x \in (a,b)} |f^{(n)}(x)|$, then

$$\left| f(x) - P_{n-1,x_0}(x) \right| \le M \frac{|x - x_0|^n}{n!}.$$

This *Lagrange error bound* provides a uniform estimate of the error implied by approximating the function with the Taylor polynomial. For any fixed $x - x_0$, this expression becomes smaller as M decreases and as n increases. Therefore, if the derivatives of increasing order of the function under consideration have a magnitude that is constant or decreasing in n, such as, for example, in the case of exponential, sine, or cosine functions, the Taylor polynomial represents an approximation that is as good as one wishes for a sufficiently high value of n.

---

[3] Other forms of the remainder are known, such as Cauchy or Peano, but for our purposes, the Lagrange form is enough.

▶ **Example 6.14** *(Taylor approximation of the exponential function)* Consider the exponential function $f(x) = e^x$. It is $f^{(k)}(x) = f(x)$, so $f^{(k)}(0) = 1$. The Taylor polynomial around $x = 0$ of order n reads $P(n, 0) = \sum_{k=0}^{n} x^k/k!$. Consider the interval $(0, \delta)$. In this interval, the maximum of $f^{(n+1)}(x)$ is achieved for $x = \delta$, so that

$$\left| f(x) - P_{n,0}(x) \right| \le e^\delta \frac{x^{n+1}}{(n+1)!}, \quad \forall x \in (0, \delta).$$

Analogously, in the interval $(-\delta, 0)$, the maximum of $f^{(n+1)}(x)$ is achieved for $x = 0$ so that

$$\left| f(x) - P_{n,0}(x) \right| \le \frac{x^{n+1}}{(n+1)!}, \quad \forall x \in (0, \delta).$$

Hence, because $e^\delta > 1$,

$$\left| f(x) - P_{n,0}(x) \right| \le e^\delta \frac{x^{n+1}}{(n+1)!}, \quad \forall x \in (-\delta, \delta).$$

Since $\forall x$, the expression $x^n/n!$ goes to zero as n increases (see Example 5.5.2), the error above goes to zero when an increasing number of terms is considered in the Taylor polynomial. For example, if we want a Taylor polynomial that approximates the function in the interval $(-1, 1)$ to a precision of 10^{-3}, we can choose an n that satisfies the condition $(n + 1)! \ge e10^3$. Setting $n = 6$ turns out to be enough. The precision of the polynomial approximation is rapidly increasing; see Fig. 6.4. With ten terms, $n = 10$, the error in the interval $(-1, 1)$ is less than one millionth.

In the previous example, we have seen that the Taylor polynomial in $x = 0$ can be used to obtain a good, in fact as good as one desires, approximation of the function

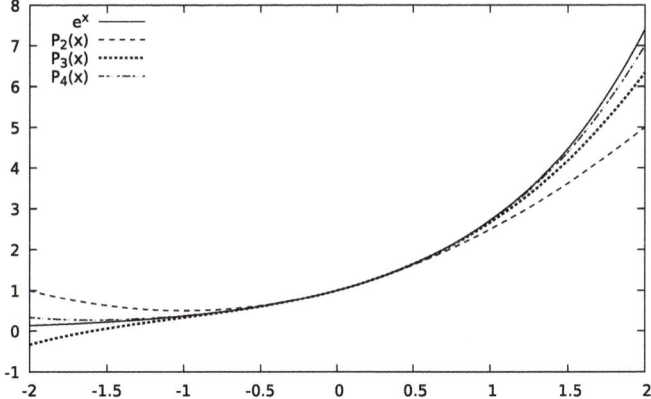

Fig. 6.4 The function e^x is progressively better approximated by polynomials of higher orders

e^x in any interval, that is, in practice, at any point of its domain. This is not true for a generic function.

▶ **Example 6.15** (*Limited precision of Taylor approximation*) The global upper bound introduced above is not always useful. Consider the function $f(x) = 1/(1-x)$. Its kth derivative is $f^{(k)}(x) = k!/(1-x)^{k+1}$, so that its Taylor polynomial around $x = 0$ of order n can be easily computed to be $P_{n,0}(x) = \sum_{k=0}^{n} x^k$, that is, the truncated geometric series. When $x \in (0, 1)$, the remainder of order n is $R_0^n(x) = x^{n+1}/(1 - (1-\lambda)x)^{n+2}$ for some $\lambda \in (0, x)$. Using the same reasoning of Example 6.14, we find that $|R_0^n(x)| < |x^{n+1}/(1 - x)^{n+2}|$. This expression converges to zero when n increases only if $|x| < 1/2$, despite the fact that the geometric series converges also when $1/2 \le |x| < 1$. However, for this particular case, we are able to explicitly compute the remainder of the Taylor polynomial,

$$\frac{1}{1-x} - P_{n,0}(x) = \frac{x^{n+1}}{1-x}.$$

As expected, the limit for $n \to \infty$ of the right-hand side expression is zero for any $|x| < 1$. Note that while Theorem 6.14 applies in any open interval around 0 that does not contain 1, the Taylor polynomial represents a good approximation of the function only in the interval $(-1, 1)$; see Fig. 6.5.

▶ **Example 6.16** (*Taylor approximation of the logarithmic function*) Consider the function $f(x) = \log(1 + x)$. It is easy to prove by recursion that $f^{(k)}(x) = (-1)^{k+1}(k-1)!/(1+x)^k$ so that the Taylor polynomial around $x = 0$ of order

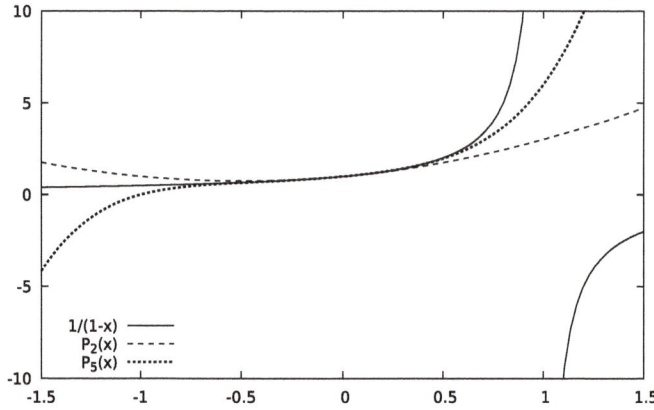

Fig. 6.5 The progressive Taylor approximations to $1/(1-x)$ are bad near the point $x = 1$, where the function has an essential discontinuity, but also for $x < 1$

$n \geq 1$ can be easily computed to be

$$P_{n,0}(x) = \sum_{k=1}^{n} \frac{(-1)^{k+1} x^k}{k}.$$

In this case, Theorem 6.14 applies in an open interval around 0 only if this interval does not contain -1. In fact, the function is not defined there.

In general, increasing the number of terms is not enough to ensure the pointwise convergence of the sequence (P_{n,x_0}) to $f(x)$ in a neighbourhood of the point x_0, as the next example shows.

▶ **Example 6.17** *(Lack of pointwise convergence of Taylor approximations)* Consider the function

$$f(x) = \begin{cases} e^{-1/x^2} & x \neq 0, \\ 0 & x = 0 \end{cases}$$

which is continuous and can be derived any number of times on the entire real axis $\mathbb{R}$ and, in particular, in $x = 0$. By using the result in Example 6.13, it is easy to show that $f^{(n)}(0) = 0$ for any n so that if we expand the function around zero, then all terms in $P_{n,0}(x) = 0$ are equal to zero. The seemingly contradiction with the expression of the error terms is reconciled once one notices that $\lim_{x \to 0} e^{-1/x^2}/|x|^n = 0$ for any n. Thus, for any n, whatever the value of M, there exists a neighbourhood of 0 for which $|f(x)| < M|x|^n/n!$.

The Taylor polynomial approximation can be adapted to obtain a *asymptotic approximation* of a function when its argument goes to plus or minus infinity. Instead of a polynomial expression in x, we are looking for a polynomial expression in $1/x$. In this way, we can obtain an approximate description of the behaviour of a function at the boundaries of its domain.[4]

▶ **Example 6.18** *(Asymptotic expansion)* Let $a \in \mathbb{R}_{>0}$ and consider the function $f(x) = \log((x + a)/x)$, defined for $x > 0$. Note that $\lim_{x \to +\infty} f(x) = 0$. Let $x = 1/y$. The behaviour of f for large values of x is equal to the behaviour of $g(y) = \log(1 + ay)$ for small and positive values of y. From Example 6.16, we know that

$$g(y) = \sum_{k=1}^{n} \frac{(-1)^{k+1} a^k y^k}{k} + o(y^n).$$

---

[4] The theory of asymptotic approximations or "expansions" of functions is more general than this. Often, one is interested to obtain expansions not in terms of powers of $1/x$, but in terms of more general sets of functions. The methods may change, but the idea remains essentially the same.

Thus, by direct substitution, we can say that when x is sufficiently large,

$$f(x) = \sum_{k=1}^{n} \frac{(-1)^{k+1}a^k}{kx^k} + o(1/x^n).$$

In particular, when $x \to +\infty$, $f(x)$ is infinitesimal of the same order as $1/x$.

In Examples 6.14, 6.15, and 6.16, we were able to obtain a good approximation of the function under consideration by using a polynomial of sufficiently high order. What happens if the order of the polynomial is allowed to become infinite? We do not know how to compute polynomials of infinite order, but we can think of them as special series.

Definition 6.8 *(Power series)* A *power series* centred at a is defined as

$$f(x) = \sum_{n=0}^{\infty} c_n(x - a)^n.$$

The power series is a function defined on the set of points at which the series on the right-hand side converges. This set is never empty, as it always contains the point $x = a$. When an infinite number of terms are considered, the Taylor approximations introduced in Example 6.14 and in Example 6.15 lead to two power series centred at zero. Let us analyse them in some detail. We start with the exponential function in Example 6.14. Formally write

$$e^x = \sum_{k=0}^{\infty} \frac{x^k}{k!}.$$

In Example 5.5.2, we prove that the series on the right-hand side is convergent for any real number x. This was precisely the definition of the exponential function. Thus, we can say that the exponential function is described by the power series in its entire domain. In other terms, we have derived a *power series expansion* of the original function valid on $\mathbb{R}$. Consider instead the hyperbolic function in Example 6.15. In this case, the Taylor expansion of infinite order leads to a geometric series,

$$\frac{1}{1 - x} = \sum_{k=0}^{\infty} x^k.$$

The equal sign in the previous expression is valid only if $|x| < 1$. Indeed, we know that the series on the right-hand side is convergent only for these points (see Example 5.5). The next theorem, credited to Cauchy and the French mathematician

Jacques Hadamard (1865–1963), shows that, in general, a power series centred on a is defined over an open interval having a as its midpoint.

Theorem 6.15 *(Cauchy–Hadamard) Define the* radius of convergence[5] *R of the power series $f(x) = \sum_{n=0}^{\infty} c_n (x - a)^n$ as*

$$\frac{1}{R} = \limsup_{n \to \infty} |c_n|^{1/n}.$$

Then the series is convergent for $x \in (a - R, a + R)$ and divergent outside this interval, that is, for $x > a + R$ and $x < a - R$.

Proof We will start by proving that the series is convergent in the interval $(a - R, a + R)$. From the definition of limit superior, $\lim_{n \to \infty} \sup_{m \geq n} |c_m|^{1/m}$. Thus, $\forall \epsilon > 0$, for sufficiently large n,

$$|c_n|^{1/n} < \left(\frac{1}{R} + \epsilon\right), \quad \text{i.e.} \quad |c_n| < \left(\frac{1}{R} + \epsilon\right)^n, \quad \text{so that}$$

$$\sum_{n=0}^{\infty} |c_n| |x - a|^n \leq \sum_{n=0}^{\infty} \left(\frac{1}{R} + \epsilon\right)^n |x - a|^n.$$

The right-hand side is a geometric series that converges if its base has a modulus less than one. In this case, it converges if $|x - a| < 1/(1/R + \epsilon)$. Since ϵ can be taken infinitesimally small, the power series is absolute convergent, and thus convergent, for $|x - a| < R$.

Next, we show that the power series is not convergent outside the above interval. First of all, given any $\epsilon > 0$, by the definition of limit superior, there are infinite terms of the series for which

$$|c_n|^{1/n} > (\frac{1}{R} - \epsilon), \quad \text{i.e.} \quad |c_n| > \left(\frac{1}{R} - \epsilon\right)^n.$$

Then, if $|x - a| > 1/(1/R - \epsilon)$, there are infinite elements of the series for which $|x - a|^n |c_n| > 1$. Thus, the elements of the series do not converge to zero, and consequently, the series cannot converge. Since ϵ can be taken infinitesimally small, we can conclude that the series does not converge for $|x - a| > R$. $\qquad\square$

The previous theorem is silent on what happens at the boundaries of the convergence interval $a \pm R$. Depending on the series considered, it might converge or not. A

---

[5] The name "radius" of convergence derives from the fact that the theorem is in general presented for power series of complex numbers. In this case, it proves that the power series converges inside a disc in the complex plane $\mathbb{C}$ of radius equal to the radius of convergence.

function represented by a power series is said to be *analytic* in the domain in which the series converges. The derivative of the function $f(x)$ in Definition 6.8 can be written as $f'(x) = \sum_{n=0}^{\infty} c_{n+1}(n+1)(x-a)^n$. By Theorem 6.15, the convergence radius of this power series is the same as the original power series that defines $f(x)$. Thus, the derivative of an analytic function is itself analytic, and an analytic function possesses derivatives of any order.

▶ **Example 6.19** *(Convergence on the boundary)* Consider the power series centred on the origin and based on the expansion in Example 6.16,

$$\sum_{k=1}^{\infty} \frac{(-1)^{k+1} x^k}{k}.$$

The radius of convergence is $\lim_{n\to\infty} \sqrt[n]{n} = 1$ so that the series converges for $x \in (-1, 1)$ and in this case its value is $\log(1+x)$. The series diverges for $x > 1$ and $x < -1$. What happens in 1 and -1? For $x = 1$, the series reduces to the harmonic series with alternating sign introduced in Example 5.5.1, which we know is convergent. Its limit is $\log 2$. Conversely, when $x = -1$, the power series becomes the harmonic series in Example 5.5.1, which we know does not converge.

If $\limsup_{n\to\infty} |c_n|^{1/n} = 0$ the radius of convergence of the power series is infinite and the series converges on the entire real axis. On the contrary, if $\limsup_{n\to\infty} |c_n|^{1/n}$ is $+\infty$, the radius of convergence of the power series is zero, and the sequence converges only for $x = a$.

If the sequence $(|c_n|^{1/n})$ converges, one can also compute the radius of convergence as $1/R = \lim_{n\to\infty} |c_{n+1}/c_n|$. For example, in the case of the exponential function, $c_{n+1}/c_n = 1/(n+1)$, so that the limit is zero and the radius of convergence of the series is infinite. In the case of the hyperbolic function, $c_{n+1}/c_n = 1$ and the radius of convergence is 1. The same applies to the function in Example 6.16, for which $c_{n+1}/c_n = n/(n+1)$.

▶ **Example 6.20** *(Basic power series expansions)* The following list of expansions around $x = 0$ summarises previous examples and might be useful to solve the exercises at the end of the chapter.

$$\frac{1}{1-x} = \sum_{n=0}^{\infty} x^n, \quad e^x = \sum_{n=0}^{\infty} \frac{x^n}{n!}, \quad \log(1+x) = \sum_{n=1}^{\infty} (-1)^{n+1} \frac{x^n}{n},$$

$$\cos x = \sum_{n=0}^{\infty} (-1)^n \frac{x^{2n}}{(2n)!}, \quad \sin x = \sum_{n=0}^{\infty} (-1)^n \frac{x^{2n+1}}{(2n+1)!}.$$

Exercises

Exercise 6.1 Consider the *p-norm* defined in Sect. 4.2.2. Prove that for any $\mathbf{x} = (x_1, x_2) \in \mathbb{R}^2$, $\lim_{p\to\infty} |\mathbf{x}|_p = \max\{|x_1|, |x_2|\}$.

Exercise 6.2 Let f be a real function defined in the interval (a, b) such that for any point $x \in (a, b)$, $\lim_{h\to 0} f(x + h) - f(x - h) = 0$. Is the function continuous in (a, b)?

Exercise 6.3 Let $f(x)$ be defined for all real x and suppose that $|f(x) - f(y)| \le (x - y)^2$ for all x and y. Prove that f is constant.

Exercise 6.4 Let f be a continuous real function on $\mathbb{R}$. Suppose that $\forall x \ne 0$, $f'(x)$ exists and $\lim_{x\to 0} f'(x) = 3$. Does it follow that $f'(0)$ exists?

Exercise 6.5 Consider the function

$$f(x) = \begin{cases} 0 & x \le 0 \\ e^{-1/x} & x > 0. \end{cases}$$

Prove $f \in C^\infty(\mathbb{R})$.

Exercise 6.6 Compute the following limits:

$$\lim_{x\to 0} \frac{\cos x - e^{-x^2/2}}{x^4}, \quad \lim_{x\to 0} \frac{\sin x - \log \cos x}{x^2}, \quad \lim_{x\to 2} \frac{e^{x-2} - x + 1}{x^2}.$$

Exercise 6.7 Suppose f is defined in a neighbourhood of x and suppose $f^{(2)}(x)$ exists. Then show that

$$\lim_{h\to 0} \frac{f(x + h) + f(x - h) - 2f(x)}{h^2} = f''(x).$$

Are you able to find an example in which the limit exists even if $f^{(2)}(x)$ does not?

Exercise 6.8 Consider the integer part function $f(x) = [x]$ defined on $\mathbb{R}_{\ge 0}$. Is it possible to find a function $g(x)$ such that $g'(x) = f(x)$?

Exercise 6.9 Consider the function

$$f(x) = \begin{cases} x^2 \sin(1/x^2) & x \ne 0, \\ 0 & x = 0. \end{cases}$$

Show that the function is differentiable in $(-1, 1)$ and its derivative is unbounded.

Exercise 6.10 Compute the limit for $n \to \infty$ of the sequences $\left(e^{-n}(1 + 1/n)^{n^2}\right)$ and $\left(e^n(1 - 1/n)^{n^2}\right)$.

Exercise 6.11 Prove that if a polynomial $P(x)$ is constant in an interval $[a, b]$, then it is constant everywhere. *Hint: Consider its derivatives.*

Exercise 6.12 Compute the Taylor polynomial of order 4 of the function $f(x) = \sqrt{\cos x}$ in $x = 0$.

Exercise 6.13 Compute $\sin 1$ with a precision of 10^{-3}.

Exercise 6.14 Consider the function $f(x) = \sqrt[3]{1 + x^3} - x$. Prove that $\lim_{x \to +\infty} f(x) = 0$. Compute the asymptotic approximation of $f(x)$ in terms of $1/x$ at $+\infty$. *Hint: Consider $y = 1/x$ and notice that $x \to +\infty$ is equivalent to $y \to 0^+$.*

Exercise 6.15 Use the Taylor polynomial approximation around an appropriate point to easily compute the following limits:

$$\lim_{x \to 0} \frac{\sin^2 x - x^2}{x^4}, \qquad \lim_{x \to 1} \frac{\log^2 x - (x - 1)^2}{(x - 1)^3}, \qquad \lim_{x \to 0} \frac{1 - \cos \sin x}{x}.$$

Exercise 6.16 Compute the Maclaurin polynomial of order n of the function

$$f(x) = \log \frac{1 + x}{1 - x}.$$

Hint: Try to use what you know about the expansion of $\log(1 + x)$.

Exercise 6.17 The functions e^x and $1/x$ have an intersection in the interval $(0, 1)$. Use the Taylor polynomial to find an approximation of the abscissa of the point. *Hint: Reduce the problem to finding the root of a suitable polynomial.*

Exercise 6.18 Let $k \in \mathbb{N}$, $k > 2$ and consider the series

$$\sum_{n=1}^{\infty} \sqrt[k]{n^k + 1} - n.$$

Use the Taylor approximation of $\sqrt[k]{1 + x}$ to prove that it is convergent. Is the series convergent for $k = 2$?

Exercise 6.19 Describe how the radius of convergence of the power series $\sum_{n=0}^{+\infty}(an + b)x^n$ depends on the value of a and b.

Exercise 6.20 Let r_a and r_b be the radius of convergence of the power series $\sum_{n=0}^{\infty} a_n x^n$ and $\sum_{n=0}^{\infty} b_n x^n$. Find the radius of convergence of $\sum_{n=0}^{\infty} a_n b_n x^n$ and $\sum_{n=0}^{\infty} a_n / b_n x^n$.

Exercise 6.21 Consider the power series $\sum_{n=1}^{\infty} x^n / (2^n + 1)$. Prove that the series is convergent for $x = 1$ and divergent for $x = 2$. Compute its radius of convergence.

Differential Calculus of Functions of Several Variables

7

7.1 Limits and Continuity in $\mathbb{R}^n$

The space $(\mathbb{R}^n, \|.\|)$ is a complete normed space. The induced topology is Hausdorff and second-countable. In this space, the Bolzano–Weierstrass theorem applies (see Sect. 2.3) and the validity of the Heine–Borel theorem was proved in Theorem 4.10. The elements of the space (points or vectors) are in bold $\mathbf{x} \in \mathbb{R}^n$. The notion of limit of a function in $\mathbf{x}$ is directly derived from Definition 2.14 using open balls (see Definition 3.5) as the neighbourhood of the point $\mathbf{x}$ and of its image $\mathbf{f}(\mathbf{x})$. The name of the function is in bold as a reminder that the image of the point is a vector with several components. Theorem 5.2 applies and all properties of the limit of sequences apply to the limit of vector-valued functions. In particular, a vector-valued function has limit in $\mathbf{x}$ if and only if all its components have limit in $\mathbf{x}$ and the components of the limit are the limits of the components. Thus, the problem of determining the limit of a vector-valued function can be broken down into separate problems of determining the limit of its components, which are real-valued functions, also called *scalar functions* or *scalar fields*, defined over a subset of $\mathbb{R}^n$. The limit of a scalar function, if it exists, is unique, and properties identical to those of the limits of functions of one variable apply.

Theorem 7.1 *Let* $f, g : \mathbb{R}^n \to \mathbb{R}$, *and assume that* $\lim_{\mathbf{x} \to \mathbf{x}_0} f(\mathbf{x}) = y_0$ *and* $\lim_{\mathbf{x} \to \mathbf{x}_0} g(\mathbf{x}) = y_0'$, *then:*

- $\lim_{\mathbf{x} \to \mathbf{x}_0} \alpha f(\mathbf{x}) + \beta g(\mathbf{x}) = \alpha y_0 + \beta y_0'$;
- $\lim_{\mathbf{x} \to \mathbf{x}_0} f(\mathbf{x}) g(\mathbf{x}) = y_0 y_0'$;
- *if* $y_0' \neq 0$, $\lim_{\mathbf{x} \to \mathbf{x}_0} f(\mathbf{x})/g(\mathbf{x}) = y_0/y_0'$.

Proof The proof is identical to the proof of Theorem 6.1, which, in turn, is based on the result of Theorem 5.2. □

G. Bottazzi, *Advanced Calculus for Economics and Finance*, Classroom Companion: Economics, https://doi.org/10.1007/978-3-031-30316-6_7

Multivariate continuous functions possess a series of important properties: their composition is continuous (see Theorem 2.24); the inverse image of an open set is open (see Theorem 2.21) and of a closed set is closed (see Theorem 2.23); the image of a connected set is connected (see Theorem 2.25) and the image of a compact set is compact (see Theorem 2.26). Moreover, note that if a real function of a real variable $g(z)$ is continuous in $z = z_0$, then the scalar function $f(x_1, \ldots, x_n) = g(x_j)$ is continuous in all points $\mathbf{x}$ with $x_j = z_0$. This property is particularly useful in practical applications. For example, it implies that any polynomial $P(\mathbf{x}) = \sum_j c_j x_1^{h_{j,1}} x_2^{h_{j,2}} \cdots x_n^{h_{j,n}}$, which is defined through the sum and product of continuous functions, is a continuous function in $\mathbb{R}^n$ and that $f(P(\mathbf{x}))$ is continuous if $f(x)$ is continuous in $P(\mathbf{x})$.

▶ **Example 7.1** *(Polar coordinates)* When studying the limit of functions defined in a subset of $\mathbb{R}^2$ it could be useful to switch to *polar coordinates*. This is a new system of coordinates for the points of the real plane $\mathbb{R}^2$ based on the two variables

$$
\begin{cases}
\rho = \sqrt{x_1^2 + x_2^2}, \\
\phi = \arctan \dfrac{x_2}{x_1},
\end{cases}
$$

where we assume that the function $\arctan(.)$ returns values in $[0, 2\pi]$. By Pythagoras' theorem,

$$
\begin{cases}
x_1 = \rho \cos \phi, \\
x_2 = \rho \sin \phi.
\end{cases}
$$

Each point $(x_1, x_2) \neq (0, 0)$ corresponds to a single couple of values (ρ, ϕ). The exception to this rule is the origin of the axis. In fact, all couples $(0, \phi)$ correspond to the point $(0, 0)$.

Polar coordinates can be useful in the study of limits and will be used in several examples. For a scalar function $g : \mathbb{R}^2 \to \mathbb{R}$, the fact that $\lim_{\mathbf{x} \to 0} g(\mathbf{x}) = l$ implies that $\lim_{\rho \to 0} g(\rho, \phi) = l$. In fact, for any neighbourhood of the origin, $N(\mathbf{0})$, there exists a sufficiently small $\delta > 0$ such that, if $\rho < \delta$, $(\rho, \phi) \in N(\mathbf{0})$. Thus, if $\forall \mathbf{x} \in N(\mathbf{0})$, $|g(\mathbf{x}) - l| < \epsilon$, then $|g(\rho, \phi) - l| < \epsilon$ for $\rho < \delta$. This is a trivial consequence of the fact that the set of open balls is a base of the Euclidean topology, see Theorem 2.2.2. Thus, if we prove that $\lim_{\rho \to 0} g(\rho, \phi)$ does not exist for some ϕ, or is different for different values of ϕ, then we can conclude that $\lim_{\mathbf{x} \to 0} g(\mathbf{x})$ does not exist (but the opposite is not true, as discussed in Example 7.7). For example, consider the function $\mathbf{f} : \mathbb{R}^2 \setminus (\mathbf{0}, \mathbf{0}) \to \mathbb{R}^2$ defined as

$$
\begin{cases}
f_1(x_1, x_2) = \dfrac{x_1 x_2}{x_1^2 + x_2^2} \\
f_2(x_1, x_2) = x_1 + x_2.
\end{cases}
$$

The second component f_2 is the sum of single variable linear functions that are continuous over $\mathbb{R}$, thus it is continuous over the entire space $\mathbb{R}^2$. The first component

f_1 is the sum and product of single variable functions which are continuous over the entire space, thus it is continuous over the entire space $\mathbb{R}^2$ apart $(0, 0)$, where it is not defined. Is it possible to make this function continuous in $(0, 0)$? We can set $f_2(0, 0) = 0$. Then we need a value c such that if we set $f_1(0, 0) = c$, the function f_1 is continuous in $(0, 0)$. For this to be the case, in polar coordinates it is necessary (but not sufficient) that $\lim_{\rho \to 0} f_1(\rho, \phi) = \lim_{\rho \to 0} \cos \phi \sin \phi = c$. However, this is impossible as the limit value is not constant and depends on the value of ϕ. Thus, the limit of the original function $\mathbf{f}$ in $(0, 0)$ does not exist.

7.2 Differential Analysis in $\mathbb{R}^n$

To introduce the notion of differential of multivariate functions, it is helpful to look at the notion of derivative of real functions from a different perspective. Consider the derivative of a function $f : \mathbb{R} \to \mathbb{R}$ in a point $x \in \mathbb{R}$, defined as

$$f'(x) = \lim_{h \to 0} \frac{f(x + h) - f(x)}{h}.$$

This definition means that the quantity $r(h) = f(x + h) - f(x) - hf'(x)$, which depends both on the function f and the point x, is such that $\lim_{h \to 0} r(h)/h = 0$. We can interpret this fact by saying that the increment $f(x + h) - f(x)$ of the function f for a small variation h of its argument is well approximated by the linear function $hf'(x)$. It is so well approximated that the approximation error $r(h)$ decreases faster than the length h over which the approximation is calculated. In this sense, the value $f'(x)$ is unique: if we consider any other real number a and define $\tilde{r}(h) = f(x + h) - f(x) - ah$,

$$\lim_{h \to 0} \frac{\tilde{r}(h)}{h} = f'(x) - a \neq 0.$$

The error $\tilde{r}(h)$ of the local linear approximation obtained using $a \neq f'(x)$ does not decrease faster than h when $h \to 0$ (Fig. 7.1).

Consider now a real vector-valued function $\mathbf{f} : \mathbb{R} \to \mathbb{R}^m$,

$$\mathbf{f}(x) = \begin{pmatrix} f_1(x) \\ \dots \\ f_m(x) \end{pmatrix}.$$

The components $f_j(x)$ with $j = 1, \dots, m$ are real functions of a single real variable. Assume that the derivatives of these components exist in x. Define the *differential* $d\mathbf{f}(x) = (f'_1(x), \dots, f'_m(x))^\mathsf{T}$ and the approximation error $\mathbf{r}(h) = \mathbf{f}(x+h) - \mathbf{f}(x) - h\,d\mathbf{f}$ such that

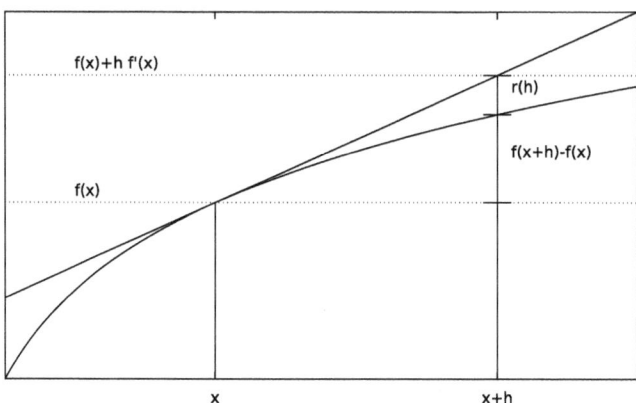

Fig. 7.1 The error $r(h)$ vanishes faster than h when $h \to 0$

$$\mathbf{f}(x+h) = \begin{pmatrix} f_1(x) + f_1'(x)h + r_1(h) \\ \cdots \\ f_m(x) + f_m'(x)h + r_m(h) \end{pmatrix}.$$

The jth component of $\mathbf{r}(h)$ is $r_j(h) = f_j(x+h) - f_j(x) - h f_j'(x)$ and, for the definition of derivative, $\lim_{h \to 0} r_j(h)/h = 0$. Therefore, $\lim_{h \to 0} \|\mathbf{r}(h)\|/h = 0$. Again, the differential $d\mathbf{f}$ constitutes a local linear approximation of the increment $\mathbf{f}(x+h) - \mathbf{f}(x)$ of the vector-valued function $\mathbf{f}$ with an error $\mathbf{r}(h)$ whose norm decreases faster than the length h over which the increment is calculated. The vector $d\mathbf{f}(x)$ is unique because the derivatives of the components of $\mathbf{f}$ are unique. In fact, if we assume that a vector $\mathbf{v}$ exists such that, defining $\tilde{\mathbf{r}}(h) = \mathbf{f}(x+h) - \mathbf{f}(x) - h\mathbf{v}$, we have $\lim_{h \to 0} \|\tilde{\mathbf{r}}(h)\|/h = 0$, then it must be true that $\lim_{h \to 0}(f_j(x+h) - f_j(x))/h = v_j, \forall j$. That is, the components of $\mathbf{v}$ must be the derivatives of the components of the function $\mathbf{f}$ with respect to x.

Next, we want to extend the idea of a local linear approximation to functions of several variables. For a function $\mathbf{f} : \mathbb{R}^n \to \mathbb{R}^m$ the linear approximation will take the form of an element of the space of $m \times n$ real matrices $M_{m,n}$ (see Sect. 4.2.3).

Definition 7.1 Let $\mathbf{f} : \mathbb{R}^n \to \mathbb{R}^m$ be defined in an open set $E \subseteq \mathbb{R}^n$ and let $\mathbf{x} \in E$. The function $\mathbf{f}$ is *differentiable* in $\mathbf{x}$ if there exists a matrix $A_{\mathbf{f}}(\mathbf{x}) \in M_{m,n}$ such that

$$\lim_{\mathbf{h} \to 0} \frac{\|\mathbf{f}(\mathbf{x}+\mathbf{h}) - \mathbf{f}(\mathbf{x}) - A_{\mathbf{f}}(\mathbf{x})\mathbf{h}\|}{\|\mathbf{h}\|} = 0.$$

The matrix $A_{\mathbf{f}}(\mathbf{x})$ is called the *differential* of $\mathbf{f}$ in $\mathbf{x}$. If $\mathbf{f}$ is differentiable in any $\mathbf{x} \in E$, then it is differentiable in E.

▶ **Example 7.2** *(Differential of a vector function)* Consider the function $\mathbf{f} : \mathbb{R}^2 \to \mathbb{R}^2$ defined as

$$\begin{cases} f_1(x_1, x_2) = x_1 x_2, \\ f_2(x_1, x_2) = x_1 + x_2. \end{cases}$$

We shall prove that the differential of $\mathbf{f}$ in $\mathbf{1} = (1, 1)$ is

$$A_\mathbf{f}(1, 1) = \begin{pmatrix} 1 & 1 \\ 1 & 1 \end{pmatrix}.$$

For an increment $\mathbf{h} = (h_1, h_2)$, the components of the approximation error $\mathbf{r}(\mathbf{h}) = \mathbf{f}(\mathbf{1} + \mathbf{h}) - \mathbf{f}(\mathbf{1}) - A_\mathbf{f}(1, 1)\mathbf{h}$ read

$$\begin{cases} r_1(x_1, x_2) = f_1(1 + h_1, 1 + h_2) - f_1(1, 1) - h_1 - h_2 = h_1 h_2, \\ r_2(x_1, x_2) = f_2(1 + h_1, 1 + h_2) - f_2(1, 1) - h_1 - h_2 = 0. \end{cases}$$

We have to prove that $\lim_{\mathbf{h} \to 0} \|\mathbf{r}(\mathbf{h})\| / \|\mathbf{h}\| = 0$. Note that

$$0 \le \frac{\|\mathbf{r}(\mathbf{h})\|}{\|\mathbf{h}\|} = \frac{|h_1 h_2|}{\sqrt{h_1^2 + h_2^2}} \le \frac{1}{2}\sqrt{h_1^2 + h_2^2} = \frac{1}{2}\|\mathbf{h}\|.$$

For the comparison theorem, the limit is actually zero and the statement is proved.

If the function $\mathbf{f}$ is differentiable in $\mathbf{x}$, for small increments $\mathbf{h}$ it is possible to approximate the increment of the components of the function $\mathbf{f}$ with a linear expression. We can write $\mathbf{f}(\mathbf{x} + \mathbf{h}) = \mathbf{f}(\mathbf{x}) + A(\mathbf{x})\mathbf{h} + \mathbf{r}(\mathbf{h})$, where $A(\mathbf{x})\mathbf{h}$ represents the usual matrix multiplication. For sufficiently small increments, the norm of the error $\mathbf{r}(\mathbf{h})$ decreases faster than the norm of the increment $\mathbf{h}$. Specifically, $\forall \epsilon > 0$, $\exists \delta_\epsilon > 0$ such that if $\|\mathbf{h}\| < \delta_\epsilon$,

$$\|\mathbf{f}(\mathbf{x} + \mathbf{h}) - \mathbf{f}(\mathbf{x})\| = \|A(\mathbf{x})\mathbf{h} + \mathbf{r}(\mathbf{h})\| \le \|A(\mathbf{x})\|_{op} \|\mathbf{h}\| + \epsilon \|\mathbf{h}\|,$$

where we have used the triangle inequality and the operator norm introduced in Definition 4.4. The previous inequality has two consequences. First, when $\|\mathbf{h}\| \to 0$, it is $\|\mathbf{f}(\mathbf{x} + \mathbf{h}) - \mathbf{f}(\mathbf{x})\| \to 0$, so we have the following.

Corollary 7.1 *If $\mathbf{f} : E \subseteq \mathbb{R}^n \to \mathbb{R}^m$ is differentiable in $\mathbf{x} \in E$, it is continuous in $\mathbf{x}$.*

The second consequence is that, for sufficiently small increments, $\|A(\mathbf{x})\|_{op}$ provides an upper bound to the incremental ratio of the function, $\|\mathbf{f}(\mathbf{x} + \mathbf{h}) - \mathbf{f}(\mathbf{x})\| / \|\mathbf{h}\|$.

▶ **Example 7.3** *(Upper bound to incremental ratio)* Consider the function $\mathbf{f}$ in Example 7.2. The operator norm of the differential $A_\mathbf{f}$ at $(1, 1)$ is the maximum of

$\|A_{\mathbf{f}}(1, 1)\mathbf{z}\|$ over all $\mathbf{z} = (z_1, z_2)$ such that $\|\mathbf{z}\| = 1$. The maximum is reached for $z_1 = z_2 = 1/\sqrt{2}$ so that $\|A_{\mathbf{f}}(1, 1)\|_{op} = 2$. Note that 2 is also the largest eigenvalue of the matrix $A_{\mathbf{f}}(1, 1)$. In conclusion, we know that for sufficiently small increments $\mathbf{h}$, it is $\|\mathbf{f}(1 + \mathbf{h}) - \mathbf{f}(1)\| \leq (2 + \epsilon)\|\mathbf{h}\|, \forall \epsilon > 0$.

Unlike the case of functions of one single variable, the differential in Definition 7.1 is not defined using the notion of the limit of the increment of the function, thus we are not automatically guaranteed about its uniqueness. However, we can easily prove it.

Theorem 7.2 *The differential is unique.*

Proof Assume that the matrices A and A' satisfy the definition of differential of a function $\mathbf{f}$ in $\mathbf{x}$. Then, for the triangle inequality,

$$\|\mathbf{f}(\mathbf{x} + \mathbf{h}) - \mathbf{f}(\mathbf{x}) - A\mathbf{h}\| + \|\mathbf{f}(\mathbf{x} + \mathbf{h}) - \mathbf{f}(\mathbf{x}) - A'\mathbf{h}\| \geq \|(A - A')\mathbf{h}\|.$$

Dividing both sides by $\|\mathbf{h}\|$ and taking the limit $\|\mathbf{h}\| \to \mathbf{0}$, the left-hand side converges to zero such that $\lim_{\mathbf{h}\to 0} \|(A-A')\mathbf{h}\|/\|\mathbf{h}\| = 0$. In particular, $\forall \mathbf{x} \in \mathbb{R}^n, \lim_{t\to 0} \|(A - A')t\mathbf{x}\|/\|t\mathbf{x}\| = 0$. Simplifying the parameter t, this implies that $\forall \mathbf{x}, \|(A-A')\mathbf{x}\| = 0$, that is, $A = A'$. $\qquad\square$

In addition to uniqueness, another similarity between the derivative of a single variable function and the differential of a function of several variables is the way in which they combine when the composition of functions is considered.

Theorem 7.3 (Chain rule) *Consider* $\mathbf{f} : E_f \subseteq \mathbb{R}^n \to \mathbb{R}^m$ *and* $\mathbf{g} : E_g \subseteq \mathbb{R}^m \to \mathbb{R}^l$. *Let* $A_{\mathbf{f}} \in M_{m,n}$ *be the differential of* $\mathbf{f}$ *in* $\mathbf{x} \in E_f$ *and* $A_{\mathbf{g}} \in M_{l,m}$ *the differential of* $\mathbf{g}$ *in* $\mathbf{f}(\mathbf{x}) \in E_g$. *Then the function* $\boldsymbol{\phi}(\mathbf{x}) = \mathbf{g}(\mathbf{f}(\mathbf{x}))$ *is differentiable in* $\mathbf{x}$ *and its differential is* $A_{\boldsymbol{\phi}} = A_{\mathbf{g}}A_{\mathbf{f}}$.

Proof Since $\mathbf{f}$ is differentiable, for $\mathbf{h} \in \mathbb{R}^n$ one has

$$\boldsymbol{\phi}(\mathbf{x} + \mathbf{h}) = \mathbf{g}(\mathbf{f}(\mathbf{x} + \mathbf{h})) = \mathbf{g}(\mathbf{f}(\mathbf{x}) + A_{\mathbf{f}}(\mathbf{x})\mathbf{h} + \mathbf{r}_f(\mathbf{h})),$$

where $\lim_{\mathbf{h}\to 0} \|\mathbf{r}_f(\mathbf{h})\|/\|\mathbf{h}\| = 0$. Since $\mathbf{g}$ is differentiable, the previous expression can be expanded as

$$\boldsymbol{\phi}(\mathbf{x} + \mathbf{h}) = \mathbf{g}(\mathbf{f}(\mathbf{x})) + A_{\mathbf{g}}(\mathbf{f}(\mathbf{x}))A_{\mathbf{f}}(\mathbf{x})\mathbf{h} + A_{\mathbf{g}}(\mathbf{f}(\mathbf{x}))\mathbf{r}_f(\mathbf{h}) + \mathbf{r}_g(A_{\mathbf{f}}(\mathbf{x})\mathbf{h} + \mathbf{r}_f(\mathbf{h})),$$

where $\lim_{\mathbf{h}'\to 0} \|\mathbf{r}_g(\mathbf{h}')\|/\|\mathbf{h}'\| = 0$ with $\mathbf{h}' \in \mathbb{R}^m$. Now notice that from the triangle inequality

$$\frac{\left\| A_{\mathbf{g}}(\mathbf{f}(\mathbf{x}))\mathbf{r}_f(\mathbf{h}) + \mathbf{r}_g(A_{\mathbf{f}}(\mathbf{x})\mathbf{h} + \mathbf{r}_f(\mathbf{h})) \right\|}{\|\mathbf{h}\|} \leq$$

$$\frac{\left\| A_{\mathbf{g}}(\mathbf{f}(\mathbf{x}))\mathbf{r}_f(\mathbf{h}) \right\|}{\|\mathbf{h}\|} + \frac{\left\| \mathbf{r}_g(A_{\mathbf{f}}(\mathbf{x})\mathbf{h} + \mathbf{r}_f(\mathbf{h})) \right\|}{\|\mathbf{h}\|}.$$

Since $\| A_{\mathbf{g}}(\mathbf{f}(\mathbf{x}))\mathbf{r}_f(\mathbf{h}) \| \leq \| A_{\mathbf{g}}(\mathbf{f}(\mathbf{x})) \|_{op} \|\mathbf{r}_f(\mathbf{h})\|$, the first ratio on the right-hand side goes to zero as $\mathbf{h} \to \mathbf{0}$. For the second ratio, multiply and divide by the same amount to obtain

$$\frac{\left\| \mathbf{r}_g(A_{\mathbf{f}}(\mathbf{x})\mathbf{h} + \mathbf{r}_f(\mathbf{h})) \right\|}{\|\mathbf{h}\|} = \frac{\left\| \mathbf{r}_g(A_{\mathbf{f}}(\mathbf{x})\mathbf{h} + \mathbf{r}_f(\mathbf{h})) \right\|}{\left\| A_{\mathbf{f}}(\mathbf{x})\mathbf{h} + \mathbf{r}_f(\mathbf{h}) \right\|} \frac{\left\| A_{\mathbf{f}}(\mathbf{x})\mathbf{h} + \mathbf{r}_f(\mathbf{h}) \right\|}{\|\mathbf{h}\|}.$$

The first ratio goes to zero when $\mathbf{h} \to \mathbf{0}$ while the second remain bounded (remember that $\| A_{\mathbf{f}}(\mathbf{x})\mathbf{h}\|/\|\mathbf{h}\| \leq \| A_{\mathbf{f}}(\mathbf{x})\|_{op}$). In conclusion,

$$\lim_{\mathbf{h} \to \mathbf{0}} \frac{\left\| A_{\mathbf{g}}(\mathbf{f}(\mathbf{x}))\mathbf{r}_f(\mathbf{h}) + \mathbf{r}_g(A_{\mathbf{f}}(\mathbf{x})\mathbf{h} + \mathbf{r}_f(\mathbf{h})) \right\|}{\|\mathbf{h}\|} = 0,$$

and the assertion is proved. □

The differential of a real-valued (or scalar) function f in $\mathbf{x}$ is often denoted by $df(\mathbf{x})$. It is a $1 \times n$ matrix, that is, a row vector. The fact that the convergence of the norm of a vector implies the convergence of all its components (and vice versa), means that if a function $\mathbf{f} : \mathbb{R}^n \to \mathbb{R}^m$ is differentiable, then all its component functions $f_j(\mathbf{x})$ from $\mathbb{R}^n$ to $\mathbb{R}$ with $j = 1, \ldots, m$ are differentiable. The differential of the jth component is $df_j(\mathbf{x})$, the jth row of the differential of $\mathbf{f}$. Thus, the problem of proving the differentiability of a vector function or finding its differential is equivalent to the problem of proving the differentiability or finding the differential of all its component functions. It follows that all the results concerning the existence of the differential and its computation can be derived considering functions from $\mathbb{R}^n$ to $\mathbb{R}$. This is indeed the kind of function that appears in the examples. Limiting the direct investigation of special cases to functions with an image in $\mathbb{R}$ will simplify the analysis. There is no reason, however, to introduce this simplification in the general statements.

Definition 7.1 introduces the notion of differential as a linear approximation of the local behaviour of a function but is totally silent about how to compute it. In the case of a function of a single variable, we know that the differential at a given point is simply the derivative computed at that point. So, the existence of a differential, that is, the existence of a local linear approximation, and the existence of the derivative are the same thing. For a multivariate function, the situation is more complicated. We will see that, in general, if we know that the function is differentiable in one point, we can compute the differential using the *partial derivatives*, quantities obtained with methods similar to the derivation of a function of one variable. However, the existence of partial derivatives does not guarantee the existence of the differential. Denote by $(\mathbf{e}_1, \mathbf{e}_2, ..., \mathbf{e}_n)$ the canonical basis of $\mathbb{R}^n$.

Definition 7.2 *(Partial derivative)* Let $\mathbf{f} : E \subseteq \mathbb{R}^n \to \mathbb{R}^m$, E open, and $\mathbf{x} \in E$. Consider the vector-valued function of the real variable t, $\mathbf{f}(\mathbf{x} + t\mathbf{e}_j)$ defined in a neighbourhood of $t = 0$. The *partial derivative* of the function $\mathbf{f}$ in $\mathbf{x}$ along the jth direction is defined as

$$\partial_j \mathbf{f}(\mathbf{x}) = \lim_{t \to 0} \frac{\mathbf{f}(\mathbf{x} + t\mathbf{e}_j) - \mathbf{f}(\mathbf{x})}{t}.$$

Other widely used notations for partial derivatives are $\partial \mathbf{f}(\mathbf{x})/\partial x_j$, $\partial_{x_j} \mathbf{f}(\mathbf{x})$ or $D_j \mathbf{f}(\mathbf{x})$. The limit in Definition 7.2 should be intended component by component: the partial derivative is a m-dimensional vector whose components are the partial derivatives of the component functions. In the case of a real-valued function $f : \mathbb{R}^n \to \mathbb{R}$, partial derivatives are real numbers. They are the derivatives in $t = 0$ of the function $f(\mathbf{x} + t\mathbf{e}_j)$ defined as the *restriction* of the original function along the specific direction in which the derivative is computed. Alternatively, considering the functional form $f(x_1, x_2, ..., x_n)$, the partial derivative is the derivative with respect to one argument, keeping the other arguments fixed.

▶ **Example 7.4** *(Partial derivatives)* Consider the function $\mathbf{f}$ in Example 7.2. In $\mathbf{x} = (x_1, x_2)$, $f_1(x_1 + t, x_2) - f_1(x_1, x_2) = tx_2$ and $f_2(x_1 + t, x_2) - f_2(x_1, x_2) = 1$, so that

$$\partial_1 \mathbf{f} = \lim_{t \to 0} \frac{1}{t} \begin{pmatrix} tx_2 \\ t \end{pmatrix} = \begin{pmatrix} x_2 \\ 1 \end{pmatrix}.$$

Analogously, $f_1(x_1, x_2 + t) - f_1(x_1, x_2) = tx_1$ and $f_2(x_1, x_2 + t) - f_2(x_1, x_2) = t$, so that

$$\partial_2 \mathbf{f} = \lim_{t \to 0} \frac{1}{t} \begin{pmatrix} tx_1 \\ t \end{pmatrix} = \begin{pmatrix} x_1 \\ 1 \end{pmatrix}.$$

Note that in $\mathbf{x} = (1, 1)$, these vectors correspond to the columns of the differential in Example 7.2.

The previous definition can be generalised to vectors other than those in the canonical basis. Identify the directions in $\mathbb{R}^n$ with the unit vector $\mathbf{u} \in \mathbb{R}^n$, such that $\|\mathbf{u}\| = 1$. If $\mathbf{u} = \sum_j c_j \mathbf{e}_j$, then $\sum_j c_j^2 = 1$.

Definition 7.3 *(Directional derivative)* Let $\mathbf{f} : E \subseteq \mathbb{R}^n \to \mathbb{R}^m$, E open, and $\mathbf{x} \in E$. The *directional derivative* of the function $\mathbf{f}$ along the direction $\mathbf{u}$ ($\|\mathbf{u}\| = 1$) is defined as

$$D_{\mathbf{u}} \mathbf{f}(\mathbf{x}) = \lim_{t \to 0} \frac{\mathbf{f}(\mathbf{x} + t\mathbf{u}) - \mathbf{f}(\mathbf{x})}{t}.$$

Again, the limit in Definition 7.3 is intended component by component, and the directional derivative is a vector of dimensions m, whose components are the directional derivatives of the component functions. When the direction $\mathbf{u}$ takes the value of a vector of the canonical basis, the directional derivative reduces to a partial derivative.

▶ **Example 7.5** *(Directional derivatives)* Consider the function $\mathbf{f}$ in Example 7.2. At $\mathbf{x} = (x_1, x_2)$,

$$(f_1(x_1 + tu_1, x_1 + tu_2) - f_1(x_1, x_2))/t = u_1 x_2 + u_2 x_1 + tu_1 u_2,$$
$$(f_2(x_1 + tu_1, x_1 + tu_2) - f_2(x_1, x_2))/t = u_1 + u_2.$$

Thus, we conclude that

$$D_{\mathbf{u}}\mathbf{f}(x_1, x_2) = \begin{pmatrix} u_1 x_2 + u_2 x_1 \\ u_1 + u_2 \end{pmatrix}.$$

When $(u_1, u_2) = (1, 0)$ we recover the partial derivative $\partial_1 \mathbf{f}(\mathbf{x})$, and when $(u_1, u_2) = (0, 1)$ the partial derivative $\partial_2 \mathbf{f}(\mathbf{x})$.

If a function is differentiable at a given point in its domain, then the partial derivatives and the directional derivatives at that point exist and are linked by a strong relationship.

Theorem 7.4 *Let* $\mathbf{f} : E \subseteq \mathbb{R}^n \to \mathbb{R}^m$ *and* $\mathbf{x} \in E$. *If* $\mathbf{f}$ *is differentiable in* $\mathbf{x}$, *then there are all partial and directional derivatives. If* $A_{\mathbf{f}}(\mathbf{x})$ *is the differential of the function* $\mathbf{f}$ *in* $\mathbf{x}$, *then for* $j = 1, \ldots, n$,

$$\partial_j \mathbf{f}(\mathbf{x}) = A_{\mathbf{f}}(\mathbf{x})_j,$$

where $A_{\mathbf{f}}(\mathbf{x})_j$ *is the* j^{th} *column of the matrix* $A_{\mathbf{f}}(\mathbf{x})$, *and* $\forall \mathbf{u}$,

$$D_{\mathbf{u}}\mathbf{f}(\mathbf{x}) = A_{\mathbf{f}}(\mathbf{x})\mathbf{u}.$$

Proof For the definition of partial derivative and differential,

$$\lim_{t \to 0} \frac{\mathbf{f}(\mathbf{x} + t\mathbf{e}_j) - \mathbf{f}(\mathbf{x})}{t} = \lim_{t \to 0} \frac{t A_{\mathbf{f}}(\mathbf{x})\mathbf{e}_j + \mathbf{r}(t\mathbf{e}_j)}{t} = A_{\mathbf{f}}(\mathbf{x})\mathbf{e}_j.$$

For the directional derivative, the proof is analogous.

If a vector function $\mathbf{f}$ admits partial derivatives, we define its *Jacobian* matrix in $\mathbf{x}$ as $J_f(\mathbf{x})_{i,j} = \partial_j f_i(\mathbf{x})$. Theorem 7.4 states that if the function is differentiable in $\mathbf{x}$, its differential is equal to the Jacobian. If the function f is real-valued, the vector of partial derivatives is denoted with ∇f and is called the *gradient* of the function,

$$\nabla f(\mathbf{x}) = \left(\frac{\partial f(\mathbf{x})}{\partial x_1}, \frac{\partial f(\mathbf{x})}{\partial x_2}, \ldots, \frac{\partial f(\mathbf{x})}{\partial x_n} \right).$$

This can be represented by writing the infinitesimal increment of the function f as $df = \partial_1 f dx_1 + \ldots + \partial_n f dx_n$. The Jacobian of a function with values in $\mathbb{R}^m$ has, in its rows, the gradient of the component functions. Note that the implication of Theorem 7.4 cannot be reversed. The existence of the Jacobian or of the gradient of the function $\mathbf{f}$ in $\mathbf{x}$ does not, in general, imply that $\mathbf{f}$ is differentiable in $\mathbf{x}$.

▶ **Example 7.6** *(Non-differentiable function with partial derivatives)* Consider the function $f : \mathbb{R}^2 \to \mathbb{R}$ defined by

$$f(x_1, x_1) = \begin{cases} 0 & \text{if } (x_1, x_1) = (0, 0), \\ x_1 x_2/(x_1^2 + x_2^2) & \text{otherwise.} \end{cases}$$

Since f is the composition of continuous functions, it is continuous in every point, apart, possibly, the origin. In polar coordinates, the function reads $f(\rho, \phi) = \cos \phi \sin \phi$. Thus, $\lim_{\rho \to 0} f(\rho, \phi)$ does not exist and the function is not continuous in $(0, 0)$. A way of realising it is by noticing that $f(\rho, \phi)$ takes any value in $[-1/2, 1/2]$ whatever the value of ρ, however small.

Concerning the partial derivatives, consider a point $(x_1, x_2) \neq (0, 0)$, then

$$\frac{\partial f}{\partial x_1} = \frac{x_1}{x_1^2 + x_2^2} - 2\frac{x_1^2 x_2}{(x_1^2 + x_2^2)^2}, \quad \frac{\partial f}{\partial x_2} = \frac{x_1}{x_1^2 + x_2^2} - 2\frac{x_1 x_2^2}{(x_1^2 + x_2^2)^2}.$$

At $(0, 0)$,

$$\lim_{h \to 0} \frac{f(h, 0) - f(0, 0)}{h} = \lim_{h \to 0} \frac{f(0, h) - f(0, 0)}{h} = 0,$$

so that the partial derivatives of the function exist on the whole $\mathbb{R}^2$, origin included. Consider the directional derivative at the origin along the direction (u_1, u_2),

$$\lim_{t \to 0} \frac{f(tu_1, tu_2) - f(0, 0)}{t} = \lim_{t \to 0} \frac{u_1 u_2}{t},$$

where we have used the fact that $u_1^2 + u_2^2 = 1$. When u_1 and u_2 are both different from zero, this limit does not exist.

In the previous example, we have seen that the existence of partial derivatives at one point does not imply the continuity of the function in that point and, consequently, the existence of the differential. However, the function lacked directional derivatives in almost all directions. As the following example clarifies, the existence of the latter is not sufficient to guarantee continuity.

▶ **Example 7.7** (*Non-differentiable function with partial and directional derivatives*)
Consider the function $f : \mathbb{R}^2 \to \mathbb{R}$ defined by

$$
f(x, y) = \begin{cases} 0 & \text{if } (x_1, x_2) = (0, 0), \\ x_1 x_2^2 / (x_1^2 + x_2^4) & \text{otherwise.} \end{cases}
$$

Being the composition of continuous and differentiable functions, f is continuous
and differentiable at every point apart, possibly, from the origin. Its partial derivatives
in $(0, 0)$ are both zero:

$$
\lim_{h \to 0} \frac{f(h, 0) - f(0, 0)}{h} = \lim_{h \to 0} \frac{f(0, h) - f(0, 0)}{h} = 0.
$$

The derivative at $(0, 0)$ along the direction (u_1, u_2), with $u_1^2 + u_2^2 = 1$, can be
computed as

$$
\lim_{t \to 0} \frac{f(tu_1, tu_2) - f(0, 0)}{t} = \lim_{t \to 0} \frac{u_1 u_2^2}{u_1^2 + u_2^4 t^2}.
$$

If $u_1 = 0$ the limit is 0 while if $u_1 \neq 0$ the limit is u_2^2/u_1. Thus, partial derivatives
and all directional derivatives of the function exist in $(0, 0)$. Can we conclude that
the function is continuous in $(0, 0)$? First of all, notice that the function is bounded.
Indeed, in polar coordinates,

$$
f(\rho, \phi) = \frac{\rho^3 \cos \phi \sin^2 \phi}{\rho^2 \cos^2 \phi + \rho^4 \sin^4 \phi} = \rho \frac{\cos \phi \sin^2 \phi}{\cos^2 \phi + \rho^2 \sin^4 \phi}
$$

and the denominator is always different from zero. We can use the polar coordinates
to find the image of the function. The function takes value zero for some ϕ, regardless
of the value of ρ. Put $f(\rho, \phi) = \alpha \neq 0$ and solve the equation with ρ as unknown
to obtain

$$
\rho = \frac{\cos \phi \pm |\cos \phi| \sqrt{1 - 4\alpha^2}}{2\alpha \sin^2 \phi}.
$$

The previous equation admits real solutions only if the argument of the square root
is real. Then we conclude that the image of the function is $[-1/2, 1/2]$. By direct
substitution of the values of $\alpha = \pm 1/2$ in the previous equation, it is immediate
to see that the function reaches its maximum value when $\rho \sin^2 \phi = \cos \phi$ and its
minimum value when $\rho \sin^2 \phi = -\cos \phi$. Recalling that ρ cannot be negative, we
can solve the previous equations in terms of ρ to find for which values of ϕ the
function reaches its maximum and minimum values. In both cases, only one root of
the second degree polynomial in $\cos \phi$ is admissible and after little algebra we obtain
that the extremal values $\pm 1/2$, for a given ρ, are achieved respectively in

$$
\phi_\pm(\rho) = \cos^{-1}\left(\pm \frac{\sqrt{1 + 4\rho^2} - 1}{2\rho} \right)
$$

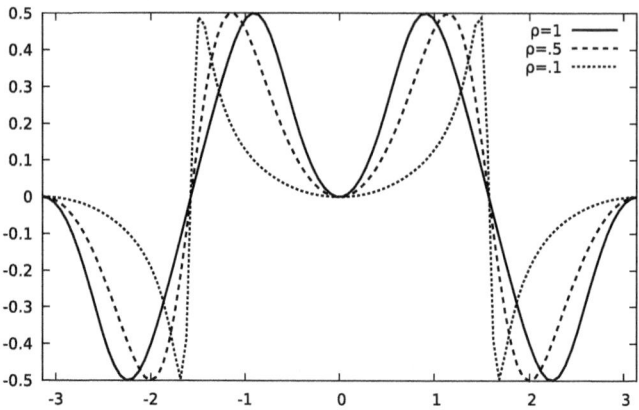

Fig. 7.2 The function f of Example 7.7 as function of $\phi \in [-\pi, \pi]$ for several values of ρ

and, for symmetry $-\phi_\pm(\rho)$. This implies that, on any circle with centre the origin, whatever small its radius ρ is, the function takes values $1/2$ and $-1/2$. An illustration of this phenomenon is shown in Fig. 7.2. Obviously, this implies that the function does not have any limit for $(x_1, x_2) \to (0, 0)$ and, consequently, the function is not continuous. However, note that $\lim_{\rho \to 0} f(\rho, \phi) = 0$ for any ϕ.

If a function $\mathbf{f} : E \subset \mathbb{R}^n \to \mathbb{R}^m$ is differentiable in E, we can assign to each point $\mathbf{x} \in E$ the differential $A_\mathbf{f}(\mathbf{x})$ of the function at that point, defining a new function from E to $M_{m,n}$.

Definition 7.4 *(Continuously differentiable)* The function $\mathbf{f} : E \subset \mathbb{R}^n \to \mathbb{R}^m$ is *continuously differentiable* in the open set E if $A_\mathbf{f} : E \subset \mathbb{R}^n \to M_{m,n}$ exists and is continuous on E. In this case we write $\mathbf{f} \in C^1(E)$.

Examples 7.7 show that the existence of the partial derivatives and all directional derivatives in a point does not guarantee the continuity of the function in that point and, consequently, the existence of the differential. However, if we also require the partial derivatives to be continuous functions in a neighbourhood of the point, then the existence of a differential at that point is guaranteed.

Theorem 7.5 *Let $\mathbf{f} : E \subseteq \mathbb{R}^n \to \mathbb{R}^m$, E open. Then $\mathbf{f} \in C^1(E)$ if and only if all partial derivatives exist and are continuous $\forall \mathbf{x} \in E$.*

Proof If the differential $A_\mathbf{f}(\mathbf{x})$ exists and is continuous then all partial derivatives exist and are continuous as they are the components of a continuous function from $\mathbb{R}^n$ to $M_{m,n}$.

Assume instead that all the partial derivatives $\partial_j f_i(\mathbf{x})$ exist and are continuous $\forall \mathbf{x} \in E$. We will show that in this case, the Jacobian $J_f(\mathbf{x})_{i,j} = \partial_j f_i(\mathbf{x})$ is the differential of $\mathbf{f}$ in $\mathbf{x}$. If this is true, then the differential is continuous because it has

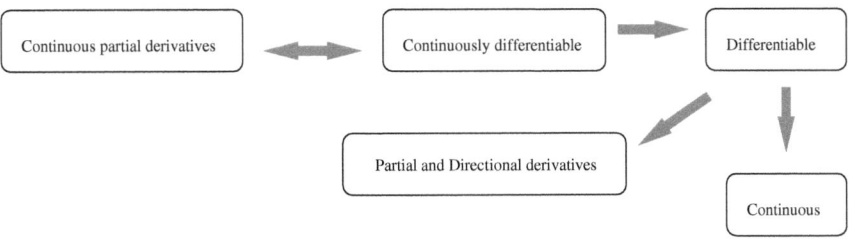

Fig. 7.3 The existence of continuous partial derivatives implies the existence of the differential and its continuity. In turn the existence of the differential implies continuity and the existence of partial derivatives. The direction of the arrows in the picture cannot in general be reversed

continuous components. To prove the assertion it is sufficient to prove that $\forall i = 1, \ldots, m,$

$$\lim_{\mathbf{h} \to 0} \frac{f_i(\mathbf{x} + \mathbf{h}) - f_i(\mathbf{x}) - \sum_{j=1}^n \partial_j f_i(\mathbf{x}) h_j}{\|\mathbf{h}\|} = 0,$$

where $\sum_{j=1}^n \partial_j f_i(\mathbf{x}) h_j$ is the ith component of $J_f(\mathbf{x})\mathbf{h}$. Consider the vectors $\mathbf{v}_k = \sum_{j=1}^k \mathbf{e}_j h_j$ with $k = 1, \ldots, n$ and $\mathbf{v}_0 = \mathbf{0}$. The expression above can be written as

$$\sum_{j=1}^n \frac{f_i(\mathbf{x} + \mathbf{v}_j) - f_i(\mathbf{x} + \mathbf{v}_{j-1}) - \partial_j f_i(\mathbf{x}) h_j}{\|\mathbf{h}\|}.$$

Since partial derivatives exist, and are continuous, applying Theorem 6.10 in a neighbourhood of $\mathbf{x}$,

$$f_i(\mathbf{x} + \mathbf{v}_j) - f_i(\mathbf{x} + \mathbf{v}_{j-1}) = f_i(\mathbf{x} + \mathbf{v}_{j-1} + \mathbf{e}_j h_j) - f_i(\mathbf{x} + \mathbf{v}_{j-1}) = \\ \partial_j f_i(\mathbf{x} + \mathbf{v}_{j-1} + \eta_j \mathbf{v}_j) h_j,$$

where $\eta_j \in (0, h_j)$. Thus, the jth element of the previous sum can be rewritten as

$$\frac{\left(\partial_j f_i(\mathbf{x} + \mathbf{v}_{j-1} + \eta_j \mathbf{v}_j) - \partial_j f_i(\mathbf{x}) \right) h_j}{\|\mathbf{h}\|}.$$

When $\|\mathbf{h}\| \to 0$, $h_j/\|\mathbf{h}\|$ is a bounded quantity. Thus, because of the continuity of partial derivatives, the previous expression goes to zero. This means that all elements of the sum go to zero, and the assertion is proved.   □

Figure 7.3 graphically illustrates the relationship between continuity, differentiability, and the existence of partial derivatives. The arrows point along the direction of logical implication. Note that the existence of the differential does not guarantee that the partial derivatives are continuous.

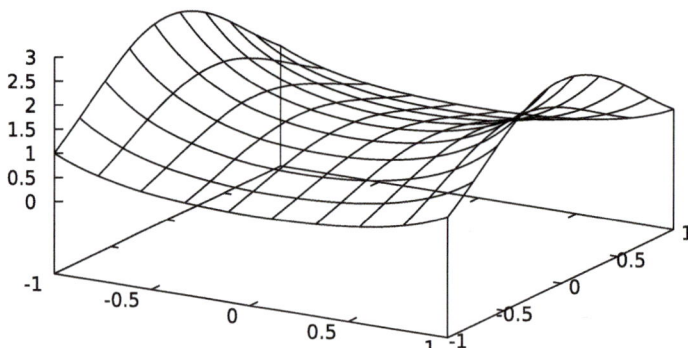

Fig. 7.4 The function of Example 7.9. The origin is a critical point but neither a minimum nor a maximum point

▶ **Example 7.8** *(Differentiable function with discontinuous partial derivatives)* Consider the function $f : \mathbb{R}^2 \to \mathbb{R}$ defined by

$$f(x_1, x_2) = \begin{cases} 0 & \text{if } (x_1, x_2) = (0, 0), \\ (x^2 + x_2^2) \sin \frac{1}{x_1^2 + x_2^2} & \text{otherwise.} \end{cases}$$

The partial derivatives in $(0, 0)$ can be easily computed

$$\partial_1 f(0, 0) = \lim_{h \to 0} \frac{f(h, 0) - f(0, 0)}{h} = \lim_{h \to 0} \frac{h^2}{h} \sin \frac{1}{h^2} = 0,$$

$$\partial_2 f(0, 0) = \lim_{h \to 0} \frac{f(0, h) - f(0, 0)}{h} = \lim_{h \to 0} \frac{h^2}{h} \sin \frac{1}{h^2} = 0.$$

The function is differentiable in the origin and $df(0, 0) = \nabla f(0, 0) = (0, 0)$. In fact, because the sine is bounded above by 1,

$$\lim_{(h_1, h_2) \to (0,0)} \left| \frac{f(h_1, h_2)}{\sqrt{h_1^2 + h_2^2}} \right| \le \lim_{(h_1, h_2) \to (0,0)} \sqrt{h_1^2 + h_2^2} = 0.$$

On the other hand, the partial derivatives at $(x, y) \neq (0, 0)$ read

$$\partial_1 f(x_1, x_2) = 2x_1 \sin \frac{1}{x_1^2 + x_2^2} - \frac{2x_1}{x_1^2 + x_2^2} \cos \frac{1}{x_1^2 + x_2^2},$$

$$\partial_2 f(x_1, x_2) = 2x_2 \sin \frac{1}{x_1^2 + x_2^2} - \frac{2x_2}{x_1^2 + x_2^2} \cos \frac{1}{x_1^2 + x_2^2}.$$

Thus, $\lim_{(x_1,x_2)\to(0,0)} \partial_1 f(x_1, x_2)$ and $\lim_{(x_1,x_2)\to(0,0)} \partial_2 f(x_1, x_2)$ do not exist.

We conclude this section with some considerations about real-valued functions that are particularly useful in applications. The local increments of a differentiable scalar function $f : \mathbb{R}^n \to \mathbb{R}$ in $\mathbf{x}$ can be expressed as $f(\mathbf{x}+\mathbf{h}) = f(\mathbf{x})+\mathbf{h}\cdot\nabla f+r(\mathbf{h})$, where the term $r(\mathbf{h})$ is negligible when $\|\mathbf{h}\|$ is sufficiently small. The function f increases along the direction $\mathbf{h}$ if the scalar product $\mathbf{h} \cdot \nabla f$ is positive and decreases along that direction if the scalar product is negative. Because $-\|\mathbf{h}\|\|\nabla f\| \le \mathbf{h}\cdot\nabla f \le \|\mathbf{h}\|\|\nabla f\|$, the direction along which the function increases the most, that is, the direction along which $f(\mathbf{x} + \mathbf{h}) - f(\mathbf{x})$ is greater while keeping the length of the increment $\|\mathbf{h}\|$ fixed, is precisely along the gradient, that is $\mathbf{h} = c\nabla f$ with $c > 0$. The direction of faster decrease is, conversely, opposite to the gradient, that is when $c < 0$. Along a direction $\mathbf{h}$ orthogonal to the gradient, the function increases less than linearly. It is useful to introduce the following.

Definition 7.5 *(Critical point)* Consider a real function $f : E \subseteq \mathbb{R}^n \to \mathbb{R}$. A point $\mathbf{x} \in \text{int } E$ is a *critical point* or *stationary point* of f if the function is differentiable in $\mathbf{x}$ and $\nabla f(\mathbf{x}) = 0$.

If $\nabla f(\mathbf{x})$ is different from zero, in any neighbourhood of $\mathbf{x}$ there are points in which the value of the function is greater than $f(\mathbf{x})$ and points in which it is smaller.

Corollary 7.2 *Consider a real function $f : E \subseteq \mathbb{R}^n \to \mathbb{R}$, differentiable in $\mathbf{x} \in$ int E. If the function has a local maximum or a local minimum in $\mathbf{x}$, then $\mathbf{x}$ is a critical point.*

However, a critical point is not always a maximum or a minimum.

▶ **Example 7.9** *(A non-extremal critical point)* Consider the function $f : \mathbb{R}^2 \to \mathbb{R}$ defined as $f(x_1, x_2) = \exp(x_1^2 - x_2^2)$. Notice that $\partial_1 f(0, 0) = \partial_2 f(0, 0) = 0$. But the point $(0, 0)$ is neither a local maximum nor a local minimum of the function. In fact, while $f(0, 0) = 1$, $f(z, 0) > 1$, and $f(0, z) < 1 \ \forall z \in \mathbb{R} \setminus \{0\}$, see Fig. 7.4.

To identify the maximum and minimum points of a function, it is common practise to impose the so-called *first-order conditions* (FOC), that is, look for its critical points by solving the system of equations $\nabla f = \mathbf{0}$. In general, if nothing is known about the function, this procedure is not guaranteed to provide the extrema we are looking for. The derivation of sufficient conditions for the identification of maxima and minima, when the function admits higher-order derivatives, is based on the extension of the Taylor approximation introduced in Theorem 6.14 to functions of several variables. This extension requires a series of results on the local behaviour of functions of several variables. They are introduced in the next section.

7.3 Mean Value Theorems

We can use the existence of the differential of the function at a point to build some bound on the variation of the function in its neighbourhood. This is a generalisation of Theorem 6.10, although the results are often weaker. We will see results of increasing generality. Let us start with the case of a vector function of a single real variable.

Theorem 7.6 (Mean value theorem) *Let* $\mathbf{f} : [a, b] \to \mathbb{R}^m$ *be differentiable in* (a, b). *Then* $\exists x \in (a, b)$ *such that*

$$\|\mathbf{f}(b) - \mathbf{f}(a)\| \le (b - a)\|\mathbf{f}'(x)\|.$$

Proof Define the real-valued function of one real variable $\phi(t) = (\mathbf{f}(b) - \mathbf{f}(a)) \cdot \mathbf{f}(t)$ with $t \in [a, b]$. The function ϕ is differentiable in (a, b) and, by Theorem 6.10, $\exists x \in [a, b]$ such that

$$\phi(b) - \phi(a) = (b - a)\phi'(x) = (b - a)(\mathbf{f}(b) - \mathbf{f}(a)) \cdot \mathbf{f}'(x).$$

For the Cauchy–Schwarz inequality,

$$(\mathbf{f}(b) - \mathbf{f}(a)) \cdot \mathbf{f}'(x) \le \|(\mathbf{f}(b) - \mathbf{f}(a)) \cdot \mathbf{f}'(x)\| \le \|\mathbf{f}(b) - \mathbf{f}(a)\|\|\mathbf{f}'(x)\|.$$

The statement is proved by substituting the last inequality in the previous equation and noticing that by substituting a and b into the initial definition of ϕ, we have $\phi(b) - \phi(a) = \|\mathbf{f}(b) - \mathbf{f}(a)\|^2$. $\square$

For real multivariate functions defined over a convex domain (see Definition 1.22), we have the following.

Theorem 7.7 (Mean value theorem) *Consider a map* $f : E \subseteq \mathbb{R}^n \to \mathbb{R}$ *with* E *convex and let* $\mathbf{f}$ *be differentiable in* int E. *Then, for any pair of points* $\mathbf{a}, \mathbf{b} \in E$, $\exists \lambda \in [0, 1]$ *such that*

$$f(\mathbf{b}) - f(\mathbf{a}) = df(\lambda\mathbf{b} + (1 - \lambda)\mathbf{a}) \cdot (\mathbf{b} - \mathbf{a}).$$

Proof Consider the function $\boldsymbol{\gamma} : [0, 1] \to \mathbb{R}^n$, $\boldsymbol{\gamma}(t) = (1 - t)\mathbf{a} + t\mathbf{b}$. Since E is convex, $\gamma(t) \subseteq E$. Then define the function $\phi(t) = f(\boldsymbol{\gamma}(t)), \phi(t) : [0, 1] \to \mathbb{R}$. This function is continuous and differentiable. According to Theorem 6.10, $\exists \lambda \in [0, 1]$ such that $\phi(1) - \phi(0) = \phi'(\lambda)$. From chain rule $\phi'(\lambda) = df(\lambda\mathbf{b} + (1 - \lambda)\mathbf{a}) \cdot (\mathbf{b} - \mathbf{a})$. Since $\phi(1) = f(\mathbf{b})$ and $\phi(0) = f(\mathbf{a})$, the statement is proved. $\square$

The previous Theorem has a straightforward implication.

Corollary 7.3 *If* $\forall \mathbf{x} \in E$, $df(\mathbf{x}) = \mathbf{0}$, *then the function* f *is constant in* E.

For vector-valued functions of several variables, we can obtain a similar, albeit weaker, result for the increment of the norm.

Theorem 7.8 (Mean value theorem) *Consider a map* $\mathbf{f} : E \subseteq \mathbb{R}^n \to \mathbb{R}^m$ *with* E *convex and let* $\mathbf{f}$ *be differentiable in* int E. *Assume that there is a real number* M *such that* $\|A_{\mathbf{f}}(\mathbf{x})\|_{op} \leq M$, $\forall \mathbf{x} \in E$. *Then, for any two points* $\mathbf{a}, \mathbf{b} \in E$,

$$\|\mathbf{f}(\mathbf{a}) - \mathbf{f}(\mathbf{b})\| \leq M\|\mathbf{b} - \mathbf{a}\|.$$

Proof Consider the function $\boldsymbol{\gamma} : [0, 1] \to \mathbb{R}^n$, $\boldsymbol{\gamma}(t) = (1 - t)\mathbf{a} + t\mathbf{b}$. Since E is convex, $\boldsymbol{\gamma}(t) \subseteq E$. Then define the vector function $\boldsymbol{\phi}(t) = \mathbf{f}(\boldsymbol{\gamma}(t))$, $\boldsymbol{\phi}(t) : [0, 1] \to \mathbb{R}^m$. By Theorem 7.6, $\exists t^*$ such that $\|\boldsymbol{\phi}(1) - \boldsymbol{\phi}(0)\| \leq |\boldsymbol{\phi}'(t^*)|$. From chain rule,

$$\boldsymbol{\phi}'(t^*) = A_{\mathbf{f}}(\boldsymbol{\gamma}(t^*))\boldsymbol{\gamma}'(t^*) = A_{\mathbf{f}}(\boldsymbol{\gamma}(t^*))(\mathbf{b} - \mathbf{a}),$$

and, for the definition of the operator norm,

$$\|A_{\mathbf{f}}(\boldsymbol{\gamma}(t))(\mathbf{b} - \mathbf{a})\| \leq \|A_{\mathbf{f}}(\boldsymbol{\gamma}(t))\|_{op}\|\mathbf{b} - \mathbf{a}\| \leq M\|\mathbf{b} - \mathbf{a}\|.$$

The statement follows by noticing that $\|\boldsymbol{\phi}(1) - \boldsymbol{\phi}(0)\| = \|\mathbf{f}(\mathbf{a}) - \mathbf{f}(\mathbf{b})\|$. $\qquad\square$

If E is compact and $f \in C^1(E)$, then M can be the maximum of the continuous function $\|A_{\mathbf{f}}\|_{op}$ in E.

7.4 Higher-Order Derivatives and Taylor Polynomial

If a function $\mathbf{f} : E \subseteq \mathbb{R}^n \to \mathbb{R}^m$ has partial derivatives in all the points of an open set E, then there is the possibility that the partial derivative functions, that is the functions that assign the value of the partial derivatived of $\mathbf{f}$ to each point of E, have themselves partial derivatives.

▶ **Example 7.10** *(Symmetric second-order derivatives)* Consider the two-variable differentiable function $f(x_1, x_2) = \exp(x_1^2 + x_2^2)$. Its partial derivatives are $\partial_1 f(x_1, x_2) = 2x_1 \exp(x_1^2 + x_2^2)$ and $\partial_2 f(x, y) = 2x_2 \exp(x_1^2 + x_2^2)$. These are differentiable functions of their argument and they can be partially derived. In particular, notice that $\partial_1 \partial_2 f(x_1, x_2) = \partial_2 \partial_1 f(x_1, x_2) = 4x_1 x_2 \exp(x_1^2 + x_2^2)$.

Analogously to what happens with the derivative of functions of one single variable, the operation of taking the partial derivative can be iterated if the functional expression allows it. In the previous example, the order in which the derivatives are taken does not seem to be relevant. However, this is not a universal property, and, in general, one should keep track of their order.

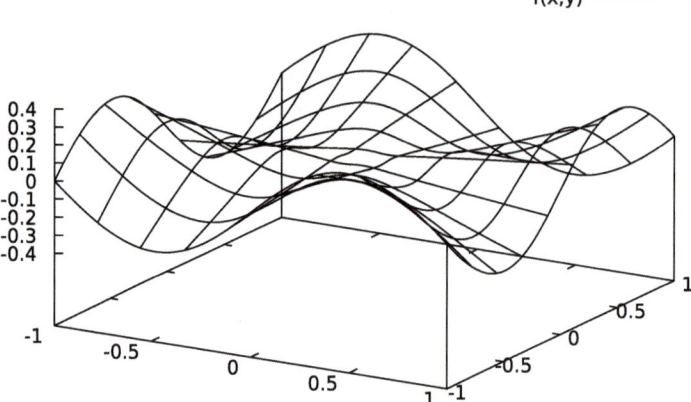

f(x,y) ——

Fig. 7.5 The function of Example 7.11. The second-order derivatives $\partial_1 \partial_2 f$ and $\partial_2 \partial_1 f$ exist in the origin but are not continuous

▶ **Example 7.11** *(Asymmetric second-order derivatives)* Consider the function $f :$ $\mathbb{R}^2 \to \mathbb{R}$ defined as

$$f(x_1, x_2) = \begin{cases} 0 & \text{if } (x_1, x_2) = (0, 0), \\ \dfrac{x_1 x_2 (x_1^2 - x_2^2)}{x_1^2 + x_2^2} & \text{otherwise.} \end{cases}$$

The function f and its partial derivatives

$$\partial_1 f = \frac{y(3x_1^2 - x_2^2)}{x_1^2 + x_2^2} - \frac{2x_1^2 x_2 (x_1^2 - x_2^2)}{(x_1^2 + x_2^2)^2}$$

$$\partial_2 f = \frac{x(x_1^2 - 3x_2^2)}{x_1^2 + x_2^2} - \frac{2x_2 x_2^2 (x_1^2 - x_2^2)}{(x_1^2 + x_2^2)^2}$$

are continuous in $\mathbb{R}^2$ with $\partial_1 f(0, 0) = \partial_2 f(0, 0) = 0$. Let's compute the second-order partial cross-derivatives in the origin. By direct substitution of the previous expression, it is immediate to see that

$$\partial_2 \partial_1 f(0, 0) = \lim_{h \to 0} \frac{\partial_1 f(0, h) - \partial_1 f(0, 0)}{h} = -1$$

while

$$\partial_1 \partial_2 f(0, 0) = \lim_{h \to 0} \frac{\partial_2 f(h, 0) - \partial_2 f(0, 0)}{h} = 1.$$

In this case, the order in which the derivatives are taken matters (Fig. 7.5).

In the previous example, the order in which the derivatives are taken matters because the second-order derivatives are not continuous. In fact, one can prove that if the second-order partial derivatives exist and are continuous, then the order in which they are taken is irrelevant. We start by proving this statement for a real function of two variables.[1]

Theorem 7.9 (Schwarz) *Assume that the function* $f : E \subseteq \mathbb{R}^2 \to \mathbb{R}$ *admits second-order partial derivatives in the open set E and they are continuous. Then* $\forall \mathbf{x} = (x, y) \in E$, $\partial_x \partial_y f(\mathbf{x}) = \partial_y \partial_x f(\mathbf{x})$.

Proof Consider a point $(x_0, y_0) \in E$ and two sufficiently small positive real numbers h and k. Define

$$u(y) = \frac{f(x_0 + h, y) - f(x_0, y)}{h}.$$

This function is differentiable in a neighbourhood of y_0. According to the mean value theorem, $\exists \theta_y \in [0, 1]$ such that

$$u(y_0 + k) - u(y_0) = k u'(y_0 + \theta_y k),$$

that is

$$\frac{u(y_0 + k) - u(y_0)}{k} = \frac{\partial_y f(x_0 + h, y_0 + \theta_y k) - \partial_y f(x_0, y_0 + \theta_y k)}{h}.$$

Because partial derivatives are differentiable, by the mean value theorem, $\exists \theta_x \in [0, 1]$ such that

$$\partial_y f(x_0 + h, y_0 + \theta_y k) = \partial_y f(x_0, y_0 + \theta_y k) + h \partial_x \partial_y f(x_0 + \theta_x h, y_0 + \theta_y k).$$

Substituting in the previous equation we obtain $u(y_0 + k) - u(y_0) = k \partial_x \partial_y f(x_0 + \theta_x h, y_0 + \theta_y k)$. Let us repeat the same procedure for the function

$$v(x) = \frac{f(x, y_0 + k) - f(x, y_0)}{k}.$$

By the mean value theorem $\exists \theta'_x \in [0, 1]$ such that $v(x_0 + h) - v(x_0) = h v'(x_0 + \theta'_x h)$. For the definition of v and again by the mean value theorem, $\exists \theta'_y \in [0, 1]$ such that $v'(x_0 + \theta'_x h) = \partial_y \partial_x f(x_0 + \theta'_x h, y_0 + \theta'_y k)$. By direct substitution it is easy to see that

$$\frac{u(y_0 + k) - u(y_0)}{k} = \frac{v(x_0 + h) - v(x_0)}{h}.$$

---

[1] The hypothesis of the following theorem can be relaxed in several respects but this is outside the scope of the present treatment.

Thus, we have proved that for sufficiently small values of h and k, there are four numbers $\theta_x, \theta_y, \theta'_x$ and θ'_y, in $[0, 1]$, which generally depend on h and k, such that

$$\partial_x \partial_y f(x_0 + \theta_x h, y_0 + \theta_y k) = \partial_y \partial_x f(x_0 + \theta'_x h, y_0 + \theta'_y k).$$

Taking the limit for $h, k \to 0$ and using the continuity of the second-order derivatives proves the assertion. $\square$

This theorem is named after the German mathematician Hermann Schwarz (1843–1921). The extension to the general case of functions of more than two variables is immediate. In fact, one is studying the restriction of the function to the two-dimensional space spanned by the two variables considered in the cross-derivatives. For vector-valued functions, the theorem applies component by component. The same result is obviously also valid for higher-order derivatives.

Definition 7.6 Consider an open domain $E \subseteq \mathbb{R}^n$. We denote by $C^K(E)$ the set of all functions $f : E \to \mathbb{R}$ that have continuous partial derivatives of order K.

If $f \in C^K(E)$, the order in which its partial derivatives are taken is irrelevant, at least until their order does not exceed K. In this case, the partial derivative of order $k \le K$ of the function f can be denoted by $\partial_1^{h_1} \ldots \partial_n^{h_n} f$ where $0 \le h_i \le k$ and $\sum_{i=1}^n h_i = k$. Another common notation is $\frac{\partial^k}{\partial x_1^{h_1} \ldots \partial x_n^{h_n}}$. The meaning of these expressions is that the function has been derived (partially) h_i times with respect to the ith variable. If $h_i = 0$, that is, if the function has not been derived with respect to the ith variable, the partial derivative symbol ∂x_i^0 or ∂_i^0 is generally omitted from the expression. The derivatives of order k are many. Specifically, their number is equal to the number of ways in which k elements can be assigned to n groups, that is, the binomial coefficient. Define the general *multinomial coefficient* $\binom{k}{h_1,\ldots,h_n} = k!/\prod_{i=1}^n h_i!$. We are now ready to introduce the Taylor approximation of multivariate functions.

Theorem 7.10 *Consider a function $f : E \subseteq \mathbb{R}^n \to \mathbb{R}$ with E convex and suppose that $f \in C^{K-1}(E)$ and all the partial derivatives of order K exist in* int E. *Let $\mathbf{x} \in E$ and define the polynomial*

$$P_{K-1,\mathbf{x}}^f(\mathbf{z}) = \sum_{k=0}^{K-1} \frac{1}{k!} \sum_{\sum h_i = k} \binom{k}{h_1 \ldots h_n} (z_1 - x_1)^{h_1} \ldots (z_n - x_n)^{h_n} \partial_1^{h_1} \ldots \partial_n^{h_n} f(\mathbf{x}),$$

where the inner summation is on all the n-tuples of nonnegative integers $(h_1, \ldots, h_n)$ such that $\sum_{i=1}^n h_i = k$. Then $\exists \lambda \in [0, 1]$ such that

$$f(\mathbf{z}) = P_{K-1,\mathbf{x}}^f(\mathbf{z}) + R_{K,\mathbf{x}}^f(\lambda \mathbf{x} + (1 - \lambda)\mathbf{z}),$$

where

$$R_{K,\mathbf{x}}^f(\mathbf{x}) = \frac{1}{K!} \sum_{\sum h_i = K} \binom{K}{h_1, \ldots, h_n} (z_1 - x_1)^{h_1} \ldots (z_n - x_n)^{h_n} \partial_1^{h_1} \ldots \partial_n^{h_n} f(\mathbf{x}).$$

Proof The proof is based on the Taylor polynomial of a real function defined in Theorem 6.14. Consider the function $\boldsymbol{\gamma} : [0, 1] \to \mathbb{R}^n$, $\boldsymbol{\gamma}(t) = (1 - t)\mathbf{x} + t\mathbf{z}$. Since E is convex, $\gamma(t) \subseteq E$. Then define the function $\phi(t) : [0, 1] \to \mathbb{R}$, $\phi(t) = f(\boldsymbol{\gamma}(t))$. This function is continuous and can be derived K times. By the chain rule,

$$\frac{d^k}{dt^k}\phi(t) = \sum_{\sum h_i = k} \binom{k}{h_1 \ldots h_n}(z_1 - x_1)^{h_1} \ldots (z_n - x_n)^{h_n} \partial_1^{h_1} \ldots \partial_n^{h_n} f((1-t)\mathbf{x}+t\mathbf{z}).$$

The polynomial $P_{K-1,\mathbf{x}}(\mathbf{z})$ is then just the Taylor polynomial of order $K - 1$ of the function $\phi(t)$, $P_{k-1,0}^\phi(t)$, as defined in Theorem 6.14, computed in $t = 0$. For this theorem, there is a $\lambda \in [0, 1]$ such that

$$f(\mathbf{z}) = \phi(1) = P_{k-1,0}^\phi(0) + \frac{1}{K!}\frac{d^K}{dt^K}\phi(\lambda).$$

By the chain rule, the last term on the right-hand side is exactly the remainder $R_{K,\mathbf{x}}^f(\lambda\mathbf{x} + (1 - \lambda)\mathbf{z})$ and the assertion is proved. □

Note that $\lim_{\mathbf{z} \to \mathbf{x}} R_{K,\mathbf{x}}^f(\mathbf{z})/\|\mathbf{z} - \mathbf{x}\|^{K-1} = 0$, so that the remainder is, in general, an infinitesimal of order higher than $K - 1$.

7.4.1 Local Maxima and Minima

The previous theorem allows for a simple characterisation of the extremal points of a scalar function. The result is better stated in terms of the next object.

Definition 7.7 *(Hessian matrix)* Consider a map $f : E \subseteq \mathbb{R}^n \to \mathbb{R}$ with E convex and $f \in C^2(E)$. Then the *Hessian matrix* at $\mathbf{x} \in \text{int } E$ is the $n \times n$ symmetric real matrix $H_{i,j}(\mathbf{x}) = \partial_i \partial_j f(\mathbf{x})$, $i, j = 1, \ldots, n$.

This matrix takes its name from the German mathematician Otto Hesse (1811–1874). The hypothesis of continuity of the second partial derivatives is sufficient, given Theorem 7.9, to guarantee that the matrix H is, in fact, symmetric. Consider a function $f : E \subseteq \mathbb{R}^n \to \mathbb{R}$ and suppose that $f \in C^2(E)$ and that it has all third-order partial derivatives in a neighbourhood of $\mathbf{x} \in E$. According to Theorem 7.10, in that neighbourhood, the function can be approximated by the second-order Taylor polynomial,

$$f(\mathbf{z}) = f(\mathbf{x}) + \nabla f(\mathbf{x}) \cdot (\mathbf{z} - \mathbf{x}) + (\mathbf{z} - \mathbf{x})^{\mathsf{T}} H(\mathbf{z} - \mathbf{x}) + o(\|\mathbf{z} - \mathbf{x}\|^2).$$

Assume that $\mathbf{x} \in E$ is a critical point, so that

$$f(\mathbf{z}) = f(\mathbf{x}) + (\mathbf{z} - \mathbf{x})^{\mathsf{T}} H(\mathbf{z} - \mathbf{x}) + o(\|\mathbf{z} - \mathbf{x}\|^2).$$

Because the last terms $o(\|\mathbf{z} - \mathbf{x}\|^2)$ can be made as small as desired in a suitable neighbourhood of $\mathbf{x}$, if $\mathbf{x}$ is a local maximum, then $\forall \mathbf{v} \in \mathbb{R}^n$, $\mathbf{v}^{\mathsf{T}} H \mathbf{v} \le 0$. This condition is equivalent to saying that the Hessian matrix is *negative semi-definite*.[2] If $\mathbf{x}$ is a local minimum, then $\forall \mathbf{v} \in \mathbb{R}^n$, $\mathbf{v}^{\mathsf{T}} H \mathbf{v} \ge 0$, that is, the Hessian matrix is *positive semi-definite*. If for any nonzero vector $\mathbf{v} \in \mathbb{R}^n$, $\mathbf{v}^{\mathsf{T}} H \mathbf{v} < 0$, then $\mathbf{x}$ is a strict local maximum and the Hessian matrix *negative definite*. Analogously, if for any nonzero vector $\mathbf{v} \in \mathbb{R}^n$, $\mathbf{v}^{\mathsf{T}} H \mathbf{v} > 0$, then $\mathbf{x}$ is a strict local minimum, and the Hessian matrix *positive definite*. Finally, if there exist vectors $\mathbf{v}, \mathbf{w} \in \mathbb{R}^n$ such that $\mathbf{v}^{\mathsf{T}} H \mathbf{v} > 0$ and $\mathbf{w}^{\mathsf{T}} H \mathbf{w} < 0$, then $\mathbf{x}$ cannot be a maximum or a minimum: In this case, there are directions along which the function increases and directions along which it decreases, and the point $\mathbf{x}$ is a *saddle point* of the function. An example is provided in Fig. 7.4.

A common criterion for discovering whether a matrix is positive or negative definite, known as *Sylvester's criterion*, after the English mathematician James Sylvester (1814–1897), is to look at the leading principal minors (the determinants of the top-left principal submatrices). If they are all positive, the matrix is positive definite. If their signs oscillate between negative and positive, the matrix is negative definite.

▶ **Example 7.12** *(A saddle point)* Consider the function of Example 7.9. We have $\partial_1^2 f(0, 0) = 2$, $\partial_2^2 f(0, 0) = -2$, and $\partial_1 \partial_2 f(0, 0) = 0$, so that the Hessian matrix is

$$H = \begin{pmatrix} 2 & 0 \\ 0 & -2 \end{pmatrix}.$$

The principal minors are 2 and -4. Thus, the Hessian matrix is neither positive nor negative definite. It is immediate to verify that the origin is a saddle point of the function.

Our findings are summarised in the following.

---

[2] The definition of positive or negative (semi-) definite symmetric matrices (or quadratic forms) is part of standard courses in linear algebra. As already mentioned, this topic is outside the scope of the present book. For those who already know these things: a symmetric matrix is positive (negative) definite if all its eigenvalues are positive (negative) and positive (negative) semi-definite if all its eigenvalues are nonnegative (nonpositive).

Corollary 7.4 (Second-order conditions) *Consider a function* $f : E \subseteq \mathbb{R}^n \to \mathbb{R}$, $f \in C^2(E)$. *Suppose that* $\mathbf{x} \in \text{int } E$ *is a critical point and that* f *has all partial derivatives of order 3 in a neighbourhood of* $\mathbf{x}$. *Let* H *be the Hessian matrix of* f *in* $\mathbf{x}$. *Then*

1. *if* $\mathbf{x}$ *is a local minimum, then* H *is positive semi-definite;*
2. *if* $\mathbf{x}$ *is a local maximum, then* H *is negative semi-definite;*
3. *if* H *is positive definite, then* $\mathbf{x}$ *is a strict local minimum;*
4. *if* H *is negative definite, then* $\mathbf{x}$ *is a strict local maximum;*
5. *if there exist* $\mathbf{v}, \mathbf{w} \in \mathbb{R}^n$ *such that* $\mathbf{v}^\mathsf{T} H \mathbf{v} > 0$ *and* $\mathbf{w}^\mathsf{T} H \mathbf{w} < 0$, *then* $\mathbf{x}$ *cannot be a maximum or a minimum.*

Some comments are required. The first two items represent necessary conditions that the extremal points must meet. However, they are not sufficient. Conversely, the following two items are sufficient, but not necessary, conditions for the characterisation of critical points. Indeed, a function can have a maximum at one point without being partially derivable there (or elsewhere). Note that even if the Hessian matrix exists, the theorem can be inconclusive. For example, it could be that all elements of H are zero. In this case, the previous theorem is totally silent on the behaviour of the function in a neighbourhood of the point. With respect to the sufficient conditions, the hypotheses of the theorem can be weakened. It is not necessary that the function has all third-order partial derivatives. In fact, it is enough that the partial derivatives of order 2 exist and are continuous. The proof of the weaker form of the theorem is not difficult, but, for space constraints, I prefer to omit it. Finally, for two variable functions, there is a simple rule: the Hessian matrix is positive (negative) definite if its determinant is positive and its trace is positive (negative). In the case where $n = 1$, we return to the case of a real function of one real variable, and the Hessian is just the second-order derivative computed at the point (see Corollary 6.4).

In addition to local extremal points, a common problem is finding the extreme values of a smooth scalar function on a compact subset E, which we know exist from the Weierstrass extreme value theorem.[3] In particular, we did not discuss how to identify extremal points on the boundary of E. The general analysis is postponed to Sect. 7.7. If the boundaries of the subset can be translated into simple restrictions on the values taken by the variables, the analysis is easier.

▶ **Example 7.13** (*Global minimum in a simple domain*) Find the minimum of the function $f(x, y) = x^2 + y^2 + 3xy$ in the set $E = \{(x, y) \mid |x| \le 1, |y| \le 1\} \subset \mathbb{R}^2$. The gradient of the function reads $(2x + 3y, 2y + 3x)$ and the Hessian matrix

$$H = \begin{pmatrix} 2 & 3 \\ 3 & 2 \end{pmatrix}.$$

---

[3] The expression "smooth" stands for "sufficiently differentiable". Depending on the context, derivatives of a sufficiently high order are assumed to exist and be continuous.

It is immediate to see that the only critical point is the origin of the axis. The Hessian matrix there is neither positive nor negative definite: the determinants of the first and second principal minors are 2 and -5, respectively. This means that $(1, 0)^{\mathsf{T}} H(1, 0) = 4 > 0$ and $(1, -1)^{\mathsf{T}} H(1, -1) = -5 < 0$. Thus the origin is a saddle point and because the set E is compact (closed and bounded), the minimum of f must be reached on its boundary.

The boundary of E is made up of four segments of length 2 and we can study the behaviour of the function on them separately. The function restricted to the segment $x = 1$ and $y \in [-1, 1]$ reads $f(1, y) = y^2 + 3y + 1$ and is strictly increasing in y. Thus, the minimum is in $(1, -1)$. Since the function is symmetric with respect to the bisectors of the first and third quadrants and of the second and fourth quadrants, we can conclude that the function reaches its minimum at $(-1, 1)$ and $(1, -1)$, where its value is -1.

7.5   Inverse Function Theorem

In this section, we show that some assumptions about the differential of a function in a point are enough to guarantee the existence of the inverse of the function at least in the neighbourhood of that point. The problem of inverting a scalar function of one real variable has already been addressed. As stated in Theorem 6.7, if the derivative of the function exists in a point and is different from zero, the function is strictly monotonic in that point. So, if the derivative is continuous, the monotonic behaviour is persistent in a neighbourhood of the point, and the inverse of the function exists. The behaviour of the inverse can be characterised using Corollary 6.2. In the case of functions of several variables, the solution is not so easy. Indeed, in this case, we lack the order relation in the domain of the function, which was essential in the definition of monotonic behaviour. The function can now be increasing along some directions and decreasing along others.

Instead of immediately presenting the general result, we start by restating the problem of the existence of the inverse in the case of a scalar function of one real variable, but avoiding any reliance on the order relation in $\mathbb{R}$ and using only concepts that can be later extended to functions of several variables. This approach has the advantage of illustrating the general ideas in a framework in which it is easier to obtain some geometric intuition and draw some pictures (Fig. 7.6).

Consider a differentiable function $f : E \subseteq \mathbb{R} \to \mathbb{R}$ with continuous derivative and let $f'(x_0) \neq 0$ with $x_0 \in \text{int } E$. Since the derivative is continuous, we can find a closed interval $U = [x_1, x_2]$, with $x_0 \in U$, such that $\forall x \in U, |f'(x) - f'(x_0)| < |f'(x_0)|/2$. Now consider an element of the image $y \in V = f(U)$. There exists at least one element $x_y \in U$ such that $f(x_y) = y$. We can prove that this element is unique. Define the function $\phi_y(x) : U \to \mathbb{R}$ as

$$\phi_y(x) = x + \frac{1}{f'(x_0)}(y - f(x)).$$

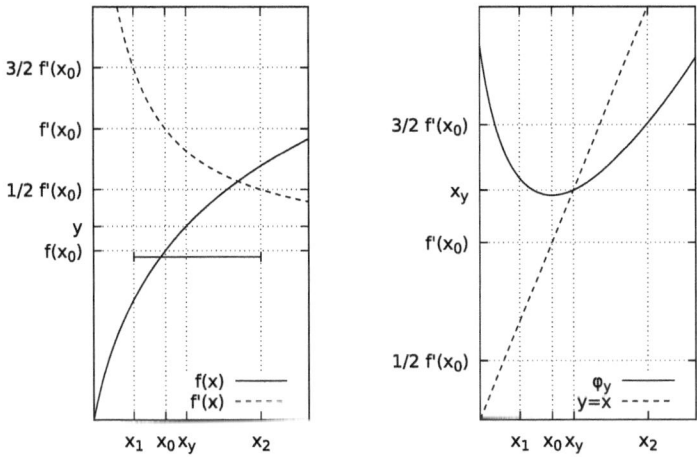

Fig. 7.6 Left panel: the function $f(x)$ and the point x_0 with $f'(x_0) > 0$. In the interval $U = [x_1, x_2]$, $|f'(x) - f'(x_0)| < f'(x_0)/2$ so that $\forall y \in f(U)$ the map ϕ_y can be used to find the unique inverse. Notice that in U the function f is monotonic and continuous and takes all the values between $f(x_1)$ and $f(x_2)$. **Right panel**: the function ϕ_y. Notice that it has an unique intersection with the curve $y(x) = x$. This intersection is the preimage of y

Any preimage x_y of y is a fixed point of ϕ_y, $\phi_y(x_y) = x_y$. Given any interval $[a, b] \subseteq U$, for the mean value theorem (see Theorem 6.10) there exists a $x \in [a, b]$ such that $\phi_y(a) - \phi_y(b) = \phi'_y(x)(a - b)$. Notice that

$$\phi'_y(x) = 1 - \frac{f'(x)}{f'(x_0)} = \frac{f'(x_0) - f'(x)}{f'(x_0)},$$

which means, by construction, that $\forall x \in U$, $|\phi'_y(x)| < 1/2$. This, in turn, implies that $|\phi_y(a) - \phi_y(b)| = |\phi'_y(x)||a - b| < |a - b|/2$, so ϕ_y is a contraction in U and, according to Theorem 5.9, it has a unique fixed point. Thus, we have proved that for any element of the image $y \in V$, there exists one and only one element $x_y \in U$ such that $f(x_y) = y$. The local restriction of the function $f : U \to V$ is one-to-one and the inverse $f^{-1} : V \to U$ is defined. According to Theorem 2.27, f^{-1} is continuous. Next, we want to prove that the inverse function f^{-1} can be derived and that its derivative in $y \in V^0$ is precisely $1/f'(x_y)$, the inverse of the derivative of the original function computed in the preimage of y. Take k small enough so that $y + k \in V$. Let x and $x + h$ in U be such that $y = f(x)$ and $y + k = f(x + h)$. Due to the fact that f and f^{-1} are continuous, when $h \to 0$, $k \to 0$, and vice versa. Thus

$$\lim_{k \to 0} \frac{f^{-1}(y + k) - f^{-1}(y)}{k} = \lim_{h \to 0} \frac{x + h - x}{f(x + h) - f(x)} = \frac{1}{f'(x)}.$$

The next theorem repeats the same construction in the case of a vector function of many variables.

Theorem 7.11 (Inverse function) *Let* $\mathbf{f} : E \subseteq \mathbb{R}^n \to \mathbb{R}^n$, *$E$ open, and assume* $\mathbf{f} \in C^1(E)$. *If the differential $A_{\mathbf{f}}(\mathbf{x}_0)$ is invertible for $\mathbf{x}_0 \in E$, then:*

- *there exists a closed ball U of centre $\mathbf{x}_0$ such that the function $\mathbf{f}$ is invertible in U and the inverse $\mathbf{g} : V = \mathbf{f}(U) \to U$ is continuous;*
- $\mathbf{g} \in C^1(V)$ *and* $A_g(\mathbf{y}) = A_{\mathbf{f}}(\mathbf{g}(\mathbf{y}))^{-1}$, $\forall \mathbf{y} \in V$.

Proof Let $\mathbf{x}_0 \in E$ and $\det A_{\mathbf{f}}(\mathbf{x}_0) \neq 0$, so that $A_{\mathbf{f}}(\mathbf{x}_0)$ is invertible. Since $A_{\mathbf{f}}$ is continuous, there exists a closed ball U centred on $\mathbf{x}_0$ such that $\forall \mathbf{x} \in U$, $\|A_{\mathbf{f}}(\mathbf{x}) - A_{\mathbf{f}}(\mathbf{x}_0)\|_{op} < 1/(2\|A_{\mathbf{f}}(\mathbf{x}_0)^{-1}\|_{op})$. Let $V = f(U)$, take $\mathbf{y} \in V$ and consider the map $\boldsymbol{\phi}_y(\mathbf{x}) = \mathbf{x} + A_{\mathbf{f}}(\mathbf{x}_0)^{-1}(\mathbf{y} - \mathbf{f}(\mathbf{x}))$. Indicating with I the identity matrix and using the chain rule to obtain the differential of $\boldsymbol{\phi}_y$,

$$\|A_{\phi_y}(\mathbf{x})\|_{op} = \|I - A_{\mathbf{f}}(\mathbf{x}_0)^{-1}A_{\mathbf{f}}(\mathbf{x})\|_{op} =$$
$$\|A_{\mathbf{f}}(\mathbf{x}_0)^{-1}(A_{\mathbf{f}}(\mathbf{x}_0) - A_{\mathbf{f}}(\mathbf{x}))\|_{op} \leq \|A_{\mathbf{f}}(\mathbf{x}_0)^{-1}\|_{op}\|A_{\mathbf{f}}(\mathbf{x}_0) - A_{\mathbf{f}}(\mathbf{x})\|_{op} < 1/2.$$

Thus, given two points $\mathbf{x}_1$ and $\mathbf{x}_2$ in the convex set U, by Theorem 7.8, $\|\boldsymbol{\phi}_y(\mathbf{x}_1) - \boldsymbol{\phi}_y(\mathbf{x}_2)\| < \|\mathbf{x}_1 - \mathbf{x}_2\|/2$, so that $\boldsymbol{\phi}_y$ is a contraction in U. Since the fixed points of $\boldsymbol{\phi}_y$ are the preimage of y, we can conclude that $\exists! \mathbf{x}_y$ such that $\mathbf{f}(\mathbf{x}_y) = \mathbf{y}$ and the inverse function $\mathbf{g}$ remains defined. According to Theorem 2.27, $\mathbf{g}$ is continuous. The first part of the statement is proved.

For the second part, assume that $\mathbf{x}$ and $\mathbf{x} + \mathbf{h}$ are both in U. Denote $\mathbf{y} = \mathbf{f}(\mathbf{x})$ and $\mathbf{y} + \mathbf{k} = \mathbf{f}(\mathbf{x} + \mathbf{h})$. The vector $\mathbf{k}$ is a function of the vector $\mathbf{h}$, but to simplify the notation I drop the explicit dependence. Since $\mathbf{f}$ and $\mathbf{g}$ are continuous and bijective, when $\|\mathbf{h}\| \to 0$, $\|\mathbf{k}\| \to 0$ and vice versa. Note that

$$\mathbf{g}(\mathbf{y} + \mathbf{k}) - \mathbf{g}(\mathbf{y}) - A_{\mathbf{f}}(\mathbf{x})^{-1}\mathbf{k} = \mathbf{h} - A_{\mathbf{f}}(\mathbf{x})^{-1}\mathbf{k} =$$
$$A_{\mathbf{f}}(\mathbf{x})^{-1}(A_{\mathbf{f}}(\mathbf{x})\mathbf{h} - \mathbf{k}) = A_{\mathbf{f}}(\mathbf{x})^{-1}(A_{\mathbf{f}}(\mathbf{x})\mathbf{h} + \mathbf{f}(\mathbf{x}) - \mathbf{f}(\mathbf{x} + \mathbf{h}))$$

so that, by the Cauchy–Schwarz inequality,

$$\|\mathbf{g}(\mathbf{y} + \mathbf{k}) - \mathbf{g}(\mathbf{y}) - A_{\mathbf{f}}(\mathbf{x})^{-1}\mathbf{k}\|/\|\mathbf{k}\| \leq$$
$$\|A_{\mathbf{f}}(\mathbf{x})^{-1}\|_{op}\|A_{\mathbf{f}}(\mathbf{x})\mathbf{h} + \mathbf{f}(\mathbf{x}) - \mathbf{f}(\mathbf{x} + \mathbf{h})\|/\|\mathbf{k}\|.$$

If we prove that the right-hand side converges to zero when $\|\mathbf{k}\| \to 0$, then the left-hand side converges to zero and the function is differentiable. Since the differential is $A_f^{-1}(\mathbf{x})$, its components are continuous so that the second part of the statement is also proved. Multiplying and dividing by $\|\mathbf{h}\|$ and rearranging the terms, the right-hand side becomes

$$\|A_{\mathbf{f}}(\mathbf{x})^{-1}\|_{op}\frac{\|\mathbf{h}\|}{\|\mathbf{f}(\mathbf{x}) - \mathbf{f}(\mathbf{x} + \mathbf{h})\|}\frac{\|A_{\mathbf{f}}(\mathbf{x})\mathbf{h} + \mathbf{f}(\mathbf{x}) - \mathbf{f}(\mathbf{x} + \mathbf{h})\|}{\|\mathbf{h}\|}.$$

The denominator of the second factor can be rewritten $\|A_{\mathbf{f}}(\mathbf{x})\mathbf{h}+\mathbf{r}(\mathbf{h})\|$. Since $A_{\mathbf{f}}(\mathbf{x})$ is invertible, $\exists \alpha > 0$ so that if $\|\mathbf{h}\|$ is sufficiently small, $\|A_{\mathbf{f}}(\mathbf{x})\| > \alpha\|\mathbf{h}\|$ (any α lower than the absolute value of all eigenvalues of A would work). Thus, if $\|\mathbf{h}\|$ is sufficiently small, the second factor is lower than $1/\alpha$. In summary, when $\|\mathbf{h}\| \to 0$, the first term is constant, the second term is bounded, and the last term converges to zero because $A_{\mathbf{f}}(\mathbf{x})$ is the differential of $\mathbf{f}$ in $\mathbf{x}$.   □

▶ **Example 7.14** *(Inverse of vector functions)* Consider the function $\mathbf{y} : \mathbb{R}^2 \to \mathbb{R}^2$ defined as

$$\begin{cases} y_1(x_1, x_2) = x_1 x_2, \\ y_2(x_1, x_2) = x_1 + x_2. \end{cases}$$

Clearly, it is $\mathbf{y} \in C^\infty(\mathbb{R}^2)$. The differential reads

$$A_{\mathbf{y}}(x_1, x_2) = \begin{pmatrix} x_2 & x_1 \\ 1 & 1 \end{pmatrix}$$

and the function is invertible in a neighbourhood of $\mathbf{x}$ if $\det A_{\mathbf{y}}(\mathbf{x}) \neq 0$, that is, if $x_1 \neq x_2$. The line $x_1 = x_2$ is sent by the function $\mathbf{y}$ to the curve $y_2^2/4 = y_1$ and the image $\mathbf{y}(\mathbb{R}^2)$ is made by all points such that $y_2^2/4 \geq y_1$. To obtain the expression of the inverse, substitute the first equation in the second, to get $y_2 = x_1 + y_1/x_1$ which solving for x_1 gives

$$x_1^{\pm}(y_1, y_2) = \frac{y_2 \pm \sqrt{y_2^2 - 4y_1}}{2}$$

and, respectively,

$$x_2^{\mp}(y_1, y_2) = y_2 - x_1^{\pm}(y_1, y_2) = \frac{y_2 \mp \sqrt{y_2^2 - 4y_1}}{2}.$$

The solutions (x_1^+, x_2^-) represent preimages such that $x_2 < x_1$ while (x_1^-, x_2^+) represent preimages such that $x_2 > x_1$.

7.6 Implicit Function Theorem

Sometimes, the relationship between variables can be defined by the fulfilment of some condition. A typical case is the couples of values (x, y) that satisfy the condition $f(x, y) = 0$ for some function f. These couples define a set in $\mathbb{R}^2$ and the question is whether we can say something about the shape of this set given the properties of f. For example, $x + y = 0$ identifies the bisector of the second and fourth quadrants, $y = -x$. In general, if f is smooth enough, then the set is a differentiable curve described by a functional relation $y = g(x)$. Before providing a general high-dimensional result, we will study a few low-dimensional cases.

7.6.1 Real Functions of Two Variables

Consider a real function of two real variables f. We want to characterise the points (x, y) for which $f(x, y) = z_0$. For this purpose, we can take advantage of the results of the previous section. Take a point $\mathbf{x}_0 = (x_0, y_0)$ such that $f(\mathbf{x}_0) = z_0$ and assume that f is continuously differentiable in $\mathbf{x}_0$ with $\partial_y f(\mathbf{x}_0) \neq 0$. Then consider the function $\mathbf{F} : E \subseteq \mathbb{R}^2 \to \mathbb{R}^2$ defined by

$$\mathbf{F}(x, y) = \begin{pmatrix} x \\ f(x, y) \end{pmatrix},$$

such that $\mathbf{F}(x_0, y_0) = (x_0, z_0)$. This function is clearly continuous and differentiable, and its differential in $\mathbf{x}$ reads

$$A_{\mathbf{F}}(\mathbf{x}) = \begin{pmatrix} 1 & 0 \\ \partial_x f & \partial_y f \end{pmatrix}.$$

Because $\det A_{\mathbf{F}}(\mathbf{x}_0) = \partial_y f(\mathbf{x}_0) \neq 0$, the differential is invertible in $\mathbf{x}_0$. Therefore, according to Theorem 7.11, there exists a closed ball B centred on $\mathbf{x}_0$ where the inverse function $\mathbf{F}^{-1} : B' \to B$ with $B' = \mathbf{F}(B)$ is defined, continuous and differentiable. In other words, for any ordered pair $(x, z) \in B'$ there exists a value of y such that $\mathbf{F}(x, y) = (x, z)$. Let Γ be the function that associates to any ordered pair (x, z) the respective y,

$$\mathbf{F}^{-1}(x, z) = \begin{pmatrix} x \\ \Gamma(x, z) \end{pmatrix}.$$

The function Γ is continuous and differentiable because it is a component of a continuous and differentiable vector function. The differential of $\mathbf{F}^{-1}$ can be obtained as

$$A_{\mathbf{F}^{-1}} = \begin{pmatrix} 1 & 0 \\ \partial_x \Gamma & \partial_z \Gamma \end{pmatrix}.$$

According to Theorem 7.11, it should be equal to the inverse of $A_{\mathbf{F}}$,

$$A_{\mathbf{F}}^{-1} = \begin{pmatrix} 1 & 0 \\ -\partial_x f / \partial_y f & 1/\partial_y f \end{pmatrix}.$$

Equating the previous matrices, we find $\partial_x \Gamma = -\partial_x f / \partial_y f$ and $\partial_z \Gamma = 1/\partial_y f$. Keeping the value of z_0 fixed, the first relation can be used to characterise the derivative of the so-called *implicit function* $g(x) = \Gamma(x, z_0)$ which describes the locus of points that satisfy the relation $f(x, y) = z_0$ in a neighbourhood of (x_0, y_0). Our findings are summarised below.

Theorem 7.12 *Let* $f : E \subseteq \mathbb{R}^2 \to \mathbb{R}$ *and* $f \in C^1(E)$. *Assume a point* $\mathbf{x}_0 = (x_0, y_0) \in$ int E *exists such that* $f(\mathbf{x}_0) = z_0$ *and* $\partial_y f(\mathbf{x}_0)$ *is different from zero. Then, there exists a neighbourhood* V *of* x_0 *and a neighbourhood* U *of* y_0 *such that* $\forall x \in V$, *there exists a unique* $y \in U$ *for which* $f(x, y) = z_0$. *Let* $g : V \to U$, $y = g(x)$ *be the function that maps* x *to the corresponding* y. *The function* g *is continuously differentiable in* U *and its differential satisfies the relation*

$$\frac{dg(x)}{dx} = -\frac{\partial_x f}{\partial_y f}.$$

Proof The neighbourhood U and V are just the projection of the closed ball B identified above on the x-axis and y-axis respectively. The function g is just the restriction of the function Γ defined above in the set $z = z_0$. ☐

In the case in which $\partial_y f(\mathbf{x}_0) = 0$ but $\partial_x f(\mathbf{x}_0) \neq 0$ one can derive the implicit function of x as a function of y, $dx(y)/dy = -\partial_y f/\partial_x f$. The only case in which the previous analysis cannot provide a differential description of the local set of points for which $f(x, y) = z_0$ is when $\nabla f(\mathbf{x}_0) = 0$, that is, when $\mathbf{x}_0$ is a critical point of the function.

▶ **Example 7.15** *(Failure of the implicit function theorem)* If the point $\mathbf{x}_0$ is a critical point of the function f, Theorem 7.12 cannot be applied. The reasons can be multiple. For example, consider the function $f(x, y) = x^2 + y^2$. In this case, the locus of points (x, y) that satisfy $f(x, y) = 0$ is made up of a single point. This is not a differentiable curve. Consider instead the function $f(x, y) = x^2 - y^2$. In this case, the points that satisfy $f(x, y) = 0$ are the lines $y = \pm x$. They intersect at the origin. There, the multiplicity of curves in a neighbourhood voids the theorem. Lastly, consider $f(x, y) = y^3 - x^2$. In this case, the locus of points that satisfy $f(x, y) = 0$ is the curve $y = |x|^{2/3}$. However, this is not differentiable at the origin.

It is useful to derive the previous result using a heuristic argument. If for any small h it is $f(x + h, g(x + h)) = f(x, g(x))$, then $d/dx f(x, g(x)) = 0$. Using the chain rule,

$$\frac{d}{dx} f(x, g(x)) = \frac{\partial}{\partial x} f(x, g(x)) + \frac{\partial}{\partial y} f(x, g(x)) g'(x) = 0$$

which is precisely the equation derived above.

▶ **Example 7.16** *(Local approximation of the implicit function)* We use Theorem 7.12 to derive a local approximation of the implicit function. Consider the function $f(x, y) = e^x - e^y + xy$. Note that $f(0, 0) = 0$. The partial derivative of the function reads $\partial_x f = e^x + y$ and $\partial_y f = -e^y + x$, so $\nabla f(0, 0) = (1, -1))$. Thus we know that there exists a differentiable function $y = g(x)$, defined in a neighbourhood of $x = 0$, that solves $f(x, g(x)) = 0$ with $g(0) = 0$. From Theorem 7.12,

$g'(x) = (e^x + y)/(e^y - x)$, whence $g'(0) = 1$. The derivative of $g'(x)$ reads

$$g''(x) = \frac{e^x + g'(x)}{e^y - x} - \frac{(e^x + y)(e^y g'(x) - 1)}{(e^y - x)^2}.$$

Substituting $x = y = 0$ and $g'(0) = 1$ we get $g''(0) = 2$. Thus $g(x) = x + x^2 + o(x^2)$. Figure 7.7 reports the function $g(x)$ and its approximation in $[-1, 1]$.

7.6.2 Real Functions of Several Variables

Consider the function $f : E \subseteq \mathbb{R}^{n+1} \to \mathbb{R}$, and assume that $f \in C^1(E)$, i.e. all first-order partial derivatives exist and are continuous. In this section, it is convenient to denote a point of E with $(\mathbf{x}, y)$, where $\mathbf{x} \in \mathbb{R}^n$, to explicitly identify the $(n+1)^{\text{th}}$ component.

Let $(\mathbf{x}_0, y_0)$ be such that $f(\mathbf{x}_0, y_0) = z_0$ and assume $\partial_y f(\mathbf{x}_0, y_0) = \partial_{n+1} f(\mathbf{x}_0, y_0) \neq 0$. Consider the function $F : E \subseteq \mathbb{R}^{n+1} \to \mathbb{R}^{n+1}$, $\mathbf{F}(\mathbf{x}, y) = (\mathbf{x}, f(\mathbf{x}, y))$. Its differential reads

$$A_{\mathbf{F}} = \begin{pmatrix} 1 & 0 & \dots & 0 & 0 \\ 0 & 1 & \dots & 0 & 0 \\ \dots & \dots & \dots & \dots & \dots \\ 0 & 0 & \dots & 1 & 0 \\ \partial_1 f & \partial_2 f & \dots & \partial_n f & \partial_y f \end{pmatrix}.$$

Note that $\det A_{\mathbf{F}} = \partial_y f \neq 0$, so that the differential of $\mathbf{F}$ is invertible in $(\mathbf{x}_0, y_0)$. Then, according to Theorem 7.11, there is a closed ball B centred in $(\mathbf{x}_0, y_0)$ where the function $\mathbf{F}$ is invertible and the inverse is continuous and differentiable. That is, $\forall (\mathbf{x}, z) \in B' = f(B), \exists (\mathbf{x}, y)$ such that $\mathbf{F}(\mathbf{x}, y) = (\mathbf{x}, z)$. Let $\Gamma : B' \to \mathbb{R}$ be the

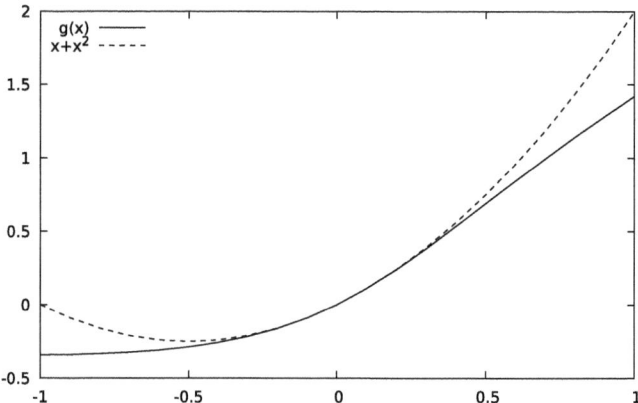

Fig. 7.7 The function $y = f(x)$ implicitly defined in Example 7.16 and its second-order Taylor approximation in a neighbourhood of $x = 0$

function that associate the y to the vector $(\mathbf{x}, z)$, that is $\mathbf{F}^{-1}(\mathbf{x}, z) = (\mathbf{x}, \Gamma(\mathbf{x}, z))$. The differential of $\mathbf{F}^{-1}$ then reads

$$
A_{\mathbf{F}^{-1}} =
\begin{pmatrix}
1 & 0 & \cdots & 0 & 0 \\
0 & 1 & \cdots & 0 & 0 \\
\cdots & \cdots & \cdots & \cdots & \cdots \\
0 & 0 & \cdots & 1 & 0 \\
\partial_1 \Gamma & \partial_2 \Gamma & \cdots & \partial_n \Gamma & \partial_z \Gamma
\end{pmatrix}.
$$

At the same time, Theorem 7.11 states that the differential of $\mathbf{F}^{-1}$ should be equal to the inverse of $A_{\mathbf{F}}$,

$$
A_{\mathbf{F}}^{-1} =
\begin{pmatrix}
1 & 0 & \cdots & 0 & 0 \\
0 & 1 & \cdots & 0 & 0 \\
\cdots & \cdots & \cdots & \cdots & \cdots \\
0 & 0 & \cdots & 1 & 0 \\
-\partial_1 f/\partial_y f & -\partial_2 f/\partial_y f & \cdots & -\partial_n f/\partial_y f & 1/\partial_y f
\end{pmatrix}.
$$

Equating term by term the previous two expressions, one gets $\partial_j \Gamma = -\partial_j f/\partial_y f$, with $j = 1, \dots, n$, and $\partial_z \Gamma = 1/\partial_y f$. All these results can be summarised as follows.

Theorem 7.13 Let $f : E \subseteq \mathbb{R}^{n+1} \to \mathbb{R}$ and $f \in C^1(E)$. Assume a point $(\mathbf{x}_0, y_0) \in$ int E exists such that $f(\mathbf{x}_0, y_0) = z_0$ and $\partial_y f(\mathbf{x}_0, y_0)$ are different from zero. Then there exists a neighbourhood V of $\mathbf{x}_0$ and a neighbourhood U of y_0 such that $\forall \mathbf{x} \in V$, $\exists! y \in U$ for which $f(\mathbf{x}, y) = z_0$. Let $g : V \to U$, $y = g(\mathbf{x})$ the function that maps $\mathbf{x}$ to the corresponding y. The function g is continuously differentiable in U and its differential satisfies the relation

$$
\partial_j g(x) = -\frac{\partial_{x_j} f}{\partial_y f} \text{ with } j = 1, \dots, n.
$$

Proof The neighbourhood U and V are just the projection of the closed ball B identified above in the n-dimensional plane containing $\mathbf{x}$ and y-axis, respectively. The function g is just the restriction of the function $\Gamma(\mathbf{x}, x)$ the set $z = z_0$ defined above. $\square$

Again, the relation of the differential of g with the differential of f can be obtained using a heuristic argument. If for small $\mathbf{h} \in \mathbb{R}^n$, $f(\mathbf{x} + \mathbf{h}, g(\mathbf{x} + \mathbf{h})) = f(\mathbf{x}, g(\mathbf{x}))$, then $\partial_j f(\mathbf{x}, g(\mathbf{x})) = 0$, $\forall j = 1, \dots, n$. Using the chain rule, $\partial_j f(\mathbf{x}, g(\mathbf{x})) = \partial_j f(\mathbf{x}, g(\mathbf{x})) + \partial_y f(\mathbf{x}, g(\mathbf{x})) \partial_j g(\mathbf{x}) = 0$, which is the expression above.

▶ **Example 7.17** (*Local approximation of the implicit function*) Consider the function $f(x, y, z) = xz + y + z^3$. Note that $f(1, 0, 0) = 0$. The partial derivatives of the function read $\partial_x f = z$, $\partial_y f = 1$, and $\partial_z f = 2z^2 + x$, so that $\nabla f(1, 0, 0) = (0, 1, 1)$. Thus, we know that there exists a differentiable function $z = g(x, y)$, defined

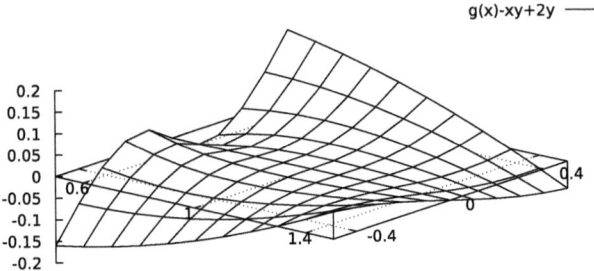

Fig. 7.8 Difference of the function $z = g(x, y)$ implicitly defined in Example 7.17 and its second-order Taylor approximation in a neighbourhood of $(x, y) = (1, 0)$

in a neighbourhood of $(x, y) = (1, 0)$, that solves $f(x, , y, g(x, y)) = 0$, with $g(1, 0) = 0$. From Theorem 7.13,

$$\partial_x g(x, y) = -\frac{z}{3z^2 + x} \quad \text{and} \quad \partial_y g(x, y) = -\frac{1}{3z^2 + x},$$

whence $\nabla g(1, 0) = (0, 1)$. Using the chain rule, the second-order partial derivative can be easily computed,

$$\partial_{xx}^2 g(x, y) = -\frac{\partial_x g}{3z^2 + x} + \frac{z}{(3z^2 + x)^2}(6z\partial_x g + 1),$$

$$\partial_{yy}^2 g(x, y) = \frac{6z\partial_y g}{(3z^2 + x)^2}, \quad \text{and} \quad \partial_{xy}^2 g(x, y) = \frac{6z\partial_y g + 1}{(3z^2 + x)^2}.$$

Using the coordinates of the point and the value of the partial derivatives at that point, $\partial_{xx}^2 g(1, 0) = \partial_{yy}^2 g(1, 0) = 0$ and $\partial_{xy}^2 g(1, 0) = 1$. The second-order Taylor approximation of the function around $(1, 0)$ is then $g(x, y) = xy - 2y + o(y^2 + (x - 1)^2)$. Figure 7.8 reports the difference of the numerically computed function $g(x, y)$ and its approximation in a neighbourhood of $(1, 0)$.

7.6.3 Vector Functions of Several Variables

The general statement pertains to the generic function $\mathbf{f} : E \subseteq \mathbb{R}^{n+k} \to \mathbb{R}^k$. A point in E will be denoted by $(\mathbf{x}, \mathbf{y})$, with $\mathbf{x} \in \mathbb{R}^n$ and $\mathbf{y} \in \mathbb{R}^k$. Given a fixed $\mathbf{z_0} \in \mathbf{f}(E)$, the problem is to characterise the locus of points $(\mathbf{x}, \mathbf{y})$ such that $f(\mathbf{x}, \mathbf{y}) = \mathbf{z_0}$. Note that if it exists, the differential of $\mathbf{f}$, $A_{\mathbf{f}}$, is a $(n + k) \times k$ matrix. In what follows, we shall make use separately of its leftmost $n \times k$ part and its rightmost $k \times k$ part, respectively, denoted by $A_{\mathbf{f},x}$ and $A_{\mathbf{f},y}$, so that $A_{\mathbf{f}} = \left(A_{\mathbf{f},x} \; A_{\mathbf{f},y}\right)$, with

$$A_{f,x}(\mathbf{x}_0, \mathbf{y}_0) = \begin{pmatrix} \partial_{x_1} f_1(\mathbf{x}_0, \mathbf{y}_0) & \cdots & \partial_{x_n} f_1(\mathbf{x}_0, \mathbf{y}_0) \\ \cdots & \cdots & \cdots \\ \partial_{x_1} f_k(\mathbf{x}_0, \mathbf{y}_0) & \cdots & \partial_{x_n} f_k(\mathbf{x}_0, \mathbf{y}_0) \end{pmatrix},$$

and

$$A_{f,y}(\mathbf{x}_0, \mathbf{y}_0) = \begin{pmatrix} \partial_{y_1} f_1(\mathbf{x}_0, \mathbf{y}_0) & \cdots & \partial_{y_k} f_1(\mathbf{x}_0, \mathbf{y}_0) \\ \cdots & \cdots & \cdots \\ \partial_{y_1} f_k(\mathbf{x}_0, \mathbf{y}_0) & \cdots & \partial_{y_k} f_k(\mathbf{x}_0, \mathbf{y}_0) \end{pmatrix}.$$

Theorem 7.14 *(Implicit function) Let* $\mathbf{f} : E \subseteq \mathbb{R}^{n+k} \to \mathbb{R}^k$ *and* $\mathbf{f} \in C^1(E)$. *Consider a* $\mathbf{z}_0 \in$ int $\mathbf{f}(E)$ *and a point* $(\mathbf{x}_0, \mathbf{y}_0) \in E$ *such that* $\mathbf{f}(\mathbf{x}_0, \mathbf{y}_0) = \mathbf{z}_0$. *Assume that the* $k \times k$ *bottom-right square submatrix of the differential* $A_{f,y}(\mathbf{x}_0, \mathbf{y}_0)$ *is invertible. Then there exist neighbourhoods* $V \subseteq \mathbb{R}^n$ *of* $\mathbf{x}_0$ *and* $U \subseteq \mathbb{R}^k$ *of* $\mathbf{y}_0$ *such that* $\forall \mathbf{x} \in V, \exists! \mathbf{y} \in U$ *for which* $\mathbf{f}(\mathbf{x}, \mathbf{y}) = \mathbf{z}_0$. *Let* $\mathbf{g} : V \to U$ *be the function that maps* $\mathbf{x}$ *to the corresponding* $\mathbf{y}$. *The function* $\mathbf{g}$ *is continuously differentiable in* U *and its differential reads* $A_{\mathbf{g}} = \big(A_{f,y}(\mathbf{x}, \mathbf{g}(\mathbf{x}))\big)^{-1} A_{f,x}(\mathbf{x}, \mathbf{g}(\mathbf{x}))$.

Proof The proof develops along the lines of the proofs of the previous sections. Consider the function $\mathbf{F} : E \subseteq \mathbb{R}^{n+k} \to \mathbb{R}^{n+k}$, $f(\mathbf{x}, \mathbf{y}) = (\mathbf{x}, \mathbf{f}(\mathbf{x}, \mathbf{y}))$. The differential of $\mathbf{F}$ reads

$$A_{\mathbf{F}}(\mathbf{x}_0, \mathbf{y}_0) = \begin{pmatrix} I_{n \times n} & 0_{n \times k} \\ A_{f,x}(\mathbf{x}_0, \mathbf{y}_0) & A_{f,y}(\mathbf{x}_0, \mathbf{y}_0) \end{pmatrix},$$

where $I_{n \times n}$ and $0_{n \times k}$ denote the identity and zero matrix, respectively. Since det $A_{A_{\mathbf{F}}} = $ det $A_{f,y}$, the differential of $\mathbf{F}$ is invertible. Then, according to Theorem 7.11, there is a closed ball B centred on $(\mathbf{x}_0, \mathbf{y}_0)$ where the function $\mathbf{F}$ is invertible and the inverse is continuous and differentiable. That is, $\forall (\mathbf{x}, \mathbf{z}) \in B' = \mathbf{F}(B)$, $\exists (\mathbf{x}, \mathbf{y})$ such that $\mathbf{F}(\mathbf{x}, \mathbf{y}) = (\mathbf{x}, \mathbf{z})$. The neighbourhoods U and V are the projection of the ball B on the space of $\mathbf{x}$ and $\mathbf{y}$, respectively. This proves the first part of the statement. Next, let $\Gamma : B' \to \mathbb{R}^k$ be the function that associates the $\mathbf{y}$ to the vector $(\mathbf{x}, \mathbf{z})$ such that $\mathbf{F}^{-1}(\mathbf{x}, \mathbf{z}) = (\mathbf{x}, \Gamma(\mathbf{x}, \mathbf{z}))$. The differential of $\mathbf{F}^{-1}$ can be written as

$$A_{\mathbf{F}^{-1}} = \begin{pmatrix} I_{n \times n} & 0_{n \times k} \\ A_{\Gamma,x} & A_{\Gamma,z} \end{pmatrix}.$$

According to Theorem 7.11 it must be equal to the inverse of the differential of $A_{\mathbf{F}}$, which can be easily derived to be

$$A_{\mathbf{F}}^{-1}(\mathbf{x}_0, \mathbf{y}_0) = \begin{pmatrix} I_{n \times n} & 0_{n \times k} \\ -A_{f,y}^{-1} A_{f,x} & A_{f,y}^{-1} \end{pmatrix}.$$

Equating the last two expressions term by term, we obtain $A_{\Gamma,x} = -A_{f,y}^{-1} A_{f,x}$ and $A_{\Gamma,z} = A_{f,y}^{-1}$. Let $\mathbf{g}$ be the restriction of the function Γ to the set $\mathbf{z} = \mathbf{z}_0$, that is

$\mathbf{g}(\mathbf{x}) = \Gamma(\mathbf{x}, \mathbf{z}_0)$. The function $\mathbf{g}$ is continuous and differentiable, and $A_{\mathbf{g}} = A_{\Gamma, x}$. By substituting in the previous equations, the statement is proved.   □

The assumption that the bottom-right square submatrix of the differential $A_{\mathbf{f}}$ is invertible implies that the gradient of the components of $\mathbf{f}$ are, in the neighbourhood in which the implicit function is considered, linearly independent vectors. This assumption has a fundamental interpretation, which will be discussed in the next section.

7.6.4 Dependent and Independent Functions

If a set is composed of points that fulfil certain functional relations, it is important to understand if the requirement of fulfilling a further relation reduces the set. In fact, if the new relation is just some combination of the previous relations, that is, it depends on them, its fulfilment does not change the original set of points. For example, the set of points $\mathbf{x} \in \mathbb{R}^n$ that satisfy the equations $f_1(\mathbf{x}) = 0$ and $f_2(\mathbf{x}) = 0$ is not modified if we also require $f_1(\mathbf{x}) + f_2(\mathbf{x}) = 0$. This consideration leads to the problem of finding an appropriate way to describe the dependence among functions.

Definition 7.8 *(Dependent functions)* Consider a function $\mathbf{f} : E \subseteq \mathbb{R}^n \to \mathbb{R}^k$, E open, with $k \leq n$ and $\mathbf{f} \in C^1(E)$. The components of $\mathbf{f}$ are said to be *dependent* on E if there exists a function $\phi : \mathbf{f}(E) \subseteq \mathbb{R}^k \to \mathbb{R}$, with $\phi \in C^1(\mathbf{f}(E))$ and $\nabla \phi \neq \mathbf{0}$, such that $\phi(\mathbf{f}(\mathbf{x})) = 0$ for each $\mathbf{x} \in E$.

If the components of $\mathbf{f}$ are not dependent, they are said to be *independent*. The previous definition might seem a bit cryptic. Its meaning is clarified in the following result.

Theorem 7.15 *If the components of the function* $\mathbf{f} : E \subseteq \mathbb{R}^n \to \mathbb{R}^k$ *are dependent in* E, *then for any* $\mathbf{x} \in E$ *there exists a neighbourhood* $N(\mathbf{x})$, *an index* $j \leq k$ *and a continuously differentiable function* $g : \mathbb{R}^{k-1} \to \mathbb{R}$ *such that* $f_j(\mathbf{z}) = g(f_1(\mathbf{z}), \ldots, f_{j-1}(\mathbf{z}), f_{j+1}(\mathbf{z}), \ldots, f_k(\mathbf{z}))$, $\forall \mathbf{z} \in N(\mathbf{x})$.

Proof Fix $\mathbf{x} \in E$ and consider $\mathbf{y} = \mathbf{f}(\mathbf{x})$. Without loss of generality, assume that the function ϕ in Definition 7.8 is such that $\partial_1 \phi(\mathbf{y}) \neq 0$ and for continuity this is true also in a neighbourhood $N(\mathbf{y})$ of $\mathbf{y}$. Then, for Theorem 7.13, there exists a continuously differentiable function g such that the set of points $\mathbf{t} \in N(\mathbf{y})$ that satisfy $\phi(\mathbf{t}) = 0$ can be defined as $t_1 = g(t_2, \ldots, t_k)$. Substituting the definition of $\mathbf{t}$ proves the assertion.   □

In other words, if the components of the function $\mathbf{f}$ are dependent, then in the neighbourhood of each point, there exists at least one component that can be written as a function ϕ of the others. If the function ϕ is linear, the components of $\mathbf{f}$ are said to be *linearly dependent*. The interesting result is that the seemingly difficult task of proving the dependence (or independence) of functions can be reduced to calculating the rank of a matrix.

Theorem 7.16 *If the rank of the differential $A_{\mathbf{f}}(\mathbf{x})$ of the function $\mathbf{f} : E \subseteq \mathbb{R}^n \to \mathbb{R}^k$ is equal to k, $\forall \mathbf{x} \in E$, then its components are independent.*

Proof We prove the opposite. Assume that the components of $\mathbf{f}$ are dependent such that $\exists \phi$, $\nabla \phi \neq \mathbf{0}$, $\phi(\mathbf{f}(\mathbf{x})) = 0$, $\forall \mathbf{x} \in E$. As the differential of a constant function is zero, $A_{\mathbf{f}}(\mathbf{x})\nabla\phi(\mathbf{f}(\mathbf{x})) = \mathbf{0}$ identically in E. The nonzero vector $\nabla\phi$ belongs to the kernel of $A_{\mathbf{f}}(\mathbf{x})$, so its rank must be lower than k. ☐

▶ **Example 7.18** *(Dependent functions)* Consider the functions $u = x + y + 2z$, $v = x - y + 2z$ and $w - x^2 + 4z^2 + 4xz + y^2$ from $\mathbb{R}^3$ to $\mathbb{R}^3$. The Jacobian matrix with their gradients in the rows reads

$$J = \begin{pmatrix} 1 & 1 & 2 \\ 1 & -1 & 2 \\ 2x + 4z & 2y & 4x + 8z \end{pmatrix}.$$

It is immediate to see that $\det J = 0$ for any point (x, y, z), since the first column is half the third column. Thus, we can conclude that the functions u, v, and w are dependent. To find the dependence, note that the upper left 2×2 minor is different from zero, so u and v are independent. Solving

$$\begin{cases} u = x + y + 2z, \\ v = x - y + 2z, \end{cases}$$

for x and y one gets $x = (u + v)/2 - 2z$ and $y = (u - v)/2$. Substituting these equations into the functional expression of w and simplifying, we finally obtain $w = u^2/2 + v^2/2$.

7.7 Constrained Optimisation

In this section, we discuss the problem of finding the maximum (or minimum) value of a function $f : \mathbb{R}^n \to \mathbb{R}$, called *objective function*, in a set E and identifying the points in the set where these values are reached, the maximum and minimum points, according to Definition 1.11. This is an example of an *optimisation problem*. Since the function we consider is not limited to be linear and the set E can have any shape, this problem is often referred to as *nonlinear programming*. The problem of finding the maximum of $f(x)$ is equivalent to the problem of finding the minimum of $-f(x)$. Therefore, the general treatment of the two problems is substantially the same. We will return to this point in Sect. 7.7.7. However, for definiteness, in the general exposition we will consider the former.

We assume that the objective function is smooth and, in any case, that $f \in C^1(E)$. We also assume that the set E is defined through the fulfilment of certain requirements or *constraints* expressed as non-strict inequalities of type $g(\mathbf{x}) \leq 0$ or equality of type $h(\mathbf{x}) = 0$, for a suitable set of functions g and h, where $g, h : \mathbb{R}^n \to \mathbb{R}$ and $g, h \in C^1$. The set E is the *feasible region* or *feasible set*. If $E = \emptyset$, the problem is said to be *unfeasible* and has no solution. The problem also lacks a solution if the function f is unbounded above in E. When they exist, the identification of the extremal points of the objective function is generally accomplished in two steps. First, all local extremal points are found according to Definition 2.3. Then, one compares the value of the function at these points to find the global maximum. In what follows, the name *candidate solutions* will be adopted for local extremal points. Candidate solutions can be further distinguished into *boundary candidate solutions*, if some of the non-strict constraint inequalities are satisfied with equality and *internal candidate solutions*, if all the non-strict constraint inequalities are satisfied as strict inequalities. Constraint inequalities that are satisfied with equality in a candidate solution are said to be *binding* or *active*.

In analogy to the discussion in Sect. 7.2, the identification of candidate solutions proceeds in two steps. Firstly, we exploit a set of necessary conditions that are essentially the equivalent of Corollary 7.2, modified when constraints have to be taken into account. These conditions are still called *first-order conditions* because of their similarity to the unconstrained problem and because they involve first-order derivatives. Secondly, we further screen these points to include only the actual interesting points, local maxima, or local minima, depending on the problem. In this step, we essentially follow considerations similar to those reported in Corollary 7.4. The following discussion will mainly emphasise the first step, presenting necessary (and, in general, not sufficient) conditions that candidate solutions must satisfy. This is the step that is often the most important and complicated. Even if the identification of extremal points is addressed here more generally than in the previous chapters, all the results we have derived so far apply. Thus, if E is compact, we know that a solution, possibly not unique, must exist. In this case, we expect to find at least one candidate solution. Moreover, if we further assume that the objective function is strictly concave, then we expect only one candidate solution associated to the maximum of the function. Before presenting the general results, it is useful to introduce the basic tools by analysing simpler cases.

7.7.1 One Dimensional Problems

Consider the function $f : [a, b] \subseteq \mathbb{R} \to \mathbb{R}$ with $f \in C^1([a, b])$. We want to solve the following problem:

$$\max_{x \in [a,b]} f(x). \tag{7.1}$$

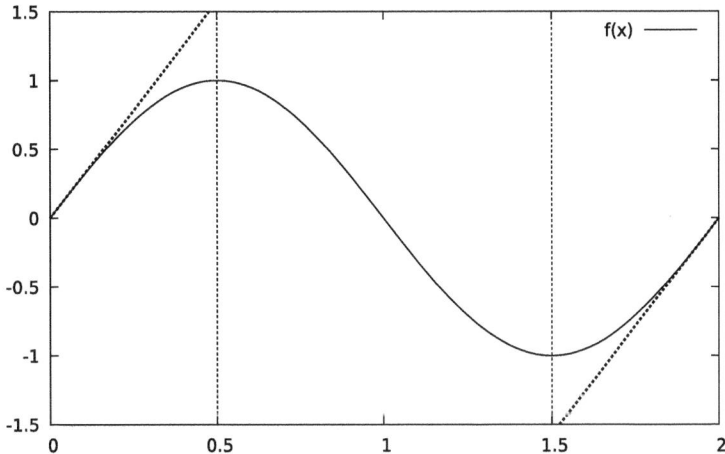

Fig. 7.9 The point $x = 0.5$ and $x = 1.5$ are critical points. The first is a candidate solution while the second corresponds to a minimum. The point $x = 2$ is also a candidate solution because $f'(2) > 0$

That is, we want to find the maximum value of the function f in the interval $[a, b]$. Refer to Fig. 7.9 for an example. The internal candidate solutions are the local maximum point x that belongs to the open interval (a, b). In these points, Corollary 7.2 dictates that $f'(x) = 0$. The points $x = a$ and $x = b$ are possible boundary candidate solutions. These two points are somehow special. In fact, $x = a$ is a local maximum if there exists a (sufficiently small) $h > 0$ such that the function does not increase in the interval $[a, a + h]$. Analogously, $x = b$ is a local maximum, if there exists a (sufficiently small) $h > 0$ such that the function does not decrease in the interval $[b - h, b]$. Since $f \in C^1$, the two requirements translate, respectively, into $f'(a) \leq 0$ and $f'(b) \geq 0$. Thus, we have the following.

Lemma 7.1 *If the point x solves the problem (7.1) then*

- *either it is an internal solution $x \in (a, b)$, in which case $f'(x) = 0$;*
- *or it is a boundary solution $x = a, b$, in which case $f'(a) \leq 0$ or $f'(b) \geq 0$, respectively.*

The reader must be aware that, in general, simply checking the conditions above is by no means sufficient for the identification of candidate solutions. As in the case of unconstrained maximisation, sufficient conditions can be obtained in some cases by looking at second-order derivatives. We will return to this problem in Sect. 7.7.5. In this case, checking second-order derivatives might not be necessary for boundary solutions $x = a, b$. If the derivative satisfies the strict inequality, then the boundary point already qualifies as a candidate solution.

▶ **Example 7.19** (*One-dimensional parametric problem*) We are interested in the maximum of the function $f(x) = x^3 + cx$ in $[-1, 1]$ as a function of c. The internal solutions must satisfy the equation $f'(x) = 3x^2 + c = 0$. This equation has two solutions at the points $x_{\pm}^* = \pm\sqrt{|c|/3}$. Note that $x_{\pm}^* \in [-1, 1]$ only if $c \in [-3, 0]$. In these critical points, the value of the function is $f(x_{\pm}^*) = \mp 2(|c|/3)^{3/2}$ so that we have only to keep the point x_-^*.

Next, consider the boundary point $x = 1$. It is a local maximum when $f'(1) = 3 + c > 0$, that is, if $c > -3$. The value of the function at this point is $f(1) = 1 + c$. Instead, the boundary point $x = -1$ is a local maximum when $f'(-1) = 3 + c < 0$, that is, $c < -3$. The value of the function at this point is $f(-1) = -1 - c$.

In conclusion, for $c < -3$ there is only one candidate solution: the maximum of the function is reached at $x = -1$ and is equal to $-1 - c$. For $c > 0$ there is only one candidate solution: the maximum of the function is reached in $x = 1$ and is equal to $1 + c$. For $c \in [-3, 0]$ there are two candidate solutions: $x = x_-^*$ and $x = 1$. The maximum of the function is then $\max\{2(|c|/3)^{3/2}, 1 + c\}$.

An alternative way to identify the points x that satisfy the necessary conditions of (7.1) is to think of them as part of the solutions $(x, \lambda_1, \lambda_2)$ of the following system of equations and inequalities:

$$\begin{cases} f'(x) + \lambda_1 - \lambda_2 = 0, \\ (x - a)\lambda_1 = 0, & a - x \le 0, & \lambda_1 \ge 0, \\ (b - x)\lambda_2 = 0, & x - b \le 0, & \lambda_2 \ge 0. \end{cases} \tag{7.2}$$

To see it, note that there are three possible types of solution. The first type is when $\lambda_1 = \lambda_2 = 0$, so that $a - x \le 0$, $x - b \le 0$ and $f'(x) = 0$. This corresponds to a critical point that is internal or at the boundary and that satisfies the necessary conditions in Lemma 7.1. The second type is when $\lambda_1 = 0$ and $\lambda_2 > 0$, implying $x = b$ and $f'(b) = \lambda_2 > 0$. This is a possible boundary solution. Analogously, if $\lambda_2 = 0$ and $\lambda_1 > 0$, then $x = a$ and $f'(a) = -\lambda_1 < 0$, which is the other possible boundary solution. The conditions in the last two rows of the system are usually called *slackness conditions*. The reason is that one of the factors on the left-hand side of the equation can be "slack", i.e. not equal to zero, but their product must always be zero. The auxiliary parameters λ_1 and λ_2 introduced in (7.2) are the Lagrangian *multipliers*. The next proposition introduces a more compact, and commonly adopted, notation for the first-order necessary conditions.

Lemma 7.2 *Given the problem in* (7.1) *consider the* Lagrangian function *or La-grangian, for short,* $L(x, \lambda_1, \lambda_2) = f(x) - \lambda_1(a - x) - \lambda_2(x - b)$. *Then the system* (7.2) *can be equivalently expressed as*

$$\begin{cases} \partial_x L = 0, \\ \lambda_i \partial_{\lambda_i} L = 0, \ \lambda_i \ge 0, \ \partial_{\lambda_i} L \ge 0, \ i = 1, 2. \end{cases}$$

It would be wrong to think of the system above as the first-order conditions of the unconstrained maximisation of the Lagrangian function. The Lagrangian function is clearly unbounded above. Rather, the conditions on λ_i are the necessary conditions for a local minimum constrained by the requirement that $\lambda_i \geq 0$: either $\lambda_i = 0$ and $\partial L/\partial \lambda_i \geq 0$ or $\lambda_i > 0$ and $\partial L/\partial \lambda_i = 0$. In this sense, the solutions of the system in Lemma 7.2 can be thought of as saddle points.

▶ **Example 7.20** (*Saddle point interpretation*) Consider the Lagrangian in Lemma 7.2 and define the function $p(x) = \min_{\lambda_1, \lambda_2 \geq 0} L$. If $x > b$ or $x < a$, then L is unbounded from below, $p(x) = -\infty$. If instead $x \in [0, 1]$, then $p(x) = f(x)$. Thus, the original maximisation problem $\max_{x \in [a,b]} f(x)$ can be restated as $\max_{x \in \mathbb{R}} p(x) = \max_{x \in \mathbb{R}} \min_{\lambda_1, \lambda_2 \geq 0} L$.

In many applications, the Lagrangian function and the multipliers' values are given a specific meaning, so that the saddle point analogy might become useful.

7.7.2 Two Dimensional Problems

Consider two functions $f, g : \mathbb{R}^2 \to \mathbb{R}$, and the problem:

$$\max_{\mathbf{x} \in E} f(\mathbf{x}), \quad E = \left\{ \mathbf{x} = (x_1, x_2) \in \mathbb{R}^2 \mid g(\mathbf{x}) \leq 0 \right\}. \tag{7.3}$$

Both f and g are assumed to be continuously differentiable. The boundary of the region E is defined by those points (x_1, x_2) for which $g(x_1, x_2) = 0$. Assume that if $g(\mathbf{x}) = 0$, then $\|\nabla g(\mathbf{x})\| > 0$, that is, the partial derivatives of g cannot both vanish on the boundary of E. Without loss of generality, we can then assume $\partial_2 g \neq 0$ so that, according to Theorem 7.13, the boundary of the region can be locally described by a continuously differentiable function $x_2 = h(x_1)$ with $h'(x_1) = -\partial_1 g/\partial_2 g$. The tangent to the function h in x_1, that is, the tangent of the curve $g(x_1, x_2) = 0$ in $(x_1, h(x_1))$, is along the direction of the vector $(1, -\partial_1 g/\partial_2 g)$. Therefore, the direction orthogonal to the surface is along the vector $\nabla g = (\partial_1 g, \partial_2 g)$. See the example in Fig. 7.10. This is a consequence of the fact that the gradient is the direction of the steepest increase of the function, that is, the direction along which the directional derivative is maximal (see Sect. 7.2). Since we consider the domain defined by $g(\mathbf{x}) \leq 0$, the gradient on the boundary points outwards with respect to the set E.

Returning to our original problem (7.3), a point $\mathbf{x}$ is an internal candidate solution if $g(\mathbf{x}) < 0$ and it is a local maximum of f, so that $\nabla f(\mathbf{x}) = 0$, or a boundary candidate solution if $g(\mathbf{x}) = 0$ and there exists a neighbourhood $N(\mathbf{x})$ of $\mathbf{x}$ such that for any $\mathbf{y} \in N(\mathbf{x}) \cap E$, $f(\mathbf{y}) \leq f(\mathbf{x})$. The latter condition implies that at the point $\mathbf{x}$, the direction of increase of the function f must point outward, towards the exterior of the set E. In other terms, $\nabla f(\mathbf{x})$ should have the same direction as $\nabla g(\mathbf{x})$. Both conditions can be written as the solutions $(\mathbf{x}, \lambda)$ of the following system:

Fig. 7.10 The vector
$(-2, -1)$ is normal to the
implicitly defined function
$g(x, y) =$
$x^2 - 12x - y + 39 = 0$ in
the point $(5, 4)$. Indeed,
according to Theorem 7.12,
the tangent of the curve in
the same point has equation
$y = 14 - 2x$

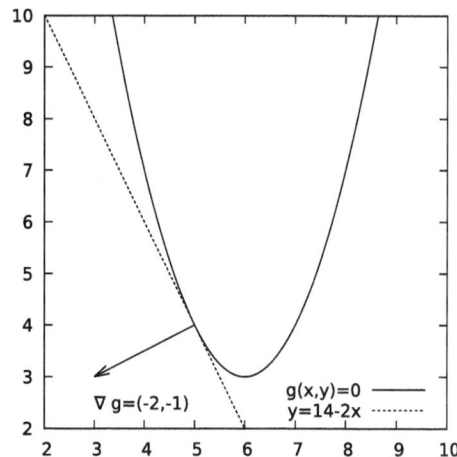

$$\begin{cases} \nabla f(\mathbf{x}) - \lambda \nabla g(\mathbf{x}) = 0, \\ g(\mathbf{x})\lambda = 0, \quad g(\mathbf{x}) \le 0, \quad \lambda \ge 0. \end{cases} \tag{7.4}$$

If $\lambda = 0$, then $g(\mathbf{x}) \le 0$ and the point is a critical point $\nabla f(\mathbf{x}) = 0$. This corresponds
to an internal or boundary candidate solution. In contrast, if $\lambda > 0$, then $g(\mathbf{x}) = 0$
and $\nabla f(\mathbf{x}) = \lambda \nabla g(\mathbf{x})$. This corresponds to a boundary solution. In analogy to the
one-dimensional problem, we can summarise our analysis in the following way.

Lemma 7.3 *Given the problem in* (7.3) *consider the* Lagrangian *function* $L(\mathbf{x}, \lambda) =$
$f(\mathbf{x}) - \lambda g(\mathbf{x})$. *Then the system* (7.4) *can be equivalently expressed as*

$$\begin{cases} \nabla_\mathbf{x} L = \mathbf{0}, \\ \lambda \partial_\lambda L = 0, \ \lambda \ge 0, \ \partial_\lambda L \ge 0. \end{cases}$$

The expression $\nabla_\mathbf{x}$ represents partial derivatives with respect to x_1 and x_2. The same
considerations about the meaning of this system of equations that were made for the
one-dimensional case still apply. In general, more than one constraint can be added
to the Lagrangian, and each constraint has its own multiplier.

▶ **Example 7.21** (*Two-dimensional problem*) Find the maximum points (x_1, x_2) of
the function $f(x_1, x_2) = (x_1 - x_2)^2$ such that $x_2 \le 4 - x_1^2$ and $x_2 \ge x_1^2 - 4$. First,
note that the set defined by the two inequalities is compact and the objective function
is continuous, so the problem has a solution. The Lagrangian function reads

$$L(x_1, x_2, \lambda_1, \lambda_2) = (x_1 - x_2)^2 - \lambda_1(x_1^2 + x_2 - 4) - \lambda_2(x_1^2 - x_2 - 4),$$

the first-order conditions with respect to x_1 and x_2,

$$x_1 - x_2 = x_1(\lambda_1 + \lambda_2), \quad x_1 - x_2 = (\lambda_2 - \lambda_1)/2,$$

and the constraints with the slackness conditions,

$$\begin{cases} \lambda_1(x_1^2 + x_2 - 4) = 0, & \lambda_1 \geq 0, x_1^2 + x_2 - 4 \leq 0, \\ \lambda_2(x_1^2 - x_2 - 4) = 0, & \lambda_2 \geq 0, x_1^2 - x_2 - 4 \leq 0. \end{cases}$$

Assuming $\lambda_1 = \lambda_2 = 0$ implies $x_1 = x_2$ and the fulfilment of the constraint inequalities $x_1^2 \pm x_1 - 4 \leq 0$ imposes $x_1 \in (1/2 - \sqrt{17/4}, -1/2 + \sqrt{17/4})$. In these points, the value of the function is 0 and they are local minima. So we can conclude that there are no internal candidate solutions. For the case $\lambda_1 = 0$ and $\lambda_2 > 0$, the solution should satisfy $x_1 - x_2 = x_1\lambda_2$, $x_1 - x_2 = \lambda_2/2$, and $x_2 = x_1^2 - 4$. The only solution is the point $\mathbf{a}_1 = (1/2, -15/4)$ with $\lambda_2 = 17/2 > 0$ and $f(\mathbf{a}_1) = (17/4)^2$. The Lagrangian function is symmetric when x_1 and x_2 are exchanged, if one also exchanges the multipliers. Thus, $\mathbf{a}_2 = (-1/2, 15/4)$ is another candidate solution with $\lambda_1 = 17/2$, $\lambda_2 = 0$ and $f(\mathbf{a}_2) = (17/4)^2$. Assuming $\lambda_1, \lambda_2 > 0$, we must have $x_2 = 4 - x_1^2 = x_1^2 - 4$, that is $x_2 = 0$ and $x_1 = \pm 2$. Substituting into the first two equations of the system, it is clear that there are no solutions of this kind. In conclusion, the function is maximised in $\mathbf{a}_1$ and $\mathbf{a}_2$ and its maximum is $(17/4)^2$.

The next section illustrates a few additional results that are necessary for the generalisation of the previous analysis to higher-dimensional problems and multiple constraints.

7.7.3 Theorems of the Alternatives

The theorems in this section offer alternatives, that is, mutually exclusive propositions. If one proposition is false, then the other is true. We start with a theorem attributed to the Hungarian mathematician and physicist Gyula Farkas (1847–1930).

Theorem 7.17 (Farkas' lemma) *Consider the convex cone C generated by the k vectors $\{\mathbf{a}_1, \ldots, \mathbf{a}_k\}$ belonging to $\mathbb{R}^n$, as in Definition 1.24, and a vector $\mathbf{a} \in \mathbb{R}^n$. Either $\mathbf{a} \in C$, or $\exists \mathbf{y} \in \mathbb{R}^n$ such that $\mathbf{y} \cdot \mathbf{a_j} \leq 0 \, \forall j = 1, \ldots, k$, and $\mathbf{y} \cdot \mathbf{a} > 0$.*

Proof First, note that if $\mathbf{a} \in C$, then there exists nonnegative $(x_1, \ldots, x_k)$ such that $\mathbf{a} = \sum_{i=1}^{k} x_i \mathbf{a}_i$. Then $\forall \mathbf{y}$ such that $\mathbf{y} \cdot \mathbf{a_j} \leq 0$, $\mathbf{y} \cdot \mathbf{a} \leq 0$. Thus, the two alternatives cannot be true at the same time.

Next, we prove that if the second alternative is false, that is, if $\forall \mathbf{y} \in \mathbb{R}^n$ such that $\mathbf{y} \cdot \mathbf{a_j} \leq 0 \, \forall j = 1, \ldots, k$, it is $\mathbf{y} \cdot \mathbf{a} < 0$, then $\mathbf{a} \in C$. By Theorem 5.19 we know that the convex cone C is closed. Thus, according to Theorem 4.12, $\exists \mathbf{b} \in C$, which is closest to $\mathbf{a}$. For $t > 0$, consider the vector $\mathbf{b} + t\mathbf{a}_j$ with $j = 1, \ldots, k$. Then $\|\mathbf{a} - \mathbf{b} - t\mathbf{a_j}\|^2 = \|\mathbf{a} - \mathbf{b}\|^2 - 2t\mathbf{a}_j \cdot (\mathbf{a} - \mathbf{b}) + t^2 \|\mathbf{a}_j\|^2$. Since this expression must be greater than $\|\mathbf{a} - \mathbf{b}\|^2$ for any positive t, it must be $\mathbf{a}_j \cdot (\mathbf{a} - \mathbf{b}) \leq 0$. For the hypothesis of the theorem, this implies that $\mathbf{a} \cdot (\mathbf{a} - \mathbf{b}) \leq 0$. Consider next the vector $(1 - t)\mathbf{b}$, which for sufficiently small t is in C. The square of the distance between $\mathbf{a}$ and $(1 - t)\mathbf{b}$ is $\|\mathbf{a} - \mathbf{b} + t\mathbf{b}\|^2 = \|\mathbf{a} - \mathbf{b}\|^2 + 2t\mathbf{b} \cdot (\mathbf{a} - \mathbf{b}) + t^2 \|\mathbf{b}\|^2$. By definition this expression must be greater than $\|\mathbf{a} - \mathbf{b}\|^2$ for any positive t, therefore we must have $\mathbf{b} \cdot (\mathbf{a} - \mathbf{b}) \geq 0$. Taking the two inequalities together, we have $(\mathbf{a} - \mathbf{b}) \cdot (\mathbf{a} - \mathbf{b}) \leq 0$. This implies $\|\mathbf{a} - \mathbf{b}\| = 0$, that is, $\mathbf{a} = \mathbf{b}$, which proves the assertion. $\square$

The previous theorem does not change if the direction of the inequalities is reversed: it is just a matter of choosing $-\mathbf{y}$ instead of $\mathbf{y}$. Theorem 7.17 is stated differently in different fields of application. We have the following equivalent statement of Farkas' lemma.

Corollary 7.5 *Given a $n \times k$ matrix A and a vector $\mathbf{a} \in \mathbb{R}^n$, the following two statements are alternatives.*

- $\exists \mathbf{x} \in \mathbb{R}^k$, $\mathbf{x} \geq \mathbf{0}$ *such that* $A\mathbf{x} = \mathbf{a}$;
- $\exists \mathbf{y} \in \mathbb{R}^n$ *such that* $\mathbf{y}^\mathsf{T} A \leq \mathbf{0}$ *and* $\mathbf{y} \cdot \mathbf{a} > 0$.

The column vectors of the matrix A in Corollary 7.5 are the vectors $\{\mathbf{a}_1, \ldots, \mathbf{a}_k\}$ of Theorem 7.17. The second theorem is named after the Israeli-American mathematician Theodore Motzkin (1908–1970).[4]

Theorem 7.18 (Motzkin's transposition theorem, reduced form) *Given a nonzero $n \times k$ matrix A and a $n \times h$ matrix C, the following two statements are alternatives.*

- $\exists \mathbf{x}_1 \in \mathbb{R}^k$, $\mathbf{x}_1 > \mathbf{0}$ *and* $\exists \mathbf{x}_2 \in \mathbb{R}^h$ *such that* $A\mathbf{x}_1 + C\mathbf{x}_2 = \mathbf{0}$;
- $\exists \mathbf{y} \in \mathbb{R}^n$ *such that* $\mathbf{y}^\mathsf{T} A \ll \mathbf{0}$ *and* $\mathbf{y}^\mathsf{T} C = \mathbf{0}$.

Proof The two statements cannot be true at the same time, otherwise we would have $\mathbf{x}_1 > \mathbf{0}$, $\mathbf{y}^\mathsf{T} A \ll \mathbf{0}$, and $\mathbf{y} A \mathbf{x}_1 = 0$. We will prove that if the first statement is false, then the second is true. Assume that there are no vectors $\mathbf{x}_1 > \mathbf{0}$ and $\mathbf{x}_2$ such that $A\mathbf{x}_1 + C\mathbf{x}_2 = \mathbf{0}$. This implies that there are no three vectors $\mathbf{z}_1 \in \mathbb{R}^k_{\geq 0}$, $\mathbf{z}_2, \mathbf{z}_3 \in \mathbb{R}^h_{\geq 0}$, such that $A\mathbf{z}_1 + B(\mathbf{z}_2 - \mathbf{z}_3) = \mathbf{0}$ and $\mathbf{1} \cdot \mathbf{z}_1 = 1$. In fact, if the first system had a solution with $\mathbf{z}_1 = \mathbf{0}$, the second condition would be false. This is equivalent to saying that there are no solutions with $\mathbf{z} \in \mathbb{R}^{k+2h}_{\geq 0}$ for the system $B\mathbf{z} = \mathbf{b}$ with

$$B = \begin{pmatrix} A & C & -C \\ 1_{1 \times k} & 0_{1 \times h} & 0_{1 \times h} \end{pmatrix}, \quad b = \begin{pmatrix} 0_{n \times 1} \\ 1 \end{pmatrix},$$

---

[4] The typical statement of Motzkin's theorem is slightly more general. For our purposes, the present reduced form is sufficient.

where 0 and 1 stand for matrices of zeros and ones, respectively, of the indicated dimensions. By Corollary 7.5, this implies that there exists a vector of dimension $n + 1$, $\mathbf{y} = (\mathbf{y}_1, y_0)$, such that $\mathbf{y}^\mathsf{T} B \leq \mathbf{0}$ and $\mathbf{y} \cdot \mathbf{b} > 0$. In terms of the component of $\mathbf{y}$, this becomes $\mathbf{y}_1^\mathsf{T} A + y_0 \mathbf{1} \leq \mathbf{0}$, $\mathbf{y}_1^\mathsf{T} C \leq \mathbf{0}$, $\mathbf{y}_1^\mathsf{T} C \geq \mathbf{0}$, and $y_0 > 0$. That is, $\mathbf{y}_1^\mathsf{T} C = \mathbf{0}$ and $\mathbf{y}_1^\mathsf{T} A \leq -y_0 \mathbf{1} \ll \mathbf{0}$, which is precisely the second statement. □

For completeness, we review a third result, attributed to the German mathematician Erich Stiemke (1892–1915), even if we are not using it in what follows.

Theorem 7.19 (Stiemke's lemma) *Given a $n \times k$ matrix A, the following two statements are alternatives.*

- $\exists \mathbf{x} \in \mathbb{R}^k$, $\mathbf{x} \gg \mathbf{0}$ *such that* $A\mathbf{x} = \mathbf{0}$;
- $\exists \mathbf{y} \in \mathbb{R}^n$ *such that* $\mathbf{y}^\mathsf{T} A < \mathbf{0}$.

Proof The two statements cannot be true at the same time, or we would have $\mathbf{x} \gg \mathbf{0}$, $\mathbf{y}^\mathsf{T} A > \mathbf{0}$, and $\mathbf{y}^\mathsf{T} A\mathbf{x} = 0$. Assume that the second statement is false. This means that there is no $\mathbf{y}$ such that $\mathbf{y}^\mathsf{T} A \leq \mathbf{0}$ and $\mathbf{y}^\mathsf{T} A\mathbf{1} < 0$. In fact, if the first inequality implies $\mathbf{y}^\mathsf{T} A = \mathbf{0}$, then the second inequality is violated. Note that the second inequality can be expressed as $\mathbf{y} \cdot (-A\mathbf{1}) > 0$. Thus, by Corollary 7.5, $\exists \mathbf{x} \geq \mathbf{0}$, such that $A\mathbf{x} = -A\mathbf{1}$, that is, $A(\mathbf{x} + \mathbf{1}) = \mathbf{0}$. Because $\mathbf{x} + \mathbf{1} \gg \mathbf{0}$, the first statement is true. □

7.7.4 First-Order Conditions

We are ready to discuss the general problem with both equality and inequality constraints. To avoid unnecessary repetitions, we anticipate here some definitions that will be commonly used in all statements. We consider an open set $D \subseteq \mathbb{R}^n$ and a continuously differentiable real-valued function $f \in C^1(D)$. Let $G_k = \{g_1, \ldots, g_k\}$ be a set of k continuous differentiable real-valued functions that represent the inequality constraints $g_i \in C^1(D)$, $\forall i = 1, \ldots, k$. $E_G = \cap_{i=1}^k \{\mathbf{x} \mid g_i(\mathbf{x}) \leq 0\}$ is the set of points in which these constraints are satisfied. Let $H_l = \{h_1, \ldots, h_l\}$ be a set of l continuously differentiable real-valued functions that represent equality constraints $h_i \in C^1(D)$, $\forall i = 1, \ldots, l$. $E_H = \cap_{j=1}^l \{\mathbf{x} \mid h_j(\mathbf{x}) = 0\}$ is set of points in which these constraints are satisfied. We assume $E = D \cap E_H \cap E_G \neq \emptyset$.

This section introduces necessary first-order conditions for the candidate solutions of the problem $\max_{\mathbf{x} \in E} f(\mathbf{x})$, expressed as relations between the gradient of the objective function and the gradient of the binding constraints. They are the constrained equivalent of Corollary 7.2. Our first result, named after the German mathematician Fritz John (1910–1994), is extremely general.

Theorem 7.20 (Fritz John) *Let* $\mathbf{x}^*$ *be a candidate solution of the problem* $\max_{\mathbf{x} \in E}$ $f(\mathbf{x})$. *Then* $\exists \boldsymbol{\lambda}^* \in \mathbb{R}_{\geq 0}^{k+1}$, *with* $\lambda_i^* = 0$ *if* $g_i(\mathbf{x}^*) < 0$, $\exists \boldsymbol{\mu}^* \in \mathbb{R}^l$, $(\boldsymbol{\lambda}^*, \boldsymbol{\mu}^*) \neq \mathbf{0}$, *such that*

$$\lambda_0^* \nabla f(\mathbf{x}^*) = \sum_{i=1}^k \lambda_i^* \nabla g_i(\mathbf{x}^*) + \sum_{j=1}^l \mu_j^* \nabla h_j(\mathbf{x}^*).$$

Proof Assume that a point $\mathbf{x}^*$ is a candidate solution and that, without loss of generality, only the first m constraints are satisfied with equality, that is, $g_i(\mathbf{x}^*) = 0$ for $i = 1, \ldots, m$ and $g_i(\mathbf{x}^*) < 0$ for $i = m + 1, \ldots, k$. If the equality constraints H_l are not independent in $\mathbf{x}^*$, then $\exists \boldsymbol{\mu}^* \neq \mathbf{0}$ such that $\sum_{j=1}^l \mu_j^* \nabla h_j(\mathbf{x}^*) = \mathbf{0}$ (see Theorem 7.16). In this case, the statement is satisfied by setting $\boldsymbol{\lambda}^* = \mathbf{0}$.

Suppose instead that the equality constraints H_l are independent in $\mathbf{x}^*$. Then, by continuity, there exists an open ball $B(\mathbf{x}^*, \epsilon) \subseteq D$, centred in $\mathbf{x}^*$ and with radius ϵ, such that $\forall \mathbf{x} \in B$, the constraints H_l are independent. We can choose ϵ small enough so that $\forall \mathbf{x} \in B$, $g_i(\mathbf{x}) < 0$, $i = m + 1, \ldots, k$. Because the functions in H_l are independent, by Theorem 7.14, l components of any point $\mathbf{x} \in B \cap E_H$ can be written as a continuous and differentiable function of the other components. Without loss of generality, we assume that they are the last l, so that $\forall \mathbf{x} \in B \cap E_H$, $\mathbf{x} = (\mathbf{z}, \mathbf{G}(\mathbf{z}))$, where $\mathbf{z}$ belongs to an appropriate open neighbourhood $N'(\mathbf{z}^*) \subseteq \mathbb{R}^{n-l}$ of $\mathbf{z}^*$, the vector of the first $n - l$ components of $\mathbf{x}^*$, and where $\mathbf{G} : \mathbb{R}^{n-l} \to \mathbb{R}^l$ is a continuously differentiable function such that $\mathbf{G}(\mathbf{z}^*)$ are the last l components of $\mathbf{x}^*$, that is, $(\mathbf{z}^*, \mathbf{G}(\mathbf{z}^*)) = \mathbf{x}^*$. The elements of $B \cap E$ correspond to the elements $\mathbf{z} \in N'(\mathbf{z}^*)$ such that $g_i(\mathbf{z}, \mathbf{G}(\mathbf{z})) \leq 0$, $i = 1, \ldots, m$.

Since $\mathbf{x}^*$ is a local maximum, there are no points $\mathbf{z} \in N'(\mathbf{z}^*)$ such that $\Delta f(\mathbf{z}) = f(\mathbf{z}, \mathbf{G}(\mathbf{z})) - f(\mathbf{x}^*) > 0$ and $g_i(\mathbf{z}, \mathbf{G}(\mathbf{z})) < 0$, $i = 1, \ldots, m$. Consider $\mathbf{u} \in \mathbb{R}^{n-l}$, $\mathbf{z}^* + \eta \mathbf{u}$ so that, by the Taylor approximation,

$$\Delta f(\mathbf{z}^* + \eta \mathbf{u}, \mathbf{G}(\mathbf{z}^* + \eta \mathbf{u})) = \eta \nabla f(\mathbf{x}^*) \cdot (\mathbf{u}, J(\mathbf{x}^*)\mathbf{u}) + o(\eta),$$
$$g_i(\mathbf{z}^* + \eta \mathbf{u}, \mathbf{G}(\mathbf{z}^* + \eta \mathbf{u})) = \eta \nabla g_i(\mathbf{x}^*) \cdot (\mathbf{u}, J(\mathbf{x}^*)\mathbf{u}) + o(\eta),$$

where $J(\mathbf{x}^*)$ is the Jacobian of the function $\mathbf{G}$ in $\mathbf{x}^*$. This must be true for any $\eta > 0$. Thus, $\nexists \mathbf{u} \in \mathbb{R}^{n-l}$ such that $\nabla f(\mathbf{x}^*) \cdot (\mathbf{u}, J(\mathbf{x}^*)\mathbf{u}) > 0$ and $\nabla g_i(\mathbf{x}^*) \cdot (\mathbf{u}, J(\mathbf{x}^*)\mathbf{u}) < 0$, $i = 1, \ldots, m$. By Theorem 7.14, the vectors $(\mathbf{u}, J(\mathbf{x}^*)\mathbf{u})$ with $\mathbf{u} \in \mathbb{R}^{n-l}$ correspond to the vectors $\mathbf{y} \in \mathbb{R}^n$ such that $\nabla h_j(\mathbf{x}^*) \cdot \mathbf{y} = 0$, $j = 1, \ldots, l$. Taking all the conditions together, we conclude that $\nexists \mathbf{y} \in \mathbb{R}^n$ such that $\mathbf{y}^\mathsf{T} A \ll 0$ and $\mathbf{y}^\mathsf{T} C = \mathbf{0}$, where $A^\mathsf{T} = (-\nabla f(\mathbf{x}^*) \nabla g_1(\mathbf{x}^*) \ldots \nabla g_m(\mathbf{x}^*))$ and $C^\mathsf{T} = (\nabla h_1(\mathbf{x}^*) \ldots \nabla h_l(\mathbf{x}^*))$. Thus, by Theorem 7.18, there exist $\boldsymbol{\lambda}^* \in \mathbb{R}^{m+1}$, $\boldsymbol{\lambda}^* > \mathbf{0}$, and $\boldsymbol{\mu}^* \in \mathbb{R}^l$, such that $A\boldsymbol{\lambda}^* + C\boldsymbol{\mu}^* = \mathbf{0}$, which reduces to the statement. □

The vector $\boldsymbol{\lambda}$ collects the multipliers associated with the inequality constraints, while $\boldsymbol{\mu}$ collects the multipliers associated with the equality constraints. The multipliers of non-binding inequality constraints are set to zero. Note that in the Fritz-John condition, there is also a multiplier λ_0 associated with the gradient of the objective function. Theorem 7.20 is so general that it may not provide significant information.

For example, if the constraints in H_l are dependent in a point $\mathbf{x}$, then the implication of Theorem 7.20 is trivially satisfied in $\mathbf{x}$ taking $\boldsymbol{\lambda} = \mathbf{0}$ and $\boldsymbol{\mu} \neq \mathbf{0}$ such that $\sum_{j=1}^{l} \mu_j \nabla h_j(\mathbf{x}) = 0$, irrespective of whether or not $\mathbf{x}$ is a candidate solution. If in $\mathbf{x}$, $g_i(\mathbf{x}) = 0$ and $\nabla g_i(\mathbf{x}) = \mathbf{0}$, the implication of Theorem 7.20 is trivially satisfied by setting $\lambda_i = 1$ and all other multipliers equal to zero. These examples reveal that to extract more information from Theorem 7.20, we need to make further assumptions about the nature of the constraints. These assumptions are commonly named *constraint qualifications* or *regularity conditions*.[5] The simplest case is when there are no binding inequality constraints and the equality constraints in the candidate solution are independent.

Theorem 7.21 (Lagrange) *If there are no binding inequality constraints and the equality constraints H_l are independent in a candidate solution $\mathbf{x}^*$, then there exists a $\boldsymbol{\mu}^* \in \mathbb{R}^l$ such that*

$$\nabla f(\mathbf{x}^*) = \sum_{j=1}^{l} \mu_j^* \nabla h_j(\mathbf{x}^*).$$

Proof Given the hypothesis, since there are no binding inequality constraints, Theorem 7.20 implies that $\exists \lambda_0 \geq 0$ and $\exists \boldsymbol{\mu}' \in \mathbb{R}^l$ such that $\lambda_0 \nabla f(\mathbf{x}^*) = \sum_{j=1}^{l} \mu_j' \nabla h_j(\mathbf{x}^*)$. Note that it must be $\lambda_0 > 0$, or the equality constraints would be dependent, which is ruled out by hypothesis. Consider $\mu_j^* = \mu_j'/\lambda_0$ to derive the statement. $\square$

In a candidate solution, the gradient of the objective function belongs to the subspace generated by the gradients of all the equality constraints. If there are constraints that are dependent on the entire set D, or in a neighbourhood of the candidate solution, then they are redundant and can be eliminated. If there are n independent equality constraints, then $E \neq \emptyset$ consists of a single point. This is the only candidate solution, and Theorem 7.21 is trivially satisfied. The previous theorem clearly applies when the original problem has no inequality constraints. If inequality constraints are present, a second simple conclusion can be obtained if all binding inequality constraints and equality constraints are independent.[6]

Theorem 7.22 (Karush–Kuhn–Tucker) *If the binding inequality constraints $g_i(\mathbf{x}^*) = 0$, $i = 1, \ldots, m$ and the equality constraints H_l are independent in a candidate solution $\mathbf{x}^*$, then there exist a $\boldsymbol{\lambda}^* \in \mathbb{R}^m$, $\boldsymbol{\lambda}^* \geq \mathbf{0}$, and a $\boldsymbol{\mu}^* \in \mathbb{R}^l$, such that*

---

[5] There are many examples of constraint qualifications encountered in applications. We cannot review but a few cases. The interested reader is referred to specific textbooks on optimisation theory.

[6] This theorem, credited to the American mathematicians William Karush (1917–1997), Harold Kuhn (1925–2014), and Albert Tucker (1905–1995), was originally derived without equality constraints. Although less historically accurate, it is convenient to present it in a more modern version.

$$\nabla f(\mathbf{x}^*) = \sum_{i=1}^{m} \lambda_i^* \nabla g_i(\mathbf{x}^*) + \sum_{j=1}^{l} \mu_j^* \nabla h_j(\mathbf{x}^*).$$

Proof The proof is similar to the proof of Theorem 7.21. The coefficient λ_0^* implied by Theorem 7.20 cannot be zero. Thus, it can be set to 1 by dividing all coefficients by its value.   □

The multipliers of the non-binding inequality constraints have been set to zero. In a candidate solution, the gradient of the objective function has a component in the convex cone generated by the gradients of all binding inequality constraints and a component in the subspace spanned by the gradients of all equality constraints. The constraint qualification in Theorem 7.22 is generally known as *linear independence constraint qualification*. To avoid dependence, it is generally convenient to express constraints (equality or inequality) as *regular* or *non-singular* curves, that is, using continuously differentiable functions g (or h) such that if $g(\mathbf{x}) = 0$, $\|\nabla g(\mathbf{x})\| > 0$.

▶ **Example 7.22** *(Mangasarian–Fromovitz constraint qualification)* Because in Theorem 7.20 the multipliers of the inequality constraints are restricted to be nonnegative, a useful result can be derived even if the constraints are dependent. Suppose that in a candidate solution $\mathbf{x}^*$ the equality constraints H_l are independent and there exists a vector $\mathbf{u} \in \mathbb{R}^n$ such that $\nabla g_i(\mathbf{x}^*) \cdot \mathbf{u} < 0$ for all binding inequality constraints, $i = 1, \ldots, m$, and $\nabla h_j(\mathbf{x}^*) \cdot \mathbf{u} = 0$, for all equality constraints, $j = 1, \ldots, l$. Consider the expression in the statement of Theorem 7.20. Because the constraints H_l are independent, it must be $\lambda^* > 0$, that is, at least some components of λ^* must be nonzero. Multiplying by $\mathbf{u}$,

$$\lambda_0^* \nabla f(\mathbf{x}^*) \cdot \mathbf{u} = \sum_{i=1}^{k} \lambda_i^* \nabla g_i(\mathbf{x}^*) \cdot \mathbf{u}.$$

If $\lambda_0^* = 0$, then $\exists \lambda_i^* > 0$, which leads to a contradiction as the left-hand side would be zero and the right-hand side would be strictly negative. Thus, $\lambda_0^* > 0$ and we can derive the same expression of Theorem 7.22.

Based on the previous results, we can define the Lagrangian function associated with the optimisation problem.

Lemma 7.4 (Lagrangian function) *Define the Lagrangian function*

$$L(\mathbf{x}, \lambda) = f(\mathbf{x}) - \sum_{i=1}^{k} \lambda_i g_i(\mathbf{x}) - \sum_{j=1}^{l} \mu_j h_j(\mathbf{x}).$$

Under the constraint qualification of Theorem 7.22 (or Example 7.22), if $\mathbf{x}^*$ *is a solution of* $\max_{\mathbf{x} \in E} f(\mathbf{x})$, *there exist a vector* $\boldsymbol{\lambda}^* \in \mathbb{R}^k_{\geq 0}$ *and a vector* $\boldsymbol{\mu}^* \in \mathbb{R}^l$ *such that* $(\mathbf{x}^*, \boldsymbol{\lambda}^*, \boldsymbol{\mu}^*)$ *solves*

$$\begin{cases} \nabla_{\mathbf{x}} L = \mathbf{0}, \\ \lambda_i \partial_{\lambda_i} L = 0, \ \lambda_i \geq 0, \ \partial_{\lambda_i} L \geq 0, \ i = 1, \ldots, k, \\ \partial_{\mu_j} L = 0, \ j = 1, \ldots, l. \end{cases}$$

The system that appears in the previous statement contains $n + k + l$ equations, relative to first-order conditions with respect to variables and multipliers, and $2k$ inequalities, relative to slackness conditions.

▶ **Example 7.23** *(Weak and strong duality)* Analogously to what we have done in Example 7.20, starting from the Lagrangian in Lemma 7.4, we can define the function $p : D \subseteq \mathbb{R}^n \to \bar{\mathbb{R}}$,

$$p(\mathbf{x}) = \min_{\boldsymbol{\lambda} \in \mathbb{R}^k_{\geq 0}, \boldsymbol{\mu} \in \mathbb{R}^h} L(\mathbf{x}, \boldsymbol{\lambda}, \boldsymbol{\mu}).$$

If $\mathbf{x} \notin E_G \cap E_H$, then the Lagrangian is unbounded from below and $p(\mathbf{x}) = -\infty$. If $\mathbf{x} \in E_G \cap E_H$, $p(\mathbf{x}) = f(\mathbf{x})$. The original maximisation problem, that is, our *primal problem*, can be restated as the unconstrained problem $\max_{\mathbf{x} \in D} p(\mathbf{x})$. Define the *Lagrangian dual* $d : \mathbb{R}^k_{\geq 0} \times \mathbb{R}^h \to \bar{\mathbb{R}}$,

$$d(\boldsymbol{\lambda}, \boldsymbol{\mu}) = \max_{\mathbf{x} \in D} L(\mathbf{x}, \boldsymbol{\lambda}, \boldsymbol{\mu}).$$

The Lagrangian *dual problem* is the problem $\min_{\boldsymbol{\lambda} \in \mathbb{R}^k_{\geq 0}, \boldsymbol{\mu} \in \mathbb{R}^h} d(\boldsymbol{\lambda}, \boldsymbol{\mu})$. According to Theorem 1.4,

$$\min_{\boldsymbol{\lambda} \in \mathbb{R}^k_{\geq 0}, \boldsymbol{\mu} \in \mathbb{R}^h} \max_{\mathbf{x} \in D} L(\mathbf{x}, \boldsymbol{\lambda}, \boldsymbol{\mu}) \geq \max_{\mathbf{x} \in D} \min_{\boldsymbol{\lambda} \in \mathbb{R}^k_{\geq 0}, \boldsymbol{\mu} \in \mathbb{R}^h} L(\mathbf{x}, \boldsymbol{\lambda}, \boldsymbol{\mu}).$$

Thus, the solution of the dual problem is an upper bound to the solution of the primal problem. This result is called *weak duality*. Because both p and d have values in the extended real number system, the inequality remains meaningful also when the functions considered are unbounded. With some further hypotheses on the objective function f and the nature of constraints, it can be proved that the solutions of the primal and dual problems are actually the same. This is called *strong duality*. Strong duality can be useful because sometimes the solution of the dual problem is more accessible than the solution of the primal problem. For the specific conditions under which strong duality applies, the reader is referred to textbooks specialised in optimisation theory.

▶ **Example 7.24** *(Intertemporal consumption-saving problem)* The vector $\mathbf{c} \in \mathbb{R}^T_{\geq 0}$, $\mathbf{c} = (c_0, \ldots, c_{T-1})$ represents the consumption plan of an economic agent over

$T - 1$ periods. The agent starts with an allocation $w_0 > 0$ and what is not consumed bears fruits, so that $w_{t+1} = \gamma w_t - c_t$, with $\gamma > 1$. The utility the agent obtains from the consumption plan $\mathbf{c}$ is $U(\mathbf{c}) = \sum_{t=0}^{T-1} \beta^t \log c_t$, with $\beta > 0$ representing the intertemporal utility discount factor. We are interested in the consumption plan that generates the highest utility, under the constraint that the final amount is nonnegative $W_T \geq 0$. The Lagrangian of the problem is

$$L = \sum_{t=0}^{T-1} \beta^t \log c_t + \sum_{t=0}^{T-1} \lambda_t c_t - \sum_{t=0}^{T-1} \mu_t (w_{t+1} - \gamma w_t + c_t) + \lambda_T w_T,$$

and the first-order conditions

$$\begin{cases} \dfrac{\beta^t}{c_t} = \mu_t, -\lambda_t, \ t = 0, \ldots, T-1 \\[2mm] \gamma \mu_t = \mu_{t-1}, \ t = 1, \ldots, T-1 \\[2mm] \lambda_T = \mu_{T-1}. \end{cases}$$

The solution requires $c_t > 0$, so that $\forall t = 1, \ldots, T-1$, $\lambda_t = 0$, $\mu_t > 0$, and $c_t = \beta^t / \mu_t$. In turn, this implies $\lambda_T > 0$, and consequently $W_T = 0$. Iterating the second equation of the system written in terms of consumption, $c_t = \beta \gamma c_{t-1}$, we get $c_t = (\beta \gamma)^t c_0$. From the intertemporal relation of the amount owned, we also get $w_0 = \sum_{t=0}^{T-1} c_t / \gamma^{t+1}$. Substituting the previous equation and summing the partial geometric series, we obtain the optimal initial consumption $c_0 = w_0 \gamma (1 - \beta)/(1 - \beta^T)$. The optimal consumption at t then reads $c_t = w_0 \gamma (\beta \gamma)^t (1 - \beta)/(1 - \beta^T)$. During $T - 1$ dates, consumption increases or decreases according to the product $\beta \gamma$. This product represents the joint effect of the returns on savings and their relative utility.

7.7.5 Second-Order Conditions

In this section we consider a point $\mathbf{x}^* \in E$ that satisfies the first-order conditions of the previous section. We want to investigate whether the point actually represents a candidate solution. For this purpose, we need a few definitions that are better anticipated here, so that the following statements will be simplified. At a point $\mathbf{x}^* \in E$, let $E_H^\perp(\mathbf{x}^*)$ be the orthogonal complement of the gradients of all equality constraints,

$$E_H^\perp(\mathbf{x}^*) = \left\{ \mathbf{v} \in \mathbb{R}^n \mid \nabla h_j(\mathbf{x}^*) \cdot \mathbf{v} = 0, \ j = 1, \ldots, l \right\},$$

$E_G^\perp(\mathbf{x}^*)$ the orthogonal complement of the gradients of all binding inequality constraints,

$$E_G^\perp(\mathbf{x}^*) = \left\{ \mathbf{v} \in \mathbb{R}^n \mid \nabla g_j(\mathbf{x}^*) \cdot \mathbf{v} = 0, \ i = 1, \ldots, m \right\},$$

$E^{\perp}_{G^+}(\mathbf{x}^*)$ the orthogonal complement of the gradients of all binding inequality constraints with positive multipliers,

$$E^{\perp}_{G^+}(\mathbf{x}^*) = \left\{ \mathbf{v} \in \mathbb{R}^n \mid \nabla g_i(\mathbf{x}^*) \cdot \mathbf{v} = 0, \mu_i > 0, i = 1, \dots, m \right\},$$

and $C^{-}_{G^0}(\mathbf{x}^*)$ the polar cone (see Definition 1.26) generated by the gradients of the inequality constraints with zero multipliers,

$$C^{-}_{G^0}(\mathbf{x}^*) = \left\{ \mathbf{v} \in \mathbb{R}^n \mid \nabla g_i(\mathbf{x}^*) \cdot \mathbf{v} \le 0, \mu_i = 0, i = 1, \dots, m \right\}.$$

Some of these sets can actually be the entire space $\mathbb{R}^n$ if some type of constraints is missing. In this section, all binding constraints are assumed to be independent. The objective function and the constraints are continuously twice differentiable. We denote by $H_f(\mathbf{x})$, $H^h_j(\mathbf{x})$, $j = 1, \dots, l$, and $H^g_i(\mathbf{x})$, $i = 1, \dots, k$, the Hessian of the objective function, the equality constraints, and the inequality constraints, respectively. Following considerations similar to those in Corollary 7.4, a set of sufficient or necessary conditions can be derived for the problem with equality constraints.

Theorem 7.23 (Second-order conditions for equality constraints) *Consider* $\mathbf{x}^* \in E = D \cap E_H$. *Assume that* $\exists \mu^* \in \mathbb{R}^l$ *such that* $\nabla f(\mathbf{x}^*) = \sum_{j=1}^{l} \mu^*_j \nabla h_j(\mathbf{x}^*)$. *Define*

$$H_L(\mathbf{x}^*) = H_f(\mathbf{x}^*) - \sum_{j=1}^{l} \mu^*_j H^h_j(\mathbf{x}^*).$$

Necessary condition: *if* $\mathbf{x}^*$ *is a constrained local maximum, then* $\forall \mathbf{v} \in E^{\perp}_H(\mathbf{x}^*)$, $\mathbf{v}^{\mathsf{T}} H_L(\mathbf{x}^*) \mathbf{v} \le 0$. Sufficient condition: *if* $\forall \mathbf{v} \in E^{\perp}_H(\mathbf{x}^*)$, $\mathbf{v}^{\mathsf{T}} H_L(\mathbf{x}^*) \mathbf{v} < 0$, *then* $\mathbf{x}^*$ *is a constrained strict local maximum.*

Proof Because the functions in H_l are independent at $\mathbf{x}^*$, by Theorem 7.14, there exists a neighbourhood B of $\mathbf{x}^*$ such that the l components of any point $\mathbf{x} \in E_H \cup B$ can be written as a continuous and differentiable function of the other components. Without loss of generality, we assume that they are the last l, so that $\forall \mathbf{x} \in B \cap E_H$, $\mathbf{x} = (\mathbf{z}, \mathbf{G}(\mathbf{z}))$, where $\mathbf{z}$ belongs to an appropriate open neighbourhood $N'(\mathbf{z}^*) \subseteq \mathbb{R}^{n-l}$ of $\mathbf{z}^*$, the vector of the first $n - l$ components of $\mathbf{x}^*$, and where $\mathbf{G} : \mathbb{R}^{n-l} \to \mathbb{R}^l$ is a continuously differentiable function such that $\mathbf{G}(\mathbf{z}^*)$ are the last l components of $\mathbf{x}^*$, that is, $(\mathbf{z}^*, \mathbf{G}(\mathbf{z}^*)) = \mathbf{x}^*$.

By Theorem 7.14, $\forall j = 1, \dots, l$, $\nabla_{\mathbf{z}} h_j + \nabla_{\mathbf{y}} h_j J = 0$ identically in $B \cap E_H$, where J is the Jacobian of $\mathbf{G}$, $\nabla_{\mathbf{z}}$ indicates the vector of partial derivatives with respect to the first $n - l$ components and $\nabla_{\mathbf{y}}$ the vector of partial derivatives with respect to the last l components. By taking a further partial derivative, using the chain rule and after some algebraic manipulation, we discover that, identically in $B \cap E_H$,

$$\left(I \ J(\mathbf{z})^{\mathsf{T}} \right) H_j(\mathbf{x}) \begin{pmatrix} I \\ J(\mathbf{z}) \end{pmatrix} + \sum_{k=1}^{l} \partial_{n-l+k} h_j(\mathbf{x}) H^G_k(\mathbf{x}) = 0,$$

where I stands for the $n - l$ dimensional identity matrix and $H_k^G(\mathbf{x})$ for the Hessian matrix of the kth component of the function G. Consider the function $F(\mathbf{z}) = f(\mathbf{z}, G(\mathbf{z}))$. Using the chain rule to obtain its Hessian,

$$H_F(\mathbf{z}) = \left(I \ J(\mathbf{z})^{\mathsf{T}}\right) H_f(\mathbf{x}) \begin{pmatrix} I \\ J(\mathbf{z}) \end{pmatrix} + \sum_{k=1}^{l} \partial_{n-l+k} f(\mathbf{x}) H_k^G(\mathbf{x}).$$

By hypotheses, $\partial_{n-l+k} f(\mathbf{x}^*) = \sum_{j=1}^{l} \mu_j^* \partial_{n-l+k} h_j(\mathbf{x}^*)$. Substituting the expression derived above, we finally obtain

$$H_F(\mathbf{z}^*) = \left(I \ J(\mathbf{z}^*)^{\mathsf{T}}\right) H_L(\mathbf{x}^*) \begin{pmatrix} I \\ J(\mathbf{z}^*) \end{pmatrix}.$$

Note that $\mathbf{x}^*$ is a constrained extremal point of f in $B \cap E_H$ if and only if $\mathbf{z}^*$ is an unconstrained extremal point of F in $N'(\mathbf{z}^*)$. From the definition of $E_H^{\perp}$, $\forall \mathbf{u} \in \mathbb{R}^{n-l}$, $\mathbf{v} = (\mathbf{u}, \mathbf{u}J) \in E_H^{\perp}$ and $\forall \mathbf{v} \in E_H^{\perp}$, $\exists \mathbf{u} \in \mathbb{R}^{n-l}$ such that $\mathbf{v} = (\mathbf{u}, \mathbf{u}J)$. Hence, the statements of the theorem come directly from Corollary 7.4 applied to the unconstrained maximum $\mathbf{z}^*$ of the function $F(\mathbf{z})$. □

According to the necessary condition, if $\exists \mathbf{v} \in E_H^{\perp}$, $\mathbf{v}^{\mathsf{T}} H_L(\mathbf{x}^*) \mathbf{v} > 0$, then $\mathbf{x}^*$ cannot be a candidate solution. The second statement is a sufficient, but not necessary, condition for identifying candidate solutions. The symmetric matrix $H_L(\mathbf{x}^*)$ is required to be negative definite with respect to all vectors orthogonal to the gradient of all constraints.

▶ **Example 7.25** (*Minimal relative entropy for discrete distributions*) Let $\mathbf{p} \in \mathbb{R}_{>0}^n$, $\sum_{i=1}^{n} p_i = 1$ be a discrete probability distributions on n possible outcomes. We want to find the defective probability distribution $\mathbf{q} \in \mathbb{R}_{\geq 0}^n$, $\sum_{i=1}^{n} q_i = a$, with $0 < a \leq 1$, that has the minimal relative entropy (or Kullback–Leibler divergence) with respect to $\mathbf{p}$, defined as $D(\mathbf{p}, \mathbf{q}) = \sum_{i=1}^{n} p_i \log p_i/q_i$. Minimising $D(\mathbf{b}, \mathbf{q})$ is equivalent to maximising $-D(\mathbf{b}, \mathbf{q})$. Thus, ignoring an ineffectual constant, we can consider the maximisation problem Lagrangian:

$$L(\mathbf{q}, \lambda) = \sum_{i=1}^{n} p_i \log q_i - \lambda(\sum_{i=1}^{n} q_i - a) + \sum_{i} \mu_i q_i$$

and obtain the first-order conditions with respect to the q's,

$$p_i/q_i = \lambda - \mu_i, \quad i = 1, \dots, n.$$

It must be $q_i > 0$, $\mu_i = 0$, and we must set $\lambda > 0$. The solution then takes the form $q_i = \lambda p_i$. By imposing the constraint, we find $\lambda = a$, so that $\mathbf{q} = a\mathbf{p}$. The Hessian $H_f(\mathbf{q})$ is a diagonal matrix with entries $(-1/(ap_1), \dots, -1/(ap_n))$ and the Hessian of the equality constraint is zero, so $H_L = H_f$. There are no binding inequality constraints. Thus, for any nonzero vector $\mathbf{u}$, it is $\mathbf{u}H_f\mathbf{u} = -\sum_{i=1}^{n} u_i^2/(2p_i) < 0$ and by the second statement of Theorem 7.23 this point is guaranteed to be a maximum. In conclusion, the probability distribution with the lowest divergence from $\mathbf{p}$ is proportional to it, or equal, if $a = 1$.

Theorem 7.24 (Second-order conditions) *Consider $\mathbf{x}^* \in E$. Assume that only the first m inequality constraints are binding in $\mathbf{x}^*$, $g_i(\mathbf{x}^*) = 0$, $i = 1, \ldots, m$ and $g_i(\mathbf{x}^*) < 0$, $i > m$. Assume further that there exist $\boldsymbol{\lambda}^* \in \mathbb{R}_{\geq 0}^m$ and $\boldsymbol{\mu}^* \in \mathbb{R}^l$ such that*

$$\nabla f(\mathbf{x}^*) = \sum_{i=1}^{m} \lambda_i^* \nabla g_i(\mathbf{x}^*) + \sum_{j=1}^{l} \mu_j^* \nabla h_j(\mathbf{x}^*).$$

Define

$$H_L(\mathbf{x}^*) = H_f(\mathbf{x}^*) - \sum_{i=1}^{m} \lambda_i^* H_i^g(\mathbf{x}^*) - \sum_{j=1}^{l} \mu_j^* H_j^h(\mathbf{x}^*).$$

Necessary condition: *if $\mathbf{x}^*$ is a constrained local maximum, then $\forall \mathbf{v} \in E_H^\perp(\mathbf{x}^*) \cap E_G^\perp(\mathbf{x}^*)$, $\mathbf{v}^\mathsf{T} H_L(\mathbf{x}^*)\mathbf{v} \leq 0$.* Sufficient condition: *if $\forall \mathbf{v} \in E_H^\perp(\mathbf{x}^*) \cap E_{G^+}^\perp(\mathbf{x}^*) \cap C_{G^0}^-(\mathbf{x}^*)$, $\mathbf{v}^\mathsf{T} H_L(\mathbf{x}^*)\mathbf{v} < 0$, then $\mathbf{x}^*$ is a constrained strict local maximum.*

Proof For the necessary condition, note that if $\mathbf{x}^*$ is a constrained local minimum, it must also remain so when all inequality constraints are replaced by equality constraints. Therefore, the first statement follows from Theorem 7.23.

Next we will prove that if the point is not a constrained strict local maximum, then the sufficient condition is violated. If $\mathbf{x}^*$ is not a constrained strict local maximum, then there exists a sequence of feasible points $(\mathbf{x}_n = \mathbf{x}^* + \delta_n \mathbf{u}_n)$, with $\|\mathbf{u}_n\| = 1$, $\delta_n > 0$, such that $\delta_n \to 0$ and $f(\mathbf{x}_n) \geq f(\mathbf{x}^*)$. By Taylor approximation, $\forall i = 1, \ldots, m$ and $\forall j = 1, \ldots, l$,

$$\delta_n \nabla g_i(\mathbf{x}^*) \cdot \mathbf{u}_n + \frac{\delta_n^2}{2} \mathbf{u}^\mathsf{T} H_i^g(\mathbf{x}^*)\mathbf{u}_n + o(\delta_n^2) \leq 0,$$

$$\delta_n \nabla h_j(\mathbf{x}^*) \cdot \mathbf{u}_n + \frac{\delta_n^2}{2} \mathbf{u}^\mathsf{T} H_j^h(\mathbf{x}^*)\mathbf{u}_n + o(\delta_n^2) = 0.$$

Consider the Taylor approximation of $\Delta f(\mathbf{x}) = f(\mathbf{x}) - f(\mathbf{x}^*)$,

$$\Delta f(\mathbf{x}) = \delta_n \nabla f(\mathbf{x}^*) \cdot \mathbf{u} + \frac{\delta_n^2}{2} \mathbf{u}^\mathsf{T} H_f(\mathbf{x}^*)\mathbf{u} + o(\delta_n^2).$$

Because $(\delta_n \mathbf{u}_n)$ is a bounded sequence, it must have a converging subsequence. So, without loss of generality we can assume that $\exists \mathbf{u}, \|\mathbf{u}\| = 1$, such that $\mathbf{u}_n \to \mathbf{u}$. Then $\forall i = 1, \ldots, m$ and $\forall j = 1, \ldots, l$, it must be $\nabla g_i(\mathbf{x}^*) \cdot \mathbf{u} \leq 0$ and $\nabla h_j(\mathbf{x}^*) \cdot \mathbf{u} = 0$. By the first-order condition, if $\lambda_i > 0$, and $\nabla g_i(\mathbf{x}^*) \cdot \mathbf{u} < 0$, then definitely $\Delta f < 0$.

Thus, it must be $\mathbf{u} \in E_H^{\perp} \cap E_{G+}^{\perp} \cap C_{G^0}^{-}$. Using the first-order condition,

$$\Delta f(\mathbf{x}) = \frac{\delta_n^2}{2} \mathbf{u}^{\mathsf{T}} H_L(\mathbf{x}^*) \mathbf{u} + o(\delta_n^2),$$

and it must be $\mathbf{u}^{\mathsf{T}} H_L(\mathbf{x}^*) \mathbf{u} \geq 0$.   □

According to the necessary condition, if $\exists \mathbf{v} \in E_H^{\perp} \cap E_G^{\perp}$, $\mathbf{v}^{\mathsf{T}} H_L \mathbf{v} > 0$, then $\mathbf{x}^*$ cannot be a candidate solution. Note that the sufficient condition requires the matrixx $H_L(\mathbf{x}^*)$ to be positive definite with respect to all vectors in the polar cone of the binding constraints with zero multipliers that are orthogonal, at the same time, to the gradient of all other binding constraints.

▶ **Example 7.26** *(Zero gradient on the boundary)* Consider the problem of finding the maximum point of the function $f(x, y) = -x^2/2 - y^2 - 3xy/2$ such that $g_1(x, y) = 2x + y \leq 0$ and $g_2(x, y) = x + 4y \leq 0$. The feasible set is unbounded, so, in principle, we do not know if a solution exists. Defining the Lagrangian function $L = f(x, y) - \lambda_1 g_1(x, y) - \lambda_2 g_2(x, y)$, the first-order conditions with respect to the variables x and y are

$$\begin{cases} -x - 3y/2 = 2\lambda_1 + \lambda_2, \\ -2y - 3x/2 = \lambda_1 + 4\lambda_2. \end{cases}$$

The gradient of the function is zero only in $(0, 0)$, thus there are no internal candidate solutions. It is easy to verify that the system has no solution for $\lambda_1 > 0$, $\lambda_2 = 0$ or for $\lambda_1 = 0$, $\lambda_2 > 0$. Hence, the only possible solution is $(x, y) = (0, 0)$ with $\lambda_1 = \lambda_2 = 0$. The Hessian matrix of the function f is constant and reads

$$H_f = \begin{pmatrix} -1 & -3/2 \\ -3/2 & -2 \end{pmatrix},$$

so that the origin is a saddle point. However, considering the vector $\mathbf{v}_{\alpha} = (1, \alpha)$, we have $\mathbf{v}_{\alpha}^{\mathsf{T}} H_f \mathbf{v}_{\alpha} = -1 - 2\alpha^2 - 3\alpha$. This quantity is negative when $\alpha < -1$ or $\alpha > -1/2$. The vectors belonging to the polar cone defined by the gradient of the constraints in $(0, 0)$, $dg_1 = (1, 2)$ and $dg_2 = (1, 4)$, are identified by the vectors $\mathbf{v}_{\alpha}$ with $\alpha < -2$ and $\alpha > -1/4$, see Fig. 7.11. Thus, the sufficient condition of Theorem 7.24 applies and the origin is an acceptable candidate solution. Being the only one, it represents the solution of the problem.

▶ **Example 7.27** *(Two-dimensional parametric problem)* Let $f(x, y) = x^2 + ay^2$ with $a \in \mathbb{R}$, $g_1(x, y) = x^2 - y - 1$, and $g_2(x, y) = y - 1$. Find the value $F(a) = \max_{x,y} f(x, y)$, such that $g_1(x, y) \leq 0$ and $g_2(x, y) \leq 0$.

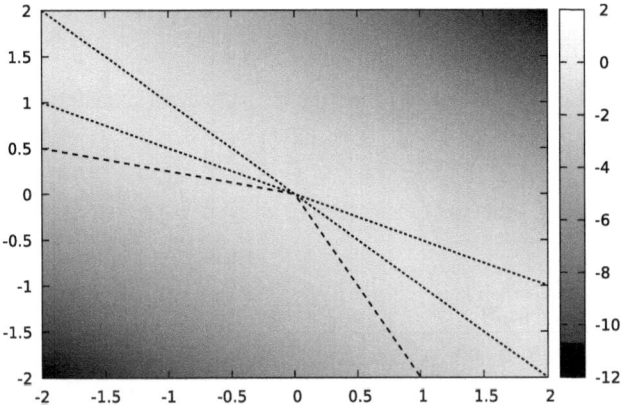

Fig. 7.11 Heat map of the function $f(x, y)$ discussed in Example 7.26. The constraints identify the bottom-left area, between the dashed lines. In the origin, the function grows in the directions between the dotted lines

From the Lagrangian function $L = f - \lambda_1 g_1 - \lambda_2 g_2$, the first-order conditions read

$$
\begin{cases}
2x(1 - \lambda_1) = 0, \\
2ay = \lambda_2 - \lambda_1, \\
\lambda_1(x^2 - y - 1) = 0, \quad \lambda_1 \geq 0, \quad x^2 - y - 1 \geq 0, \\
\lambda_2(y - 1) = 0, \quad \lambda_2 \geq 0, \quad y - 1 \geq 0.
\end{cases}
$$

Setting $\lambda_1 = \lambda_2 = 0$ we find that the origin $(0, 0)$ is the only internal critical point. The Hessian matrix of the objective function f reads

$$
H_f = \begin{pmatrix} 2 & 0 \\ 0 & 2a \end{pmatrix}.
$$

The origin is never a maximum as, for any value of a, $\mathbf{v}^\mathsf{T} H_f \mathbf{v} > 0$ for $\mathbf{v} = (1, 0)$,

Looking for candidate solutions at the boundary $y = 1$, we set $\lambda_1 = 0$ and $\lambda_2 > 0$. This condition identifies the point $(0, 1)$ with $\lambda_2 = 2a$ (point A in Fig. 7.12). The solution is acceptable only if $a > 0$. The gradient of the constraint at this point is $dg_1 = (0, 1)$. This is orthogonal to $\mathbf{v} = (1, 0)$ and $\mathbf{v} H_f \mathbf{v} > 0$, so by Theorem 7.24 this cannot be a candidate solution.

Looking for candidate solutions at the boundary $x^2 = 1 + y$, we set $\lambda_1 > 0$ and $\lambda_2 = 0$. There are two cases. If $\lambda_1 \neq 1$, the unique candidate solution becomes $(0, -1)$ with $\lambda_1 = 2a$ (point C in Fig. 7.12). At this point $dg_1 = (0, -1)$ so that the space of vectors orthogonal to the binding constraints is spanned by $\mathbf{v} = (1, 0)$. The Hessian of the Lagrangian $H_L = H_f - H_1^g$ at this point is

$$
H_L = \begin{pmatrix} 2 - 2a & 0 \\ 0 & 2a \end{pmatrix},
$$

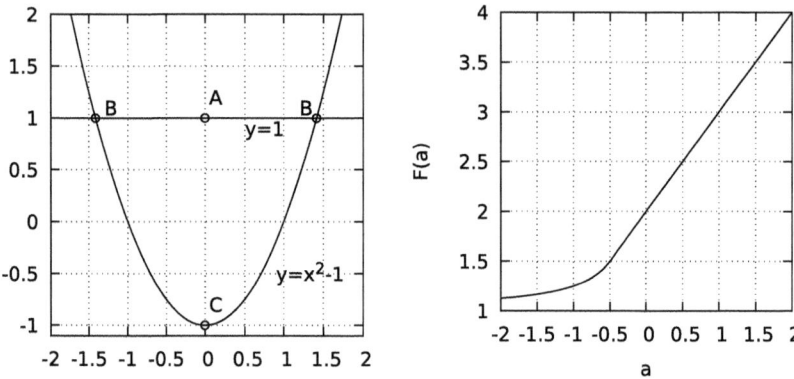

Fig. 7.12 The feasible set (left) and the behaviour of the maximum function (right) for Example 7.27

so that $\mathbf{v}^T H_L \mathbf{v} = 2 - 4a$. Thus, this is a candidate solution if $a \in (0, 1/2)$ and $F(a) = a$. If, instead, $\lambda_1 = 1$, we identify the points $(x_\pm(a), -1/(2a))$, with $x(a)_\pm = \pm\sqrt{1 - 1/(2a)}$. We must also require $-1 < 1/(2a) \le 1$, that is, $a \ge 1/2$ or $a < -1/2$. In these points, $dg_1 = (2x_\pm(a), -1)$. The orthogonal complement of the gradient is spanned by $\mathbf{v} = (1, 2x(a)_\pm)$ and

$$H_L = \begin{pmatrix} 0 & 0 \\ 0 & 2a \end{pmatrix},$$

so that $\mathbf{v}^T H_L \mathbf{v} = 4(2a - 1)$. In summary, these are candidate solutions if $a < -1/2$, with $F(a) = 1 - 1/(4a)$.

Finally, we set $\lambda_1 > 0$ and $\lambda_2 > 0$. These conditions identify the points $(\pm\sqrt{2}, 1)$ (points B in Fig. 7.12) in which $F(a) = 2 + a$. Since in this case $\lambda_2 = 1 + 2a$, these are acceptable solutions only if $a \ge -1/2$. Note that dg_1 and dg_2 are independent and have positive multipliers. There are no nonzero vectors orthogonal to both, so the sufficient condition of Theorem 7.24 is trivially satisfied.

In summary, if $a < -1/2$, the maxima are in $(x_\pm(a), -1/(2a))$ and $F(a) = 1 - 1/(4a)$, while if $a \ge -1/2$ the maxima are in $(\pm\sqrt{2}, 1)$ and $F(a) = 2 + a$.

7.7.6 Envelope Theorem

Suppose that the function to be maximised and/or the constraints depend on some parameter t. Specifically, we consider a real parameter t in an interval I and assume that the objective function $f(\mathbf{x}, t)$, the set of inequality constraints $G_k(t) = \{g_1(\mathbf{x}, t), \dots, g_k(\mathbf{x}, t)\}$, and of equality constraints $H_l(t) = \{h_1(\mathbf{x}, t), \dots, h_l(\mathbf{x}, t)\}$, can depend on t. Let $E_H(t)$ and $E_G(t)$ be the sets of points that satisfy all the equality and inequality constraints, respectively, for a given value of t. We assume that $\forall t \in I, E(t) = D \cap E_H(t) \cap E_G(t) \ne \emptyset$ and that the constraint functions are regular and independent. The next theorem shows that under mild regularity conditions, the points that satisfy the first-order condition of Theorem 7.22 belong to a differentiable curve. In addition, the theorem provides a method to compute the derivative of the value of the objective function in those points.

Theorem 7.25 (Envelope theorem) *Let* $x_0 \in E(t_0)$ *satisfy the first-order condition of Theorem 7.22 with constraints* $H_l(t_0)$ *and* $G_k(t_0)$ *and with positive multipliers for the m binding inequality constraints,* $\lambda_i > 0$ *and* $g_i(\mathbf{x}, t_0) = 0$, $i \leq m$, *and* $g_i(\mathbf{x}, t_0) < 0$ *for* $i > m$. *Assume that in a neighbourhood of* $(\mathbf{x}, t_0)$ *the objective function and the constraint functions are continuously twice differentiable. Then there exist a neighbourhood* $N(t_0) \subseteq \mathbb{R}$ *and a continuously differentiable function from* $N(t_0)$ *to* $\mathbb{R}^{n+l+m}$, $(\mathbf{x}(t), \boldsymbol{\mu}(t), \boldsymbol{\lambda}(t))$, *such that* $\mathbf{x}(t)$ *satisfies the first-order condition with constraints* $H_l(t)$ *and* $G_k(t)$, $\lambda_i(t) > 0$ *for* $i \leq m$ *and* $g_i(\mathbf{x}, t) < 0$ *for* $i > m$. *Moreover, if* $F(t) = f(\mathbf{x}(t), t)$,

$$\frac{dF(t)}{dt} = \frac{\partial L(\mathbf{x}(t), \boldsymbol{\mu}(t), \boldsymbol{\lambda}(t)))}{\partial t} =$$

$$\frac{\partial f(\mathbf{x}(t), t)}{\partial t} - \sum_{i=j}^{l} \mu_j(t) \frac{\partial h_j(\mathbf{x}(t), t)}{\partial t} - \sum_{i=1}^{m} \lambda_i(t) \frac{\partial g_i(\mathbf{x}(t), t)}{\partial t}.$$

Proof Define the gradient of the Lagrangian

$$dL(\mathbf{x}, t) = df(\mathbf{x}, t) - \sum_{i=j}^{l} \mu_j(t) dh_j(\mathbf{x}, t) - \sum_{i=1}^{m} \lambda_i(t) dg_i(\mathbf{x}(t), t)$$

and consider the function $\mathbf{G} : \mathbb{R}^{n+l+m+1} \to \mathbb{R}^{n+l+m}$,

$$\mathbf{G}(\mathbf{x}, \boldsymbol{\mu}, \boldsymbol{\lambda}, t) = \begin{pmatrix} dL(\mathbf{x}, t) \\ h_j(\mathbf{x}, t) \quad i = 1, \dots, l \\ g_i(\mathbf{x}, t) \quad i = 1, \dots, m \end{pmatrix}.$$

The solution in the statement is a point $(\mathbf{z}_0, t_0)$ with $\mathbf{z}_0 = (\mathbf{x}, \boldsymbol{\mu}, \boldsymbol{\lambda})$ such that $\mathbf{G}(\mathbf{z}_0, t_0) = \mathbf{0}, \lambda_i > 0$ for $i \leq m$ and $g_i(\mathbf{x}, t_0) < 0$ for $i > m$. Note that $\mathbf{G}$ is linear in $\boldsymbol{\mu}$ and $\boldsymbol{\lambda}$, therefore, by hypothesis, it is continuously differentiable in a neighbourhood of $(\mathbf{z}_0, t_0)$. Therefore, by the implicit function theorem, there exists a neighbourhood $N(t_0)$ and a continuously differentiable function $\mathbf{z}(t) = (\mathbf{x}(t), \boldsymbol{\mu}(t), \boldsymbol{\lambda}(t))$ such that $\mathbf{G}(\mathbf{z}(t), t) = \mathbf{0}, \forall t \in N(t_0)$. Due to the continuity of the constraint functions, one can take the neighbourhood small enough such that $\lambda_i(t) > 0$ for $i \leq m$ and $g_i(\mathbf{x}(t), t) < 0$ for $i > m$. This proves the first part of the statement. For the second part, by the chain rule,

$$\frac{dF(t)}{dt} = \frac{\partial f(\mathbf{x}(t), t)}{\partial t} + d\mathbf{f}(\mathbf{x}(t), t) \cdot \frac{d\mathbf{x}(t)}{dt}.$$

Substituting the first-order condition $dL(\mathbf{x}(t), t) = \mathbf{0}$,

$$\frac{dF(t)}{dt} = \frac{\partial f(\mathbf{x}(t), t)}{\partial t} + \sum_{i=j}^{l} \mu_j dh_j(\mathbf{x}(t), t) \cdot \frac{d\mathbf{x}(t)}{dt} + \sum_{i=1}^{m} \lambda_i dg_i(\mathbf{x}(t), t) \cdot \frac{d\mathbf{x}(t)}{dt}.$$

By the implicit function theorem, $h_j(\mathbf{x}, t) = 0$ and $g_i(\mathbf{x}, t) = 0$ imply $dh_j(\mathbf{x}(t), t) \cdot d\mathbf{x}(t)/dt = -\partial_t h_j(\mathbf{x}(t), t)$ and $dg_i(\mathbf{x}(t), t) \cdot d\mathbf{x}(t)/dt = -\partial_t g_i(\mathbf{x}(t), t)$, respectively. Substituting into the above equation proves the second part of the statement.   □

In other words, the derivative of the function $F(t)$ with respect to the parameter t is equal to the partial derivative of the Lagrangian function with respect to t, taken with the maximum point and multipliers fixed. The interpretation of the previous theorem is different in different situations. If we know that $\mathbf{x}(t)$ is the maximum point of $f(\mathbf{x}, t)$ in $E(t)$, then the theorem can be used to calculate the derivative of the *value function* $F(t) = \max_{\mathbf{x} \in E(t)} f(\mathbf{x}, t)$. For example, this is the case when the problem has a unique extremal point. More generally, Theorem 7.25 can be used to describe how the value of the objective function computed in different candidate solutions changes when the parameter t is varied. In principle, it can also be applied to curves of points that satisfy the first-order condition of Theorem 7.22 but are not local extrema. Note that the assumption of constraint independence can be replaced with the constraint qualification of Example 7.22.

▶ **Example 7.28** *(Utility maximisation)* The utility of the consumption of a bundle of commodities $\mathbf{x} \in \mathbb{R}^n_{\geq 0}$ is given by a function $U(\mathbf{x}) \in C^1(\mathbb{R}^n_{\geq 0})$ such that for $i = 1, \ldots, n$, $\partial_i U > 0$ (non-satiating hypothesis) and $\lim_{x_i \to 0^+} U = -\infty$ (Inada condition). The prices of the commodities are $\mathbf{p} \in \mathbb{R}^n_{>0}$ and the consumption $\mathbf{x}$ is affordable if $\mathbf{x} \cdot \mathbf{p} \leq w$ for some positive w (wealth). We want to maximise U while keeping consumption nonnegative and affordable. The Lagrangian function of the problem reads

$$L = U(\mathbf{x}) + \sum_{i=1}^{n} \lambda_i x_i - \mu \left(\sum_{i=1}^{n} x_i p_i - w \right).$$

The consumption levels first-order conditions read

$$\partial_i U(\mathbf{x}) = -\lambda_i + \mu p_i, \quad i = 1, \ldots, n.$$

Since the function U is strictly increasing, we must have $\mu > 0$. The Inada condition rules out any solution with zero consumption of any commodity. Thus, we must set $\lambda_i = 0, \forall i$. The value of μ is found by imposing the condition that the optimal bundle is feasible, $\mu = \sum_{i=1}^{n} x_i \partial_i U / w$. If U^* is the utility of the optimal consumption level that exists by Theorem 2.30, even if we did not solve the problem explicitly, Theorem 7.25 tells us that $dU^*/dw = \partial_w L = \mu$. In this case, the multiplier μ can be interpreted as the marginal utility of wealth.

In applying Theorem 7.25 to the problem with inequality constraints discussed in Theorem 7.22, one has to check that there are no constraints such that $\lambda_i = 0$ and $g_i(\mathbf{x}, t) = 0$. If this is the case, when the value of t changes, the constraints that were not previously binding can become binding or vice versa. In fact, even if the objective function and the constraints are continuously differentiable functions in t, the solution of a problem of the type presented in Theorem 7.22 might not be differentiable.

▶ **Example 7.29** (*Non-differentiable maximum function*) Consider the function $F(t) = \max_{x \in [-1,1]} tx^2$. This corresponds to the maximum of the function tx^2 with constraints $x - 1 \le 0$ and $-x - 1 \le 0$. The objective function and the constraints are defined by continuously twice differentiable functions. It is immediate to realise that $F(t) = 0$ if $t \le 0$ and $F(t) = t$ if $t > 0$. The function $F(t)$ is not differentiable in $t = 0$. In fact, for $t < 0$ the constraints $x - 1 \le 0$ and $-x - 1 \le 0$ are not binding. The value of their multipliers in the Lagrangian problem is zero. If $t > 0$, then one of the constraints is always binding and the value of its multiplier is $2t$.

7.7.7 Minimisation Problems

The methods introduced in the previous sections can be adapted with minimal modifications to minimisation problems. The multipliers λ associated with the inequality constraints in Theorems 7.20 and 7.22 must change sign. Now, they must be lower than or equal to zero. The requirements for equality constraints remain the same. In Lemma 7.4 we can keep the same definition of the Lagrangian function L, and modify the slackness conditions to require $\lambda_i \le 0$. Alternatively, we can change the sign of the multipliers in the definition of the Lagrangian and keep the same expression for the slackness condition. Both approaches are found in the literature. Concerning second-order conditions, the polar cone of the gradient of the binding inequality constraints with multipliers equal to zero must be replaced with the dual cone, see Definition 1.26. The necessary and sufficient conditions involving the inequality constraints in Theorem 7.24 also change sign. The symmetric matrix H_L is now required to be positive semi-definite for the necessary conditions and positive definite for the sufficient conditions, with respect to the appropriate linear spaces and cones. Apart from a possible redefinition of the Lagrangian function, Theorem 7.25 remains the same.

Exercises

Exercise 7.1 Consider the function $f : \mathbb{R}^2 \to \mathbb{R}$ defined by

$$f(x, y) = \begin{cases} 0 & \text{if } (x, y) = (0, 0), \\ x^n y^m / (x^2 + y^2) & \text{otherwise}, \end{cases}$$

with n and m nonnegative integers. For which values of n and m is the function continuous at $(0, 0)$? For which values is it differentiable? Set $n = m = 1$ and compute the supremum of the function on the set of points $(x, y) \in \mathbb{R}^2$ that satisfy the constraint $x^2 + y^2 = 1$. Does this supremum correspond to a local maximum of the function in $\mathbb{R}^2$? Does it correspond to a strict local maximum?

Exercise 7.2 Consider the function $f : \mathbb{R}^2 \to \mathbb{R}$ defined by

$$f(x, y) = \begin{cases} \alpha x^2 y & \text{if } y = x^2, \\ \beta y & \text{otherwise,} \end{cases}$$

with $\alpha, \beta \in \mathbb{R}$. For which values of α and β is the function continuous or differentiable in $\mathbb{R}^2$? For which values is it continuous/differentiable at the origin? Let $\beta = 3$ and find all values of α and all vectors $\mathbf{u}$, $\|\mathbf{u}\| = 1$, such that $D_{\mathbf{u}} f(1, 1) = 3$.

Exercise 7.3 Consider the function $f : \mathbb{R}^2 \to \mathbb{R}$ defined by

$$f(x, y) = \begin{cases} 0 & \text{if } (x, y) = (0, 0), \\ e^{-1/(x^2+y^2)} & \text{otherwise.} \end{cases}$$

Is the function continuous at $(0, 0)$? Is it differentiable? Find the maximum and minimum of the function on the set of points (x_1, x_2) such that $\max\{|x_1|, |x_2|\} \le 1$. Compute the second-order Taylor polynomial of the function around the origin.

Exercise 7.4 Consider the function $f : \mathbb{R}^n \to \mathbb{R}_{\ge 0}$, $f(\mathbf{x}) = \log(1 + \|\mathbf{x}\|^2)$. Use Theorem 7.7 and the Cauchy-Schwarz inequality to derive an upper bound to the incremental ratio $|f(\mathbf{x}) - f(\mathbf{y})|/\|\mathbf{x} - \mathbf{y}\|$ in the ball of radius r centred at the origin.

Exercise 7.5 Consider the function $f : \mathbb{R}^3 \to \mathbb{R}$ defined by

$$f(x, y, z) = \sin(x + y) + \cos(x - y) + z.$$

Given a generic point $\mathbf{w} = (w_1, w_2, w_3) \in \mathbb{R}^3$, consider the set $A(\mathbf{w}) = [w_1 - 1/2, w_1 + 1/2] \times [w_2 - 1/2, w_2 + 1/2] \times \{w_3\}$. Find a Taylor polynomial that approximates f in $A(\mathbf{w})$ to a precision of 10^{-1}. Use this polynomial around the origin to approximate $f(0.1, 0.1, 0)$.

Exercise 7.6 Find the tangent to the curve in $\mathbb{R}^2$ implicitly defined by $f(x, y) = x^3 + xy^2 + y^2 = c$ where it intersects with the axes, as a function of c.

Exercise 7.7 Consider the curve implicitly defined by the equation $\log(1 + x^2 + y^2) + \log(1 + x^2) = c$. Find the points of $\mathbb{R}^2$ where the tangent to the curve is parallel to the x axis as a function of c.

Exercise 7.8 Compute the derivative of the functions $y(x)$ and $z(x)$ implicitly defined by the system of equations

$$\begin{cases} f(x, y, z) = e^x + e - y + z = 0, \\ g(x, y, z) = e^{-x} + yz - 1 = 0, \end{cases}$$

in terms of x, y, and z.

Exercise 7.9 Consider the functions $y(x)$ and $z(x)$ implicitly defined by

$$\begin{cases} f(x, y, z) = -xy + 2xz + \beta z + x^2 = 0, \\ g(x, y, z) = \alpha y^2 + 2\gamma x + \beta z^2 = 0, \end{cases}$$

with $(\alpha, \beta, \gamma) \in \mathbb{R}^3$. For each set $S_1 = \{x = 0, y \neq 0, z \neq 0\}$, $S_2 = \{y = 0, x \neq 0, z \neq 0\}$ and $S_3 = \{z = 0, x \neq 0, y \neq 0\}$, determine the possible values of α, β, and γ such that the derivatives $y'(x)$ and $z'(x)$ exist in the whole set.

Exercise 7.10 Consider the functions u, v, and w from $\mathbb{R}^2 \to \mathbb{R}$,

$$u = x + y, \quad v = x - y, \quad w = \frac{2 \cos x \sin x}{\cos 2x} + 2 \cos x \cos y.$$

Prove that they are dependent and find their functional relation.

Exercise 7.11 Consider the function $f(x, y) = xy(x + y - 1)$. Find all its strict local maximum and minimum points. Find its minimum and maximum in the closed triangle with vertices $A = (0, 0)$, $B = (1, 0)$, and $C = (0, 2)$.

Exercise 7.12 Consider the function $f(x, y) = x^4 + 2x^2y^2 + y^4 - 2x^2$. Find all its local maximum and minimum points. Find its minimum and maximum in the closed circle with the centre in the origin and radius equal to $r > 0$, as a function of r.

Exercise 7.13 Consider the function $f(x, y) = \log(x) + \log(a + y)$ with $a \geq 0$. Solve the problem

$$\max f(x, y) \quad \text{such that} \quad x + y \leq w \quad \text{and} \quad x, y \geq 0,$$

as a function of $w > 0$.

Exercise 7.14 Let $f(x)$ be a strictly increasing real-valued function and r a positive constant. Find the maximum of the function $F : \mathbb{R}^n \to \mathbb{R}$, $F(\mathbf{x}) = \sum_{i=1}^n f(x_i)$ such that $\sum_{i=1}^n x_i^2 \leq r^2$.

Exercise 7.15 (*Dual norm*) Based on the definitions in Exercise 4.18 and Sect. 4.2.2, prove that the dual of the p-norm $\|.\|_p$ is the q-norm $\|.\|_q$ with $1/p + 1/q = 1$. In particular, the dual of the Euclidean norm is the Euclidean norm itself. How does this relate to the Holder inequality in Theorem 4.4?

Exercise 7.16 (*Failure of Karush–Kuhn–Tucker condition*) In $\mathbb{R}^2$, find the minimum point (x, y) of the function $f(x, y) = y$ such that $g(x, y) = x^2 - y^3 \leq 0$. Check that in the solution there is no $\lambda \geq 0$ such that $df = \lambda dg$. Why Theorem 7.22 does not apply? *Hint: Check the nature of the constraint that defines the feasible set.*

Exercise 7.17 (*Failure of Karush–Kuhn–Tucker condition*) In $\mathbb{R}^2$, find the maximum point (x, y) of the function $f(x, y) = x + y$ such that $g_1(x, y) = x^2 + y^2 \leq 1$ and $g_2(x, y) = (x - 2)^2 + y^2 \leq 1$. Check that in the solution there is no $\lambda_1, \lambda_2 \geq 0$ such that $df = \lambda_1 dg_1 + \lambda_2 dg_2$. Why Theorem 7.22 does not apply? *Hint: Check the nature of the constraints that define the feasible set.*

Exercise 7.18 (*Strong duality failure*) With reference to Example 7.23, consider the problem of maximising x^2 for $x \in [0, 1]$. Derive the associated Lagrangian and prove that strong duality does not apply. In other terms, the solution of the dual problem is greater than the solution of the primal problem.

Exercise 7.19 (*Strong duality success*) With reference to Example 7.23, consider the problem of maximising $\sqrt{x}$ for $x \in [0, 1]$. Derive the associated Lagrangian and prove that strong duality applies. In other terms, the solution of the dual problem is equal to the solution of the primal problem.

Exercise 7.20 (*Consumption smoothing*) An agent enjoys an utility $U(\mathbf{c}) = \sum_{t=1}^{T} \beta^t c_t^\alpha$ from the consumption stream $\mathbf{c} = (c_1, \ldots, c_T) \in \mathbb{R}_{\geq 0}^T$ over T periods. Assume that $\alpha, \beta \in (0, 1)$, so that the agent is impatient and prefers to consume sooner rather than later. The present value of the entire consumption is $v(\mathbf{c}) = \sum_{t=1}^{T} c_t$. Find the consumption stream that maximises $U(\mathbf{c})$ so that $v(\mathbf{c}) \leq w$, for a constant $w > 0$.

Exercise 7.21 (*Production problem*) Given a quantity of labour $l \geq 0$ and capital $k \geq 0$, a firm produces an output $y = k^\alpha l^\beta$ with $\alpha, \beta > 0$. If the positive constants p_y, p_k, and p_l represent the price of output good, capital, and labour respectively, find, if there exists, the production level y that maximises the profit $\pi = p_y y - p_k k - p_l l$ as a function of α, β, and prices. *Hint: Check also the second-order conditions.*

Exercise 7.22 (*Consumption-leisure problem*) With reference to Example 7.28, let the utility of consumption be $U = \prod_{i=1}^{n} x_i^{\alpha_i}$ with $\alpha_i \in (0, 1)$ $\forall i$. Assume that the agent has an endowment of H units of time. By working an amount l of time, the agent receives an income wl, with $w > 0$, and enjoys a utility from *leisure* equal to $H - l$. Maximise the total utility of the agent $U + H - l$ so that $\sum_{i=1}^{n} p_i x_i \leq wl$ and $0 \leq l \leq H$.

Integral Calculus

8

8.1 Definite Integrals

Consider a bounded function $f : \mathbb{D} \subseteq \mathbb{R} \to \mathbb{R}$ and an interval $[a, b] \subseteq \mathbb{D}$. Of paramount relevance for the following discussion is the notion of a partition of the interval.

Definition 8.1 (*Partition of the interval*) A partition $P = \{x_0 = a, x_1, ..., x_{N-1}, x_N = b\}$ of the interval $[a, b]$ is a set of increasing points belonging to the interval, $x_i \in (a, b)$ and $x_i < x_{i+1}, \forall i \in 1, ..., N - 2$.

The partition P divides the closed interval $[a, b]$ into N closed subintervals $[x_i x_{i+1}]$ with overlapping endpoints.

Definition 8.2 (*Upper and lower sum*) Consider a bounded function f over the interval $[a, b]$. For any finite partition P of $[a, b]$, the *upper sum* is

$$U_f(P) = \sum_{i=1}^{N} (x_i - x_{i-1}) \sup_{[x_{i-1}, x_i]} f,$$

and the *lower sum* is

$$L_f(P) = \sum_{i=1}^{N} (x_i - x_{i-1}) \inf_{[x_{i-1}, x_i]} f,$$

where $\sup_{[x,y]} f$ and $\inf_{[x,y]} f$ are, respectively, the supremum and infimum of the function f in the closed interval $[x, y]$.

© The Author(s), under exclusive license to Springer Nature Switzerland AG 2023
G. Bottazzi, *Advanced Calculus for Economics and Finance*, Classroom
Companion: Economics, https://doi.org/10.1007/978-3-031-30316-6_8

For any bounded function f and any partition P, $U_f(P) \geq L_f(P)$. Given two partitions P and P' of $[a, b]$, the partition $P'' = P \cup P'$ contains all points of P and P', with duplicate points duly removed. We write $P \subseteq P'$ if P' is a *refinement* of the partition P, that is, P' contains all the points of P plus possibly others. Some or all of the intervals in which P divides $[a, b]$ are further divided when considering P'. In this case $P' = P \cup P'$.

Theorem 8.1 *If $P \subseteq P'$, then $U_f(P) \geq U_f(P')$ and $L_f(P) \leq L_f(P')$.*

Proof Consider a point $y \in (x_{i-1}, x_i)$ with $x_{i-1}, x_i \in P$ and $y \in P' \setminus P$. For the property of the supremum

$$\sup_{[x_{i-1}, x_i]} f \geq \sup_{[x_{i-1}, y]} f \quad \text{and} \quad \sup_{[x_{i-1}, x_i]} f \geq \sup_{[y, x_i]} f,$$

so that

$$(x_i - x_{i-1}) \sup_{[x_{i-1}, x_i]} f \geq (y - x_{i-1}) \sup_{[x_{i-1}, y]} f + (x_i - y) \sup_{[y, x_i]} f.$$

Since

$$U_f(P \cup \{y\}) = U_f(P) +$$
$$(y - x_{i-1}) \sup_{[x_{i-1}, y]} f + (x_i - y) \sup_{[y, x_i]} f - (x_i - x_{i-1}) \sup_{[x_{i-1}, x_i]} f,$$

we have $U_f(P) \geq U_f(P \cup \{y\})$. At the same time, using a similar argument based on the properties of the infimum, it is immediate to see that $L_f(P) \leq L_f(P \cup \{y\})$. This shows that each time a new point y is added to the partition, the upper sum may decrease and the lower sum may increase. By repeating the procedure, adding any $y \in P'/P$, the assertion follows.   □

Let $\mathbb{P}([a, b])$ be the set of all possible partitions of the interval $[a, b]$. Note that, regardless of the partition considered, the upper and lower sums of f over $[a, b]$ are bounded above by $(b - a) \sup_{[a,b]} f$ and below by $(b - a) \inf_{[a,b]} f$. Then, both the set of the upper sums obtained considering any possible partition $\mathbb{U}_f = \{U_f(P) \mid P \in \mathbb{P}([a, b])\}$ and that of the lower sums obtained in the same way $\mathbb{L}_f = \{L_f(P) \mid P \in \mathbb{P}([a, b])\}$, are bounded. The next result shows that they are also separated.

Theorem 8.2 *For any $U \in \mathbb{U}_f$ and any $L \in \mathbb{L}_f$, it is $U \geq L$.*

Proof Let P be such that $U = U_f(P)$ and P' be such that $L = L_f(P')$. Consider $P'' = P \cup P'$. Then, since P'' is a refinement of both P and P', $U = U_f(P) \geq U_f(P'') \geq L_f(P'') \geq L_f(P') = L$, which proves the assertion.   □

We are ready to define when the integral of the function f over the interval $[a, b]$ exists and what its value is.

Definition 8.3 (*Riemann integral*) If $\sup \mathbb{L}_f = \inf \mathbb{U}_f$ then the function f is Riemann integrable on $[a, b]$. The *integral* of the function f on $[a, b]$, indicated by $\int_a^b dx f(x)$, is the unique element that separates the sets of lower and upper sums, that is, $\sup \mathbb{L}_f$ and $\inf \mathbb{U}_f$. The set of functions that can be integrated on $[a, b]$ is denoted by $\mathcal{R}([a, b])$.

This integral takes its name from the German mathematician Bernhard Riemann (1826–1866). The symbol used to denote integration $\int$ is a peculiarly shaped "s" letter, for "sum". The functional expression on its right is called *integrand*. The variable used within the integral, most often x in this book, has no intrinsic meaning. It is like the index in a summation, i.e. $\int_a^b dx f(x) = \int_a^b dy f(y)$.[1] The quantity $\sup \mathbb{L}_f$ corresponds to the *lower integral* and the quantity $\inf \mathbb{U}_f$ to the *upper integral* of the function over the interval, so that Definition 8.3 can be restated by saying that the function is integrable if its upper and lower integrals are equal. The Riemann integral of a function f over an interval $[a, b]$ is often called *definite*, because the integration endpoints a and b are specified.

▶ **Example 8.1** (*Integral of the constant function*) Consider the constant function $f(x) = c$ and the interval $[a, b]$. Whatever partition P one takes, $U_f(P) = L_f(P) = c(b - a)$. We can conclude that the constant function can be integrated over any interval, and the value of the integral is just the value of the function multiplied by the length of the interval.

The previous example is particularly simple. In general, one proves the integrability of a function f on a given interval by showing that for any $\epsilon > 0$, there exists a partition P such that $U_f(P) - L_f(P) < \epsilon$. In fact, this is possible only if the two sets $\mathbb{L}_f$ and $\mathbb{U}_f$ share a common boundary, that is, if f is integrable on $[a, b]$.

▶ **Example 8.2** (*Dirichlet function*) Consider a function $f(x)$ whose value is 0 for the rational and 1 for the irrational numbers,

$$f(x) = \begin{cases} 0 & \text{if } x \in \mathbb{Q}, \\ 1 & \text{if } x \in \mathbb{R} \setminus \mathbb{Q}. \end{cases}$$

Because rationals are dense in $\mathbb{R}$, in any subinterval $[x_{i-1}, x_i]$ of any partition P of any interval $[a, b]$, there are both rational and irrational numbers. Then, $\sup_{[x_{i-1}, x_i]} f = 1$

---

[1] In many textbooks the position of the integrand and the integration variable are reversed, as in $\int f(x) dx$. This is just a convention. To remind the readers that dx is nothing more than an index, it is preferable to position it near the integration symbol.

and $\inf_{[x_{i-1},x_i]} f = 0$. Hence, $U_f(P) = b - a$ and $L_f(P) = 0$ for any P. We therefore conclude that the function f cannot be integrated in any interval.

Another possible approach to proving integrability is to identify a sequence of partitions of a given interval (P_n) and show that, for a function f, $\lim_{n\to\infty} U_f(P_n) = \lim_{n\to\infty} L_f(P_n) = c$ for some constant c. This is possible only if the function f is integrable on the interval considered and the value of the integral is c. However, if for a specific sequence (P_n), $\lim_{n\to\infty} U_f(P_n) \neq \lim_{n\to\infty} L_f(P_n)$, we cannot conclude that the integral of the function does not exist. We only know that the sequences of upper and lower sums computed on the specific sequence of partitions that we considered do not converge to it.

▶ **Example 8.3** (*Integral of the identity function*) A *regular grid*, also called a lattice or an equispaced partition, over an interval $[a, b]$ is a partition $P_n = \{x_h = a + h(b - a)/n \mid h = 0, ..., n\}$. Consider the identity function $f(x) = x$. Using the regular grid, the upper sum reads

$$U_x(P_n) = \sum_{h=1}^{n}(x_h - x_{h-1}) \sup_{[x_{h-1},x_h]} x = \sum_{h=1}^{n}(x_h - x_{h-1})x_h$$

$$= \sum_{h=1}^{n}\frac{b - a}{n}(a + \frac{b - a}{n}h) = a(b - a) + \frac{(b - a)^2}{n^2}\frac{n(n + 1)}{2},$$

while the lower sum is

$$L_x(P_n) = \sum_{h=1}^{n}(x_h - x_{h-1}) \inf_{[x_{h-1},x_h]} x = \sum_{h=1}^{n}(x_h - x_{h-1})x_{h-1}$$

$$= \sum_{h=1}^{n}\frac{b - a}{n}(a + \frac{b - a}{n}(h - 1)) = a(b - a) + \frac{(b - a)^2}{n^2}\frac{n(n - 1)}{2}.$$

From the previous expressions $\lim_{n\to\infty} U_x(P_n) = \lim_{n\to\infty} L_x(P_n) = b^2/2 - a^2/2$.

▶ **Example 8.4** (*Thomae's function*) Consider the real function f defined on $[0, 1]$ as follows:

$$f(x) = \begin{cases} 0 & \text{if } x \in \mathbb{R}\backslash\mathbb{Q}, \\ 1/q & \text{if } x \in \mathbb{Q} \text{ and } x = p/q \text{ in lowest terms.} \end{cases}$$

Notice that for any equispaced partition P_n, if n is a prime number, then in any interval $[(h - 1)/n, h/n]$, $f(x) \geq h/n$. Therefore, $U_f(P_n) \geq s_n/n^2$, with $s_n = n(n + 1)/2$. Since $L_f(P_n) = 0$, it seems that using equispaced partitions, we cannot prove that the function f is integrable. We have to be slightly more smart than that.

Note that if $q > 1$, the function takes the value $1/q$ for at most $q - 1$ points. For example, $f(x) = 1/10$ for x equal to $1/10, 3/10, 7/10$ and $9/10$. Therefore, fixing n, there is a finite set of points $Q_n \subset \mathbb{Q}$ for which $f(x) > 1/n$: the rational points whose lowest terms representation p/q has $q < n$. For what said above, the number of points in Q_n is bounded above by s_n.

Consider a partition $\tilde{P}_n$ that has intervals of size $1/s(n)^2$ around the points in Q_n plus the regular grid of size n. In any interval that does not contain a point of Q_n, $f \leq 1/n$. In the intervals that contain a point of Q_n, $f \leq 1$. So,

$$U_f(\tilde{P}_n) \leq \sum_{h=1}^{n} \frac{1}{n}\frac{1}{n} + s(n)\frac{1}{s(n)^2} = \frac{1}{n} + \frac{1}{s(n)},$$

and $\lim_{n \to \infty} U_f(\tilde{P}_n) = 0$. As we have $L_f(\tilde{P}_n) = 0$, the function f is Riemann integrable in $[0, 1]$ and $\int_0^1 dx f(x) = 0$.

8.1.1 Properties of the Definite Integral

It is useful to review a series of properties that derive directly from the formal definition of the integral.

Theorem 8.3 *Consider an interval $[a, b]$ and $c \in [a, b]$. Then $f \in \mathcal{R}([a, b])$ if and only if $f \in \mathcal{R}([a, c])$ and $f \in \mathcal{R}([c, b])$. Moreover,*

$$\int_a^c dx f + \int_c^b dx f = \int_a^b dx f.$$

Proof Assume $f \in \mathcal{R}([a, b])$. Then $\forall \epsilon > 0$, take a partition P of $[a, b]$ such that $U_f(P) - L_f(P) < \epsilon$. Now consider the partition $P_1 = P \cup \{c\}$. This is a refinement of P, so that $U_f(P_1) - L_f(P_1) < \epsilon$. Now $P_1 = P_2 \cup P_3$ where P_2 contains all the points in the interval $[a, c]$ and P_3 all the points in the interval $[c, b]$. Then $U_f(P_2) - L_f(P_2) < \epsilon$ and $U_f(P_3) - L_f(P_3) < \epsilon$, which implies $f \in \mathcal{R}([a, c])$ and $f \in \mathcal{R}([b, c])$.

Let us prove the opposite implication. For any $\epsilon > 0$, let P_1 be a finite partition of $[a, c]$ and P_2 a finite partition of $[c, b]$ such that $U_f(P_1) - L_f(P_1) < \epsilon/2$ and $U_f(P_2) - L_f(P_2) < \epsilon/2$. Consider the partition $P_3 = P_1 \cup P_2$. By construction, $U_f(P_3) = U_f(P_1) + U_f(P_2), L_f(P_3) = L_f(P_1) + L_f(P_2)$, and $U_f(P_3) - L_f(P_3) < \epsilon$. This implies $f \in \mathcal{R}([a, b])$. From the definition of integral,

$$\int_a^c dx f + \int_c^b dx f - \epsilon < L_f(P_1) + L_f(P_2) = L_f(P_3) \leq \int_a^b dx f \leq$$

$$\leq U_f(P_3) = U_f(P_1) + U_f(P_2) < \int_a^c dx f + \int_c^b dx f + \epsilon.$$

Being true for any $\epsilon > 0$, this proves the second part of the assertion. $\qquad\square$

It is customary to consider integrals with reverse endpoints. If $a \geq b$ we pose $\int_b^a dxf = -\int_a^b dxf$ and obviously $\int_a^a dxf = 0$. In general, if a set of points E can be written as the union of intervals overlapping at most on the endpoints $E = \cup_{i=1}^n [a_i, b_i]$, one can write[2]

$$\int_E dxf = \sum_{i=1}^n \int_{a_i}^{b_i} dxf,$$

provided that all the integrals on the right-hand side exist.

Under suitable conditions, we can divide the integral of the sum of functions into the sum of their integrals.

Theorem 8.4 *If $f, g \in \mathcal{R}([a, b])$, then $f + g \in \mathcal{R}([a, b])$ and*

$$\int_a^b dx(f + g) = \int_a^b dxf + \int_a^b dxg.$$

Proof By hypothesis $\forall \epsilon > 0$, there exists a partition P such that $U_f(P) - L_f(P) < \epsilon/2$ and $U_g(P) - L_g(P) < \epsilon/2$. For the property of supremum and infimum $U_{f+g}(P) \leq U_f(P) + U_g(P)$ and $L_{f+g}(P) \geq L_f(P) + L_g(P)$, so that $U_{f+g}(P) - L_{f+g}(P) < \epsilon$, which proves the first part of the assertion. From the definition of an integral,

$$\int_a^b dxf + \int_a^b dxg - \epsilon < L_f(P) + L_g(P) \leq L_{f+g}(P) \leq \int_a^b dx\,(f + g)$$

$$\leq U_{f+g}(P) \leq U_f(P) + U_g(P) \leq \int_a^b dxf + \int_a^b dxg + \epsilon.$$

Being true for any $\epsilon > 0$, this proves the second part of the assertion. $\qquad\square$

If f is integrable in $[a, b]$, then cf, with c any constant, is integrable on the same interval. The value of the integral of cf is c times the integral of f, that is, $\int_a^b dxcf = c \int_a^b dxf$. Together with Theorem 8.4, this implies that the function from $\mathcal{R}([a, b])$ to $\mathbb{R}$ that assigns to each integrable function the value of its integral, $f \to \int_a^b dxf$, is a linear map. The integral also interacts in a natural way with the order relation.

---

[2] The extension of the notion of integral to more complicated sets is non-trivial and pertains to the topic of measure theory, see Chap. 9.

Theorem 8.5 *If $f, g \in \mathcal{R}([a, b])$, and $f \geq g$ in $[a, b]$, then $\int_a^b \mathrm{d}x f \geq \int_a^b \mathrm{d}x g$.*

Proof This is a trivial consequence of the fact that for any partition P, if $f \geq g$, $U_f(P) \geq U_g(P)$ and $L_f(P) \geq L_g(P)$. □

8.1.2 Riemann Integrable Functions

This section discusses criteria for the identification of integrable functions. The reader is warned that the question of which functions can be Rieman integrated on a given interval cannot be definitively resolved without the use of Lebesgue integration theory (see Sect. 9.3.1). However, some general results can be established.

Theorem 8.6 *If $f \in \mathcal{R}([a, b])$ and $g \in C^0([\inf_{[a,b]} f, \sup_{[a,b]} f])$, then $g \circ f \in \mathcal{R}([a, b])$.*

Proof If g or f are constant, the proof is trivial. Thus, we assume that they are not and set $m = \inf_{[a,b]} f$, $M = \sup_{[a,b]} f$, and $\Delta = \sup_{[a,b]} g \circ f - \inf_{[a,b]} g \circ f > 0$.

Consider any $\epsilon > 0$. We have to prove that there exists a partition P of $[a, b]$ such that $U_{g \circ f}(P) - L_{g \circ f}(P) < \epsilon$. Since g is continuous in $[m, M]$, it is uniformly continuous (see Theorem 3.5). Thus, there exists a $\delta > 0$ such that, for any $x, y \in [m, M]$, if $|x - y| < \delta$, $|g(x) - g(y)| < \epsilon/(2(b - a))$. At the same time, since f is integrable, there exists a partition P such that $U_f(P) - L_f(P) < \delta\epsilon/(2\Delta)$. We can show that this is actually the partition that we were looking for.

The subintervals in which the partition P divides the interval $[a, b]$, can be divided into two groups: A and B. Index the intervals by the index of their upper bound. If $\sup_{[x_{i-1}, x_i]} f - \inf_{[x_{i-1}, x_i]} f < \delta$, then $i \in A$, otherwise $i \in B$. Now notice that

$$\sum_{i \in B}(x_i - x_{i-1})\delta \leq \sum_{i \in B}(x_i - x_{i-1})\left(\sup_{[x_{i-1}, x_i]} f - \inf_{[x_{i-1}, x_i]} f\right) \leq$$

$$U_f(P) - L_f(P) < \frac{\delta\epsilon}{2\Delta},$$

so that $\sum_{i \in B}(x_i - x_{i-1}) < \epsilon/(2\Delta)$. In contrast, if $i \in A$, the distance of the image of the points in $[x_i, x_{i-1}]$ is always less than δ, so that by hypothesis $\delta_i = \sup_{[x_{i-1}, x_i]} g \circ f - \inf_{[x_{i-1}, x_i]} g \circ f < \epsilon/(2(b - a))$.

We compute the difference of the upper and lower sums of the composed function in the two sets of indices A and B separately,

$$D_A = \sum_{i \in A}(x_i - x_{i-1})\delta_i < \frac{\epsilon}{2(b - a)} \sum_{i \in A}(x_i - x_{i-1}) \leq \epsilon/2,$$

and

$$D_B = \sum_{i \in B}(x_i - x_{i-1})\delta_i \leq \Delta \sum_{i \in B}(x_i - x_{i-1}) < \epsilon/2.$$

Because $U_{g \circ f}(P) - L_{g \circ f}(P) = D_A + D_B < \epsilon$ the assertion is proved.   ☐

From Theorem 8.6 the following result follows directly.

Corollary 8.1 *Consider* $f, g \in \mathcal{R}([a, b])$. *Then* $fg \in \mathcal{R}([a, b])$, $|f| \in \mathcal{R}([a, b])$, *and* $\int_a^b dx |f(x)| \geq \left| \int_a^b dx f(x) \right|$.

Proof To prove the first statement, notice that $fg = \frac{1}{2}(f + g)^2 - \frac{1}{2}f^2 - \frac{1}{2}g^2$. This is the sum of continuous transformations of integrable functions, thus it is integrable. To prove the second statement, just apply Theorem 8.6 with $g(x) = |x|$. The third statement follows by noticing that $|f(x)| + f(x)$ and $|f(x)| - f(x)$ are both integrable functions and, being positive, their integral cannot be negative, so that $\int_a^b dx |f(x)| \geq \pm \int_a^b dx f(x)$.   ☐

If $|f(x)| < M$ in $[a, b]$, then $\left| \int_a^b dx f(x) \right| < M(b - a)$. The implication in Corollary 8.1 cannot be reversed. That is, if $|f| \in \mathcal{R}([a, b])$ we cannot, in general, conclude that $f \in \mathcal{R}([a, b])$. Example 8.3 shows that the identity function $f(x) = x$ is integrable. Thus, from Theorem 8.6, we can derive another result of great practical importance.

Corollary 8.2 *If* $f \in C^0([a, b])$, *then* $f \in \mathcal{R}([a, b])$.

A second important class of integrable functions is that of monotonic functions.

Theorem 8.7 *If* f *is monotonic in* $[a, b]$, *then* $f \in \mathcal{R}([a, b])$.

Proof Assume that f is increasing and define $\Delta = f(b) - f(a)$. Then $\forall \epsilon > 0$, choose a partition P such that $\forall i, x_i - x_{i-1} < \epsilon/\Delta$. So

$$U_f(P) - L_f(P) = \sum_{i=1}^N (x_i - x_{i-1}) \left(\sup_{[x_i, x_{i-1}]} f - \inf_{[x_i, x_{i-1}]} f \right) =$$

$$\sum_i (x_i - x_{i-1})(f(x_i) - f(x_{i-1})) < \frac{\epsilon}{\Delta} \sum_i f(x_i) - f(x_{i-1}) = \epsilon.$$

The proof is similar for a decreasing function.   ☐

Now imagine having a function that is monotonic or continuous in $[a, b]$ apart from a point $y \in [a, b]$. Let $\Delta = \sup_{[a,b]} f - \inf_{[a,b]} f$. Fix a $\epsilon > 0$. Then there exists a partition P_1 of $[a, y - \epsilon/(6\Delta)]$ such that $U_f(P_1) - L_f(P_1) < \epsilon/3$ and there exists a partition P_2 of $[y + \epsilon/(6\Delta), b]$ such that $U_f(P_2) - L_f(P_2) < \epsilon/3$. Let P_3 be any partition of $[y - \epsilon/(6\Delta), y + \epsilon/(6\Delta)]$. It is immediate to see that $U_f(P_3) - L_f(P_3) < \epsilon/3$. Thus, $P = P_1 \cup P_2 \cup P_3$ is a partition of $[a, b]$ such that $U_f(P) - L_f(P) < \epsilon$ and we have shown that the function f can be integrated on $[a, b]$. The essential conclusion brought about by the previous reasoning is that the behaviour of the bounded function f at an isolated point is not important to decide whether it can be integrated or not. Since we can always rewrite the integral over $[a, b]$ as the sum of the integrals over the intervals that compose a partition of $[a, b]$ (see Theorem 8.3), we can always deal with a finite number of problematic points, as we did above. Our conclusion is summarised below.

Corollary 8.3 *If f is a bounded function on $[a, b]$ with a finite number of discontinuities and/or with a finite number of points in which it is not monotonic, then $f \in \mathcal{R}([a, b])$.*

8.1.3 Improper Integrals

Under certain conditions, it is possible to extend the notion of definite integral to unbounded intervals, such as $[a, +\infty)$, $(-\infty, a]$, or the whole $\mathbb{R}$.

Definition 8.4 (*Improper unbounded integral*) If $\forall z > 0$, $f \in \mathcal{R}([a, a + z])$, the *improper integral* on $[a, +\infty)$ is defined as

$$\int_a^{+\infty} \mathrm{d}x f(x) = \lim_{z \to +\infty} \int_a^{a+z} \mathrm{d}x f(x),$$

provided that the limit exists. The set of functions (improperly) integrable on $[a, +\infty)$ is denoted by $\mathcal{R}([a, +\infty))$. Analogously, if $\forall z > 0$, $f \in \mathcal{R}([a - z, a])$, the *improper integral* on $(-\infty, a]$ is defined as

$$\int_{-\infty}^a \mathrm{d}x f(x) = \lim_{z \to +\infty} \int_{a-z}^a \mathrm{d}x f(x),$$

provided that the limit exists. The set of functions (improperly) integrable on $(-\infty, a]$ is denoted by $\mathcal{R}((-\infty, a])$. Finally, if both improper integrals on the right-hand side exist, we define

$$\int_{-\infty}^{+\infty} \mathrm{d}x f(x) = \int_0^{+\infty} \mathrm{d}x f(x) + \int_{-\infty}^0 \mathrm{d}x f(x).$$

The set of functions that can be (improperly) integrated on $\mathbb{R}$ is denoted by $\mathcal{R}(\mathbb{R})$.

For the last definition, it is important that both limits exist separately. A simple condition to determine whether a function can be integrated in an improper sense constitutes a new convergence criterion for real number series (see Chap. 5).

Theorem 8.8 *Consider a function $f(x) \geq 0$ decreasing in $[N, +\infty)$, with $N \in \mathbb{N}$. The series $\sum_{n=N}^{\infty} f(n)$ converges if and only if $f \in R([N, +\infty))$.*

Proof Let $n \geq N$. For a monotonically decreasing function, $f(n) \geq f(x) \geq f(n+1)$ for $x \in [n, n+1]$, so that

$$f(n+1) \leq \int_{n}^{n+1} dx f(x) \leq f(n).$$

Summing element by element,

$$\sum_{n=N+1}^{\infty} f(n) \leq \int_{N}^{+\infty} dx f(x) \leq \sum_{n=N}^{\infty} f(n),$$

proving the assertion. $\qquad\qquad\qquad\qquad\qquad\qquad\qquad\qquad\qquad\qquad\qquad\square$

Positive functions can be integrated over unbounded intervals if their asymptotic decrease is sufficiently fast. If a positive function f is integrable in $[a, +\infty)$, any positive function $g > 0$ that is integrable on any bounded interval $[a, b]$ and is asymptotically lower than f, is integrable on $[a, +\infty)$. Similar reasoning applies to $(-\infty, a]$.

▶ **Example 8.5** (*Improper integrals from asymptotic behaviours*) Consider a function $f \in R([a, a+z])$ for any $z > 0$. We want to show that if for some $\alpha > 1$, $\lim_{x \to +\infty} f(x)x^{\alpha} = 0$, then $f \in R([a, +\infty))$. Due to the limit, $\exists x_0 > a$ such that if $x > x_0$, then $|f(x)| < 1/x^{\alpha}$. Consider an integer $n_0 > x_0$. Notice that $\sum_{n=n_0}^{\infty} |f(n)| \leq \sum_{n=n_0}^{\infty} 1/n^{\alpha}$. The term on the right is a generalised harmonic series (see Example 5.21), and for $\alpha > 1$ it is convergent. Hence, the series $\sum_{n=n_0}^{\infty} f(n)$ is absolutely convergent, and consequently, convergent. Using Theorem 8.8 the assertion follows.

Another possible extension of the definite integral involves considering unbounded functions. Since the theory developed so far specifically applies to bounded functions, this is a second case of improper integration.

Definition 8.5 (*Improper bounded integral*) Consider an interval $[a, b]$ and a function f bounded on $[a, b)$ and such that $\lim_{x \to b^-} f(x) = \pm\infty$. If $f \in R([a, z])$ for $z \in (a, b)$ we define the *improper integral*

$$\int_{a}^{b} dx f(x) = \lim_{z \to b^-} \int_{a}^{z} dx f(x),$$

provided that the limit exists. The set of functions that can be (improperly) integrated over $[a, b)$ is denoted $\mathcal{R}([a, b))$. Analogously, if f is bounded on $(a, b]$, such that $\lim_{x \to a^+} f(x) = \pm\infty$ and $f \in \mathcal{R}([z, b])$ for all $z \in (a, b)$ we define the *improper integral*

$$\int_a^b dx f(x) = \lim_{z \to a^+} \int_z^b dx f(x),$$

provided that the limit exists. The set of functions that can be (improperly) integrated over $(a, b]$ is denoted by $\mathcal{R}((a, b])$.

If f is divergent at a point $c \in (a, b)$, it can still be (improperly) integrated over $[a, b]$ and we set

$$\int_a^b dx f(x) = \lim_{z \to c^-} \int_a^z dx f(x) + \lim_{z \to c^+} \int_z^b dx f(x),$$

provided that the two improper integrals on the right-hand side exist.

Cases that mix the situation in Definitions 8.4 and 8.5, such as an improper integral in $(a, +\infty)$ of a function diverging in a, can be reduced to sums of improper integrals of the previous kinds by splitting the integration interval into convenient pieces. What is important to understand is that all integrals in all pieces in which the interval is split must exist independently for the original integral to exist.

8.1.4 Integral of Vector-Valued Functions

The notion of Riemann integral can be easily extended to real functions with values in $\mathbb{R}^n$. The essential idea is that the integration is intended as a component-by-component operation.

Definition 8.6 A vector-valued function $\mathbf{f} : [a, b] \subset \mathbb{R} \to \mathbb{R}^n$ is integrable in $[a, b]$, $\mathbf{f} \in \mathcal{R}((a, b])$, if all its components are integrable functions, $f_i \in \mathcal{R}((a, b])$, and $\int_a^b dx \mathbf{f}(x) = (\int_a^b dx f_1(x), \ldots, \int_a^b dx f_n(x))$.

The image $\mathbf{l} = \mathbf{f}([a, b])$ of a vector-valued function identifies a set of points in the space $\mathbb{R}^n$. If $\mathbf{f} \in C^1([a, b])$, then $\mathbf{l}$ is a *differential curve*. The function $\mathbf{f}$ represents a possible *parametrisation* of the curve $\mathbf{l}$. Any continuously differentiable and invertible function $\phi : \mathbb{R} \to \mathbb{R}$ can be used to define a new parametrisation of $\mathbf{l}$ with $\mathbf{f} \circ \phi$. The theory of integration helps to derive some important quantities related to $\mathbf{l}$. From the Taylor expansion $\mathbf{f}(x + \delta x) - \mathbf{f}(x) = d\mathbf{f}\delta x + o(\delta x)$, we discover that the length of the infinitesimal arc between the points $\mathbf{f}(x)$ and $\mathbf{f}(x + \delta x)$ is proportional to the norm of the differential $\|d\mathbf{f}(x)\| |\delta x|$. From this observation, the following is derived.

Definition 8.7 (*Length of a differential curve*) Let $\mathbf{f} : [a, b] \subset \mathbb{R} \to \mathbb{R}^n$ be a differential curve. Its *length* is defined as $l = \int_a^b dx \|d\mathbf{f}(x)\|$.

Because we assumed $d\mathbf{f}$ to be continuous, Corollary 8.2 guarantees that the length of the curve is well defined.

▶ **Example 8.6** (*Length of the hypercube diagonal*) Consider the function $\mathbf{f} :$ $[0, 1] \to \mathbb{R}^n$ with $f_i(x) = ax$. When x varies from 0 to 1, this function describes a straight line between the origin and the point $(a, \ldots, a)$. This is the diagonal of the nth dimensional hypercube with side length a. In this case, $d\mathbf{f} = (a, \ldots, a)$ and applying Definition 8.7 we obtain the length of the diagonal as $l(a) = \int_0^1 dx \sqrt{na^2} = \sqrt{n}a$.

A differential curve can also be defined implicitly by a condition like $\mathbf{G}(x, \mathbf{f}(x)) = \mathbf{z}_0$, where $\mathbf{G} : E \subseteq \mathbb{R}^{n+1} \to \mathbb{R}^n$, $\mathbf{G} \in C^1(E)$, and $\mathbf{z}_0 \in \mathbb{R}^n$. Theorem 7.14 provides the expression for $d\mathbf{f}$ that can be used to compute the length of the arc of the implicitly defined curve applying Definition 8.7. The parametric description obtained through the implicit function theorem can be only local, so we might be required to use different parameters in different parts of the curve.

8.2 The Fundamental Theorem of Calculus

In practise, even for relatively simple functions, it is difficult to compute the integral using its definition and the limit of appropriately defined partitions, as was done in Examples 8.3 or 8.4. The fundamental result toward a more manageable computation of integrals is their connection to the concept of an "inverse derivative" or "anti-derivative" introduced in the following.

Definition 8.8 (*Primitive function*) Consider a function f defined over an interval $[a, b]$. The function $F(x)$ is a *primitive function* of $f(x)$ in $[a, b]$ if $dF(x)/dx = f(x)$, for any $x \in [a, b]$.

The previous definition does not clarify which functions actually possess a primitive. However, it is clear that if $F_1(x)$ and $F_2(x)$ are primitive functions of f in $[a, b]$, their difference $F_1(x) - F_2(x)$, having derivative zero, is constant in this interval. In other terms, if $F(x)$ is a primitive function of $f(x)$, then $F(x) + c$, with c any real constant, is also a primitive function. A second key notion for the present analysis is the following.

Definition 8.9 (*Integral function*) Let $f \in \mathcal{R}([a, b])$, then for any $x \in [a, b]$ the *integral function* is defined as $F(x) = \int_a^x dz f(z)$.

We can immediately derive an important result.

Corollary 8.4 *The integral function is continuous.*

Proof For any couple of points x and $x + h$ in $[a, b]$ it is

$$|F(x+h) - F(x)| = \left| \int_x^{x+h} dz f(z) \right| \leq h \sup_{[x,x+h]} |f(x)|$$

so that $\lim_{h \to 0} |F(x+h) - F(x)| = 0$.   □

The connection between the primitive function of Definition 8.8 and the integral function of Definition 8.9 is not immediately obvious. The first is defined starting from the notion of derivative and the second, being an integral, depends on the relative positions of the sets of upper and lower sums. The following theorem, which is considered "fundamental" for its importance in the development of modern calculus, clarifies their relationship.

Theorem 8.9 (Fundamental theorem of calculus) *If $f \in C^0([a, b])$, then the integral function $F(x)$ of $f(x)$ in $[a, b]$ is differentiable and is a primitive function of $f(x)$, that is, $dF(x)/dx = f(x)$ for any $x \in [a, b]$.*

Proof We have to show that $\lim_{h \to 0}(F(x+h) - F(x))/h = f(x)$. One has

$$h \inf_{[x,x+h]} f \leq F(x+h) - F(x) = \int_x^{x+h} dz f(z) \leq h \sup_{[x,x+h]} f$$

so that

$$\inf_{[x,x+h]} f \leq \frac{F(x+h) - F(x)}{h} \leq \sup_{[x,x+h]} f.$$

Since f is continuous in x, $\forall \epsilon > 0$, $\exists \delta > 0$ such that $|f(x') - f(x)| < \epsilon$ if $|x' - x| < \delta$. If $h < \delta$, then $f(x) - \epsilon < \inf_{[x,x+h]} f$ and $\sup_{[x,x+h]} f < f(x) + \epsilon$, from which the assertion follows.   □

Due to the previous theorem, the primitive function $F(x)$ is often named the *indefinite integral* of f and is denoted with $\int dx f(x)$. The difference with respect to the definite integral is the lack of any endpoint specification on the integral sign. It is important to note that the integral function is, in general, not differentiable. In fact, to obtain a differentiable integral function, Theorem 8.9 assumes that the function f is continuous.

▶ **Example 8.7** (*Non-differentiable integral function*) Consider the left continuous step function $I : \mathbb{R} \to \mathbb{R}$, with $I(x) = 0$ if $x \leq 0$ and $I(x) = 1$ if $x > 0$. The function I is monotonic and therefore integrable in any interval. Its integral function is

$$F(x) = \int_0^x dz I(z) = \begin{cases} 0 & x < 0, \\ x & x \geq 0. \end{cases}$$

For any interval $[a, b]$, $\int_a^b dz I(z) = F(b) - F(a)$, but F is not differentiable in $x = 0$. The step function is also known as *Heaviside* function, named after the English mathematician Oliver Heaviside (1850–1925), and often denoted by the symbol $\theta(x)$. We will revisit the step function in Example 8.14.

An integrable function that is not continuous might not have a primitive. However, if the function has a finite number of discontinuities, a primitive can be defined in all the intervals between the discontinuities. At the same time, the fact that a function has a primitive in a given interval does not imply that it is Riemann integrable. For example, the function of Example 6.11 is differentiable in $[-1, 1]$, but its derivative is unbounded and consequently not Riemann integrable.

▶ **Example 8.8** (*Differentiable and Lipschitz continuous functions*) We can use the theory of integration to prove that a continuously differentiable function is Lipschitz continuous. Consider an interval (a, b) and assume $f \in C^1((a, b))$. Then according to Theorem 8.9,

$$f(y) = f(x) + \int_x^y dz f'(z), \quad \forall x, y \in (a, b).$$

Define $M = \sup_{x \in (a,b)} |f'(x)|$. This quantity exists because f' is continuous. Then $\forall x, y \in (a, b)$,

$$|f(y) - f(x)| = \left| \int_x^y dz f'(z) \right| \leq \int_x^y dz |f'(z)| \leq M |y - x|.$$

We can conclude that the function f is Lipschitz continuous in (a, b).

The practical relevance of Theorem 8.9 is that knowing a primitive function, we are able to compute the integral on any domain in which the function is defined. If $F(x)$ is the primitive function of $f(x)$ in $[a, b]$, then for any couple of points $x_1 \leq x_2$ in $[a, b]$,

$$\int_{x_1}^{x_2} dz f(z) = F(x_2) - F(x_1).$$

At least for the large class of piecewise continuous functions, the problem of computing the integral is reduced to the problem of identifying a primitive.

▶ **Example 8.9** (*Basic indefinite integrals*) The following list is obtained starting from the expression of the derivative of well-known functions (see Examples 6.5, 6.6, and 6.8). The primitive functions are defined at the points in which the original functions are continuous. Primitive functions are often reported adding a $+c$, to remind the reader that they are not unique. Let us consider this point understood and omit $+c$.

$$\int dx\, x^\alpha = \frac{1}{\alpha+1}x^{\alpha+1}, \alpha \neq -1 \qquad \int dx\frac{1}{x} = \log|x| \qquad \int dx\, e^x = e^x$$

$$\int dx\, a^x = \frac{1}{\log a}a^x \qquad \int dx\, \log x = x\log x - x \qquad \int dx\, \sin x = -\cos x$$

$$\int dx\, \cos x = \sin x \qquad \int dx\, \tan x = \log|\sec x| \qquad \int dx\, \sinh x = \cosh x$$

$$\int dx\, \cosh x = \sinh x \qquad \int dx\frac{a}{a^2+x^2} = \arctan\frac{x}{a} \qquad \int dx\frac{a}{a^2-x^2} = \frac{1}{2}\log\left|\frac{x+a}{x-a}\right|$$

$$\int dx\frac{1}{\sqrt{a^2-x^2}} = \arcsin\frac{x}{a} \qquad \int dx\frac{1}{\sqrt{x^2-a^2}} = \mathrm{arccosh}\frac{x}{a} = \log(x+\sqrt{x^2-a^2})$$

$$\int dx\frac{a}{x\sqrt{x^2-a^2}} = \mathrm{arcsec}\frac{x}{a} \qquad \int dx\frac{1}{\sqrt{x^2+a^2}} = \mathrm{arcsinh}\frac{x}{a} = \log(x+\sqrt{x^2+a^2}).$$

There are a number of rules that can be applied to the identification of a primitive function and to the integration of complex expressions, but two are the most widely used. The first has to do with the chain rule derived in Theorem 6.6.

Theorem 8.10 (Integration by substitution) *Let $h(x) \in C^1([a,b])$ be strictly increasing. Then if f is continuous in $[h(a), h(b)]$,*

$$\int_{h(a)}^{h(b)} dx\, f(x) = \int_a^b dx\, h'(x) f(h(x)).$$

Proof Since $f(x)$ is continuous, it has a primitive function $F(x)$. Then the left-hand side can be written as $F(h(b)) - F(h(a))$. From the chain rule, $dF(h(x))/dx = h'(x) f(h(x))$. Thus, $F(h(x))$ is the primitive function of the integrand on the right-hand side and the statement follows. □

To exploit Theorem 8.10 one tries to change the variable of integration through a transformation $h(x)$ that makes the expression of the integrand similar to a known derivative.

▶ **Example 8.10** (*Finding a primitive function by substitution*) Given two constants, a and b, we want to compute the indefinite integral $\int dx\, \log(ax+b)$ when $x > -b/a$. Consider $y = h(x) = ax + b$ so that $x = (y - b)/a$. According to Theorem 8.10, replace dx with dy/a so that the integral is reduced to $1/a \int dy \log y$ which, checking the table in Example 8.9, is just $y(\log y - 1)/a$. Substituting the definition of y, one recovers the primitive we are interested into

$$\int dx \log(ax + b) = \left(x + \frac{b}{a}\right)(\log(ax + b) - 1).$$

▶ **Example 8.11** (*Line integral of a scalar field*) Consider a differential curve $\mathbf{l} \subset \mathbb{R}^n$, parameterised by a function $\mathbf{f} : [a, b] \subset \mathbb{R} \to \mathbb{R}^n$, and a scalar field $F : \mathbb{R}^n \to \mathbb{R}$. The composition $F \circ \mathbf{f}$ is a real-valued function. In principle, we can integrate this function over the curve. If $F \circ \mathbf{f} \in \mathcal{R}((a, b])$ the *line integral* of F over $\mathbf{l}$ is $\int_{\mathbf{l}} F = \int_a^b dx \|d\mathbf{f}\| F(\mathbf{f}(x))$. Note that this definition does not depend on the specific parameter that is adopted to describe the curve. In fact, if $\phi : \mathbb{R} \to \mathbb{R}$ is a continuously differentiable and invertible function, according to Theorem 8.10, $\int_{\mathbf{l}} F = \int_{\phi^{-1}(a)}^{\phi^{-1}(b)} dx \phi'(x) \|d\mathbf{f}\| F(\mathbf{f}(\phi(x)))$, and, from the chain rule, $\phi'(x)\|d\mathbf{f}\| = \|d\mathbf{f} \circ \phi\|$.

The second rule is related to the formula for the derivative of a product of two functions, presented in Theorem 6.5.

Theorem 8.11 (Integration by part) *Consider $f, g \in \mathcal{R}([a, b])$ and assume that their primitive functions F and G exist, then*

$$\int_a^b dx f(x)G(x) + \int_a^b dx F(x)g(x) = F(b)G(b) - F(a)G(a).$$

Proof The proof follows directly from observing that $F(x)G(x)$ is a primitive function of $f(x)G(x) + F(x)g(x)$. ☐

▶ **Example 8.12** (*Reduction formulae*) Let n be a natural number, we want to compute the integral $I_n(x) = \int_0^x dz z^n e^{-z}$. In the table of Example 8.9 we read that z^n is a primitive function of nz^{n-1} and, with a simple change of variables, $-e^{-z}$ is a primitive function of e^{-z}. Applying Theorem 8.11 with $F(z) = z^n$ and $G(z) = -e^{-z}$,

$$I_n(x) - n I_{n-1}(x) = F(x)G(x) - F(0)G(0) = -x^n e^{-x},$$

which provides a recursive rule to compute the desired integral for any n, once we observe that $I_0(x) = G(x) - G(0) = 1 - e^{-x}$. This kind of recursive rule is common in integral calculus. They are known as *reduction formulae*.

We conclude this section by providing a few results on the limit and derivative of the integral function that can often be encountered in applications. If a function is continuous, the limit of the function is the function computed at the limit. This is also true for the integral if the integrand is continuous.

Lemma 8.1 *Consider a closed interval $I \subset \mathbb{R}$ and $t_0 \in I$. Let $f(x, t) \in C^0([a, b] \times I)$. Then $\lim_{t \to t_0} \int_a^b dx f(x, t) = \int_a^b dx f(x, t_0)$.*

Proof Since the function $f(x, t)$ is continuous in the compact set $[a, b] \times I$, it is uniformly continuous. Hence, $\forall \epsilon > 0$, $\exists \delta > 0$ such that if $\mathbf{x_1} = (x_1, t_1)$ and $\mathbf{x_2} = (x_2, t_2)$ are points in $[a, b] \times I$ such that $\|\mathbf{x_1} - \mathbf{x_2}\| < \delta$, $|f(x_1, t_2) - f(x_2, t_2)| < \epsilon$. In particular, this means that if $|t - t_0| < \delta$, then $|f(x, t) - f(x, t_0)| < \epsilon/(b - a)$, $\forall x$. Then, the statement follows by noting that

$$\left| \int_a^b dx f(x, t) - \int_a^b dt f(x, t_0) \right| \le$$
$$\int_a^b dx \, |f(x, t) - f(x, t_0)| \le \frac{\epsilon}{b - a} \int_a^b dx - \epsilon.$$

In Lemma 8.1, continuity is important because it guarantees that $f(x, t)$ is integrable in x for any value of t, including t_0. Another operation which is often necessary is the computation of the derivative of a function which is defined through an integral. The variable with respect to which the derivative is required might be the value of a parameter in the integrand or in the endpoints of integration (or both). If the involved functions are sufficiently smooth, one can pass the derivative under the integral and exploit the functional dependence at the endpoints.

Theorem 8.12 (Leibnitz integral rule) *Consider $f(x, t) \in C^1(I \times J)$ with I and J closed real intervals and two functions $a(x), b(x) \in C^1(I)$ with images in J. Then if $F(x) = \int_{a(x)}^{b(x)} dt f(x, t)$,*

$$\frac{d}{dx} F(x) = b'(x) f(x, b(x)) - a'(x) f(x, a(x)) + \int_{a(x)}^{b(x)} dt \partial_x f(x, t).$$

Proof Consider a sufficiently small $h > 0$. Without loss of generality we can assume $a(x) < b(x)$. By adding and removing $f(x + h, t)$ in the second integral and rearranging terms,

$$\frac{F(x + h) - F(x)}{h} = \frac{1}{h} \int_{b(x)}^{b(x+h)} dt f(x + h, t) +$$
$$- \frac{1}{h} \int_{a(x)}^{a(x+h)} dt f(x + h, t) + \int_{a(x)}^{b(x)} dt \frac{f(x + h, t) - f(x, t)}{h}.$$

Consider the first integral on the right-hand side. For the mean value theorem there exists a $\lambda_x \in [0, 1]$ such that $b(x+h) - b(x) = hb'(x + \lambda_x h)$. Assuming $b(x+h) > b(x)$, consider the interval $K = [b(x), b(x + h)]$ and note that

$$b'(x + \lambda_x h) \inf_{t \in K} f(x{+}h, t) \le$$

$$\frac{1}{h} \int_{b(x)}^{b(x+h)} dt f(x + h, t) \le b'(x + \lambda_x h) \sup_{t \in K} f(x + h, t).$$

If instead $b(x + h) < b(x)$, just exchange the order of the endpoints in the definition of K. For the continuity of f, when $h \to 0$, the interval K reduces to the single point $b(x)$. Thus, both the infimum and the supremum converge to $f(x, b(x))$, and the integral expression converges to the first term of the sum in the statement. For the second term, we can apply precisely the same consideration. For the third and last term, by the mean value theorem, $\exists \theta_x \in [0, 1]$ such that

$$\int_{a(x)}^{b(x)} dt \, \frac{f(x + h, t) - f(x, t)}{h} = \int_{a(x)}^{b(x)} dt \, \partial_x f(x + \theta_x h, t).$$

For the continuity of the partial derivative, and by applying Lemma 8.1, the expression in the statement is recovered when $h \to 0$. ☐

The previous result is named after one of the fathers of modern calculus, the German mathematician Gottfried Leibniz (1646–1716).

8.3 Riemann–Stieltjes Integral

In this section, we study how to extend the notion of integral to include the possibility of "weighting" the points in the integration interval by a given measure or density expressed through a nondecreasing real function α. Let us start by introducing a suitable definition for the lower and upper sums.

Definition 8.10 (*Upper and lower sum*) Consider a nondecreasing function α in the interval $[a, b]$. Given a bounded function f in the same interval, for any finite partition P of $[a, b]$, the *upper sum* of the function f with respect to the *Stieltjes measure* α is defined as

$$U_f(P, \alpha) = \sum_{i=1}^{N} (\alpha(x_i) - \alpha(x_{i-1})) \sup_{[x_{i-1}, x_i]} f,$$

while the *lower sum* is

$$L_f(P, \alpha) = \sum_{i=1}^{N} (\alpha(x_i) - \alpha(x_{i-1})) \inf_{[x_{i-1}, x_i]} f,$$

where $\sup_{[x,y]} f$ and $\inf_{[x,y]} f$ represent the supremum and infimum of the function f in the interval $[x, y]$.

Since the function α is nondecreasing, the quantities $\alpha(x_i) - \alpha(x_{i-1})$ that appear in the lower and upper sums are nonnegative. While the Riemann definition of the upper and lower sum depends on the partition P of the interval $[a, b]$, the Stieltjes definition, named after the Dutch mathematician Thomas Stieltjes (1856–1894), depends on both the partition and the weighting function α. Changing the latter, in general, changes the value of the sum. If $\alpha(x) = x$, then we return to the Riemann case. Similarly to the Riemann case, the upper sum cannot increase and the lower sum cannot decrease when the partition is refined.

Theorem 8.13 *Let P' be a refinement of partition P, $P \subseteq P'$, then $U_f(P, \alpha) > U_f(P', \alpha)$ and $L_f(P, \alpha) \leq L_f(P', \alpha)$.*

Proof Consider a point $y \in (x_{i-1}, x_i)$ with $x_{i-1}, x_i \in P$ and $y \in P'/P$. For any interval $[x, y]$, define $\Delta_\alpha(x, y) = \alpha(y) - \alpha(x)$. In analogy to the Riemann case,

$$\Delta_\alpha(x_{i-1}, x_i) \sup_{[x_{i-1}, x_i]} f \geq \Delta_\alpha(x_{i-1}, y) \sup_{[x_{i-1}, y]} f + \Delta_\alpha(y, x_i) \sup_{[y, x_i]} f,$$

and because

$$U_f(P \cup \{y\}, \alpha) = U_f(P, \alpha) + \Delta_\alpha(y, x_i) \sup_{[x_{i-1}, y]} f +$$
$$+ \Delta_\alpha(x_{i-1}, y) \sup_{[y, x_i]} f - \Delta_\alpha(x_{i-1}, x_i) \sup_{[x_{i-1}, x_i]} f,$$

we have $U_f(P, \alpha) \geq U_f(P \cup \{y\}, \alpha)$. Using a similar argument based on the properties of the infimum, it is immediate to see that $L_f(P, \alpha) \leq L_f(P \cup \{y\}, \alpha)$. Each time a new y is added to the partition, the upper sum may decrease and the lower sum may increase. By repeating the procedure, adding any $y \in P'/P$, the assertion follows. □

Let $\Delta_\alpha = \alpha(b) - \alpha(a)$. For any partition P, the upper and lower sums of f in $[a, b]$ are bounded above by $\Delta_\alpha \sup_{[a,b]} f$ and below by $\Delta_\alpha \inf_{[a,b]} f$. Hence, $\mathbb{U}_f(\alpha) = \{U_f(P, \alpha) \mid P \in \mathbb{P}([a, b])\}$ and $\mathbb{L}_f(\alpha) = \{L_f(P, \alpha) \mid P \in \mathbb{P}([a, b])\}$, are bounded sets. In analogy to the Riemann case, they are also separated.

Theorem 8.14 *For any $x \in \mathbb{U}_f(\alpha)$ and any $y \in \mathbb{L}_f(\alpha)$, $x \geq y$.*

Proof The proof is identical to the proof of Theorem 8.2. □

Definition 8.11 (*Stieltjes integral*) If $\sup \mathbb{L}_f(\alpha) = \inf \mathbb{U}_f(\alpha)$ then the function f is Stieltjes integrable on $[a, b]$ with respect to the measure α. The *integral* of the function f, with respect to α, on $[a, b]$, indicated by $\int_a^b d\alpha(x) f(x)$, is the unique element that separates the sets of lower and upper sums, that is, sup $\mathbb{L}_f(\alpha)$ and inf $\mathbb{U}_f(\alpha)$. The set of integrable functions on $[a, b]$ with respect to α is denoted by $\mathcal{R}_\alpha([a, b])$.

We can avoid the direct specification of the variable x (recall that it is just an index) and write $\int_a^b d\alpha f$. In general, sup $\mathbb{L}_f$ is the *lower integral* and inf $\mathbb{U}_f$ is the *upper integral* of the function F in the interval $[a, b]$ with respect to α. The previous definition can be restated as saying that the function is integrable if its upper and lower integrals are equal.

▶ **Example 8.13** (*Stieltjes integral of the constant function*) Consider the constant function $f(x) = c$. It is immediate to see that $f \in \mathcal{R}_\alpha([a, b])$ for any nondecreasing α and any interval $[a, b]$ and that $\int_a^b d\alpha c = (\alpha(b) - \alpha(a))c$. The constant function can be integrated with respect to any measure in any interval. At the same time, if the measure is constant $\alpha = c$, $\int_a^b d\alpha f = 0$ for any bounded function f. Any bounded function can be integrated with respect to the constant Stieltjes measure in any interval.

8.3.1 Stieltjes Integrable Functions

Analogously to what has been done with the Riemann integral, it is possible to identify a few classes of functions that can be integrated. We start by exploring the relationship between continuity and integrability.

Theorem 8.15 *If f is bounded in $[a, b]$ and continuous except in a finite number of points in which α is continuous, then $f \in \mathcal{R}_\alpha([a, b])$.*

Proof If α or f are constant, the theorem is trivial. We assume that they are increasing and that f has N discontinuities. For any interval $[x, y]$, define $\Delta_f(x, y) = \sup_{[x,y]} f - \inf_{[x,y]} f$, $\Delta_\alpha(x, y) = \alpha(y) - \alpha(x)$, $\Delta_f = \Delta_f(a, b) > 0$, and $\Delta_\alpha = \Delta_\alpha(a, b) > 0$. We have to prove that $\forall \epsilon > 0$ there exists a partition P of $[a, b]$ such that $U_f(P, \alpha) - L_f(P, \alpha) < \epsilon$.

We can build a partition P such that a single discontinuity of f is at most contained in any subinterval. Since in a neighbourhood of any discontinuity of f, α is continuous, we can always reduce the width of the intervals that contain one discontinuity of f such that $\Delta_\alpha(x_{i-1}, x_i) < \epsilon/(2\Delta_f N)$. In these intervals

$$\Delta_\alpha(x_{i-1}, x_i)\Delta_f(x_{i-1}, x_i) < \epsilon/(2N).$$

In all other intervals, the function f is uniformly continuous, so we can always find a sufficiently fine partition for which $\Delta_f(x_{i-1}, x_i) < \epsilon/(2\Delta_\alpha)$. In these intervals

$$\Delta_\alpha(x_{i-1}, x_i)\Delta_f(x_{i-1}, x_i) < \epsilon\Delta_\alpha(x_{i-1}, x_i)/(2\Delta_\alpha).$$

Let A be the set of indexes relative to the N subintervals that contain one discontinuity of f and B be the set of indexes relative to the intervals in which f is continuous. Consider the two quantities

$$D_A = \sum_{i \in A} \Delta_\alpha (x_{i-1}, x_i) \Delta_f (x_{i-1}, x_i) < \sum_{i \in I} \frac{\epsilon}{2N} = \frac{\epsilon}{2},$$

and

$$D_B = \sum_{i \in B} \Delta_\alpha (x_{i-1}, x_i) \Delta_f (x_{i-1}, x_i) < \sum_{i=1}^{N} \frac{\epsilon \Delta_\alpha (x_{i-1}, x_i)}{2 \Delta_\alpha} = \frac{\epsilon}{2}.$$

Because $U_f(P, \alpha) - L_f(P, \alpha) = D_A + D_B < \epsilon$, the assertion follows. $\qquad\square$

In particular, if $f \in C^0([a, b])$, then f is integrable on $[a, b]$ with respect to any measure α.

▶ **Example 8.14** (*Step function as measure*) Consider the left continuous *step function* of Example 8.7. We want to prove that if $f \in C^0([a, b])$ and $z \in [a, b]$, then $\int_a^b dI(x - z) f(x) = f(z)$. For any partition P of $[a, b]$, if $[x_{i-1}, x_i]$ is the interval that contains z, it is

$$U_f(P, I(x - z)) - L_f(P, I(x - z)) = \sup_{[x_{i-1}, x_i]} f - \inf_{[x_{i-1}, x_i]} f.$$

In all other intervals, the measure is constant, so that their contribution to the lower and upper sums is zero. Since f is continuous, it is uniformly continuous, so that for any $\epsilon > 0$ we can find a partition P with a sufficiently small interval $[x_{i-1}, x_i]$ such that

$$\sup_{[x_{i-1}, x_i]} f - \inf_{[x_{i-1}, x_i]} f = \max_{[x_{i-1}, x_i]} f - \min_{[x_{i-1}, x_i]} f < \epsilon.$$

In this partition, $\max_{[x_{i-1}, x_i]} f - f(z) < \epsilon$, that is $U_f(P, I(x - z)) - f(z) < \epsilon$. This proves that the integral exists and its value is precisely $f(z)$.

Analogously to the Riemann case, the composition of a continuous function with an integrable function is integrable.

Theorem 8.16 *If* $f \in R_\alpha([a, b])$ *and* $g \in C^0([\inf_{[a,b]} f, \sup_{[a,b]} f])$ *then* $g \circ f \in R_\alpha([a, b])$.

Proof The following proof is essentially identical to the Riemann case but we repeat it for completeness. If g, f, or α are constant, the proof is trivial. We assume that they are not constant. For any interval $[x, y]$, define $\Delta_f(x, y) = \sup_{[x,y]} f - \inf_{[x,y]} f$, $\Delta_{g \circ f}(x, y) = \sup_{[x,y]} g \circ f - \inf_{[x,y]} g \circ f$, and $\Delta_\alpha(x, y) = \alpha(y) - \alpha(x)$. Set $m = \inf_{[a,b]} f$, $M = \sup_{[a,b]} f$, $\Delta_{g \circ f} = \Delta_{g \circ f}(a, b) > 0$, and $\Delta_\alpha = \Delta_\alpha(a, b) > 0$.

Consider any $\epsilon > 0$. We have to prove that there exists a partition P of $[a, b]$ such that $U_{g \circ f}(P, \alpha) - L_{g \circ f}(P, \alpha) < \epsilon$. Since g is continuous in $[m, M]$, it is uniformly continuous. Thus, there exists a $\delta > 0$ such that for any $x, y \in [m, M]$, if $|x - y| < \delta$, $|g(x) - g(y)| < \epsilon/(2\Delta_\alpha)$. At the same time, since f is integrable, there exists a partition P such that $U_f(P, \alpha) - L_f(P, \alpha) < \delta\epsilon/(2\Delta_{g \circ f})$. We can show that this is actually the desired partition.

The intervals of P can be divided into two groups: A and B. If $\Delta_f(x_{i-1}, x_i) < \delta$, then $i \in A$, otherwise $i \in B$. Note that

$$\sum_{i \in B} \Delta_\alpha(x_{i-1}, x_i)\delta \le \sum_{i \in B} \Delta_\alpha(x_{i-1}, x_i)\Delta_f(x_{i-1}, x_i) \le \frac{\delta\epsilon}{2\Delta_{g \circ f}},$$

so that $\sum_{i \in B} \Delta_\alpha(x_{i-1}, x_i) \le \epsilon/(2\Delta_{g \circ f})$. At the same time, if $i \in A$, the distance of the image of the points in $[x_i, x_{i-1}]$ is always less than δ, so that $\Delta_{g \circ f}(x_{i-1}, x_i) < \epsilon/(2\Delta_\alpha)$. Considering separately the summation on the two groups of indexes, define

$$D_A = \sum_{i \in A} \Delta_\alpha(x_{i-1}, x_i)\Delta_{g \circ f}(x_{i-1}, x_i) \le \frac{\epsilon}{2\Delta_\alpha} \sum_{i \in A} \Delta_\alpha(x_{i-1}, x_i) \le \frac{\epsilon}{2},$$

and

$$D_B = \sum_{i \in B} \Delta_\alpha(x_{i-1}, x_i)\Delta_{g \circ f}(x_{i-1}, x_i) \le \Delta_{g \circ f} \sum_{i \in B} \Delta_\alpha(x_{i-1}, x_i) \le \frac{\epsilon}{2}.$$

Summing both inequalities, we obtain $U_{g \circ f}(P, \alpha) - L_{g \circ f}(P, \alpha) = D_A + D_B < \epsilon$, which proves the assertion. $\qquad\square$

▶ **Example 8.15** (*Integral of a continuous measure with respect to itself*) Assume that the measure α is continuous. We want to compute $\int_a^b d\alpha\alpha$. Since α is continuous, it can be integrated with respect to any measure (c.f Theorem 8.15) and, in particular, with respect to α. To compute the value of the integral notice that for any partition P it is

$$U_\alpha(P, \alpha) + L_\alpha(P, \alpha) = \sum_{i=1}^{n}(\alpha(x_i) - \alpha(x_{i-1}))(\alpha(x_i) + \alpha(x_{i-1})) =$$

$$\sum_{i=1}^{n} \alpha(x_i)^2 - \sum_{i=1}^{n} \alpha(x_{i-1})^2 = \alpha(b)^2 - \alpha(a)^2.$$

For any $\epsilon > 0$ there exists a partition P such that $U_\alpha(P, \alpha) - L_\alpha(P, \alpha) < \epsilon$, so

$$\frac{U_\alpha(P, \alpha) + L_\alpha(P, \alpha)}{2} - \frac{\epsilon}{2} \le L_\alpha(P, \alpha) \le \int_a^b d\alpha\alpha$$

$$\le U_\alpha(P, \alpha) \le \frac{U_\alpha(P, \alpha) + L_\alpha(P, \alpha)}{2} + \frac{\epsilon}{2},$$

which implies that

$$\int_a^b d\alpha\alpha = \frac{U_\alpha(P,\alpha) + L_\alpha(P,\alpha)}{2} = \frac{1}{2}\alpha(b)^2 - \frac{1}{2}\alpha(a)^2.$$

In general, this result cannot be extended to discontinuous measures. To see it, consider the step function $I(x)$ of Example 8.14. For any partition P of $[-1, 1]$ let x_i be the lowest point greater than 0. Outside the interval $[x_{i-1}, x_i]$ the function I is constant, thus $\forall j \neq i$, $I(x_j) - I(x_{j-1}) = 0$. Hence, we have

$$U_I(P, I) - L_I(P, I) = (I(x_i) - I(x_{i-1}))^2 = 1,$$

which shows that, in this case, the set of upper sums and the set of lower sums cannot be contiguous.

Next, we move to the study of the relation between monotonicity and integrability, extending Theorem 8.7. As already seen in Theorem 8.15, the discontinuities in the measure function are special points that should be treated with particular care.

Theorem 8.17 *If f is monotonic in $[a, b]$ and continuous in the points in which α is discontinuous, then $f \in \mathcal{R}_\alpha([a, b])$.*

Proof If f or α are constant, the result has already been proved. Assume that both functions are not constant and f is increasing. For any interval $[x, y]$, define $\Delta_f(x, y) = \sup_{[x,y]} f - \inf_{[x,y]} f$, $\Delta_\alpha(x, y) = \alpha(y) - \alpha(x)$, $\Delta_f = \Delta_f(a, b) > 0$, and $\Delta_\alpha = \Delta_\alpha(a, b) > 0$. Consider $\epsilon > 0$ and start with a partition P such that, in any interval $[x_{i-1}, x_i]$, at most one between α and f is discontinuous. If f is continuous in a given interval, the partition can be refined so that $\Delta_f(x_{i-1}, x_i) < \epsilon/(2\Delta_\alpha)$. Using the refined partition,

$$\Delta_\alpha(x_{i-1}, x_i)\Delta_f(x_{i-1}, x_i) < \Delta_\alpha(x_{i-1}, x_i)\epsilon/(2\Delta_\alpha).$$

If α is continuous in a given interval, the partition can be further refined such that $\Delta_\alpha(x_{i-1}, x_i) < \epsilon/(2\Delta_f)$. Using the refined partition,

$$\Delta_\alpha(x_{i-1}, x_i)\Delta_f(x_{i-1}, x_i) < \Delta_f(x_{i-1}, x_i)\epsilon/(2\Delta_f).$$

Let A be the set of indexes relative to the subintervals in which α is discontinuous and B the set of indexes relative to the intervals in which α is continuous in the refined partition $\tilde{P}$. Consider the difference of the upper and lower sums computed on these sets separately:

$$D_A = \sum_{i \in A} \Delta_\alpha(x_{i-1}, x_i)\Delta_f(x_{i-1}, x_i) \leq \sum_{i \in I} \frac{\epsilon \Delta_\alpha(x_{i-1}, x_i)}{2\Delta_\alpha} < \frac{\epsilon}{2},$$

and

$$D_B = \sum_{i \in B} \Delta_\alpha(x_{i-1}, x_i)\Delta_f(x_{i-1}, x_i) \le \sum_{i \in B} \frac{\epsilon \Delta_f(x_{i-1}, x_i)}{2\Delta_f} < \frac{\epsilon}{2}.$$

Because $U_f(\tilde{P}, \alpha) - L_f(\tilde{P}, \alpha) = D_A + D_B < \epsilon$, the assertion is proved. The proof is similar for a decreasing function f. $\square$

The number of discontinuities of α and f can be infinite, as long as they belong to a subinterval of $[a, b]$ in which the other function is continuous. This might seem peculiar compared to Theorem 8.15. However, notice that in Theorem 8.17 the function f is required to be monotonic. This is a powerful assumption. In fact, notice that if f is a monotonically increasing function in $[a, b]$, for any partition P

$$U_f(P, \alpha) - L_f(P, \alpha) = U_\alpha(P, f) - L_\alpha(P, f).$$

That is, f is integrable with respect to α if and only if α is integrable with respect to f. But don't be mistaken: the value of the two integrals is in general not the same.

If the measure function is differentiable and its derivative can be Riemann integrated, the Stieltjes integral reduces to an integral of Riemann type.

Theorem 8.18 *Assume $\alpha(x)$ is derivable in $[a, b]$ and its derivable is Riemann integrable, $\alpha' \in \mathcal{R}([a, b])$. Then $\int_a^b d\alpha f$ exists if and only if $\int_a^b dx\alpha'(x)f(x)$ exists and they are equal.*

Proof We will prove that the integrals in the statement share the same upper and lower integrals. Let $M = \sup_{[a,b]} |f|$ and for any partition P, $\Delta x_i = x_i - x_{i-1}$, $\Delta\alpha_i = \alpha(x_i) - \alpha(x_{i-1})$. Because $\alpha' \in \mathcal{R}([a, b])$, $\forall \epsilon > 0$ there exists a partition P such that

$$U_{\alpha'}(P) - L_{\alpha'}(P) = \sum_{i=1}^N \Delta x_i \left(\sup_{[x_{i-1}, x_i]} \alpha' - \inf_{[x_{i-1}, x_i]} \alpha' \right) < \frac{\epsilon}{M}.$$

Moreover, because $\alpha(x)$ is derivable in $[a, b]$, according to Theorem 6.10, for any interval $[x_{i-1}, x_i]$, $\exists z_i \in [x_{i-1}, x_i]$ such that $\Delta\alpha_i = \Delta x_i \alpha'(z_i)$. Thus, on P, using the properties of the supremum,

$$\left| U_f(P, \alpha) - U_{\alpha'f}(P) \right| = \left| \sum_{i=1}^N \Delta x_i \left(\alpha'(z_i) \sup_{[x_{i-1}, x_i]} f - \sup_{[x_{i-1}, x_i]} \alpha'f \right) \right|$$

$$\le \sum_{i=1}^N \Delta x_i \left| \alpha'(z_i) \sup_{[x_{i-1}, x_i]} f - \sup_{[x_{i-1}, x_i]} \alpha'f \right| \le$$

$$\leq \sum_{i=1}^{N} \Delta x_i \sup_{[x_{i-1}, x_i]} \left| (\alpha'(z_i) - \alpha') f \right| \leq \sum_{i=1}^{N} \Delta x_i \sup_{[x_{i-1}, x_i]} \left| \alpha'(z_i) - \alpha' \right| M \leq$$

$$\leq \sum_{i=1}^{N} \Delta x_i \left(\sup_{[x_{i-1}, x_i]} \alpha' - \inf_{[x_{i-1}, x_i]} \alpha' \right) M < \epsilon.$$

In the same way it is possible to show that $\forall \epsilon > 0$, there exists a partition P such that $\left| L_f(P, \alpha) - L_{\alpha' f}(P) \right| < \epsilon$. The statement is thus proved. $\square$

The hypotheses of Theorem 8.18 are valid if $\alpha(x) \in C^1([a, b])$. The next example is essentially a restatement of Theorem 8.10.

▶ **Example 8.16** (*Integral of a function of the measure*) Assume $\alpha \in C^1([a, b])$. We want to compute $\int_a^b d\alpha f(\alpha)$, where f is a continuous function with a primitive F. By Theorem 8.18, $\int_a^b d\alpha f(\alpha) = \int_a^b dx \alpha'(x) f(\alpha(x))$. The integrand is a product of continuous and integrable function, so it is integrable. Note that $F(\alpha(x))$ is a primitive of the integrand, so we easily obtain $\int_a^b d\alpha f(\alpha) = F(\alpha(b)) - F(\alpha(a))$.

8.3.2 Properties of the Stieltjes Integral

A series of useful properties of the Stieltjes integral are collected below. Their proofs are trivial or essentially identical to the proofs of the same properties for the Riemann integral and are left to the reader.

Theorem 8.19 *Let* $f, g \in \mathcal{R}_\alpha([a, b])$ *then*

1. $f + g \in \mathcal{R}_\alpha([a, b])$;
2. $fg \in \mathcal{R}_\alpha([a, b])$;
3. $cf \in \mathcal{R}_\alpha([a, b])$ *with* c *any real constant;*
4. $f \in \mathcal{R}_{c\alpha}([a, b])$ *with* c *any real constant and* $\int_a^b d(c\alpha) f = c \int_a^b d\alpha f$;
5. $|f| \in \mathcal{R}_\alpha([a, b])$ *and* $\left| \int_a^b d\alpha f \right| \leq \int_a^b d\alpha |f|$;
6. *if* $f(x) \leq g(x) \ \forall x \in [a, b]$, *then* $\int_a^b d\alpha f \leq \int_a^b d\alpha g$;
7. *if* $c \in [a, b]$ *then* $f \in \mathcal{R}_\alpha([a, c])$ *and* $f \in \mathcal{R}_\alpha([c, b])$ *and* $\int_a^c d\alpha f + \int_c^b d\alpha f = \int_a^b d\alpha f$;
8. *if* $f \in \mathcal{R}_\beta([a, b])$, *then* $f \in \mathcal{R}_{\alpha+\beta}([a, b])$ *and* $\int_a^b d(\alpha+\beta) f = \int_a^b d\alpha f + \int_a^b d\beta f$.

▶ **Example 8.17** (*Distribution function*) The *distribution function* $F(x)$ of a (continuous or discrete) *random variable* (r.v.) X is defined as $F(x) = \text{Prob}\{X \leq x\}$. The distribution function is nonnegative and nondecreasing. If x_0 and x_1 are respectively the smaller and larger value that the r.v. X can take, it is $F(x) = 0$ for $x < x_0$ and

$F(x) = 1$ for $x \geq x_1$. The set $[x_0, x_1]$ is called the *support* of the r.v. Note that, in general, $x_0, x_1 \in \bar{\mathbb{R}}$ (see Sect. 5.3). The kth *moment* of the random variable X is defined as $M_k = \int dF x^k$, where the interval of integration covers the r.v. support.

As an example of a discrete random variable, consider a lottery that pays the amount $(x_1, \ldots, x_n)$ with probabilities $(p_1, \ldots, p_n)$. With reference to Example 8.14, it is immediate to see that the distribution function can be written as $F(x) = \sum_{i=1}^{n} p_i I(x - x_i)$. Using the last property of Theorem 8.19,

$$M_k = \sum_{i=1}^{n} p_i \int dI(x - x_i) x^k = \sum_{i=1}^{n} p_i x_i^k.$$

Analogously to the Riemann integral, we can also introduce the notion of improper integral for the Stieltjes case. The definition is identical to Definition 8.4, where the limit of proper Riemann integrals is replaced with the limit of proper Stieltjes integrals.

▶ **Example 8.18** (*Exponential distribution*) A random variable X is exponentially distributed in $[0, +\infty)$ if its distribution function is

$$F(x) = \begin{cases} 0 & x < 0, \\ 1 - e^{-\lambda x} & x \geq 0. \end{cases}$$

In this case,case, the central moments introduced in Example 8.17 are improper Stieltjes integral $M_k = \int_0^\infty dF x^k$. Let us see how to compute them. According to the definition of an improper integral, we have to compute $M_k(L) = \int_0^L dF x^k$ for a (large) real number L, and then consider the limit when L becomes infinite. First, notice that the Stieltjes measure $F(x)$ in $[0, L)$ is derivable with continuous derivative $F'(x) = \lambda e^{-\lambda x}$. According to Theorem 8.18 the (proper) Stieltjes integral above can be reduced to a (proper) Riemann integral

$$M_k(L) = \int_0^L dx\, x^k \lambda e^{-\lambda x}.$$

Now note that $-e^{-\lambda x}$ is a primitive of $\lambda e^{-\lambda x}$. Thus, $M_0(L) = 1 - e^{-\lambda L}$. For $k \geq 1$, applying Theorem 8.11,

$$M_k(L) = k \int_0^L dx\, x^{k-1} e^{-\lambda x} - L^k e^{-\lambda L} = \frac{k}{\lambda} M_{k-1}(L) - L^k e^{-\lambda L}.$$

Since $\lim_{L \to +\infty} L^k e^{-\lambda L} = 0$ for any k, we finally obtain the recursive formula:

$$M_k = \lim_{L \to +\infty} M_k(L) = \frac{k}{\lambda} \lim_{L \to +\infty} M_{k-1}(L) = \frac{k}{\lambda} M_{k-1},$$

with $M_0 = 1$. Thus, by recursion, $M_k = k!/\lambda^k$. In particular, for the first moment, also known as the *mean* of the r.v., $M_1 = 1/\lambda$. The *variance* of a r.v. is defined as $V = M_2 - M_1^2$. In this case, $V = 1/\lambda^2$.

We conclude this Section with a useful generalisation of the Jensen inequality in Corollary 1.1 from finite summations to integrals.

Corollary 8.5 (Jensen's inequality, integral form) *Consider an interval $I = [a, b]$, a Stieltjes measure α such that $\Delta\alpha = \alpha(b) - \alpha(a) > 0$ and two bounded functions f and g such that $g, f \circ g \in \mathcal{R}_\alpha(I)$. Then if f is concave in I,*

$$f\left(\int_I d\alpha g/\Delta\alpha\right) \geq \int_I d\alpha f \circ g/\Delta\alpha,$$

while if f is convex in I,

$$f\left(\int_I d\alpha g/\Delta\alpha\right) \leq \int_I d\alpha f \circ g/\Delta\alpha.$$

Proof A direct proof is easy to produce. I omit it because a more general result for Lebesgue integrals will be offered in Corollary 9.5. □

Even if the integrals of the functions g and $f \circ g$ do not exist, the inequalities above apply to the upper and lower integrals separately.

Exercises

Exercise 8.1 Use the regular grid in Example 8.3 to compute $\int_a^b dx e^x$ and $\int_a^b dx x^2$ in any closed interval $[a, b]$.

Exercise 8.2 Prove that if $|f| \in \mathcal{R}([a, b])$, we cannot in general conclude that $f \in \mathcal{R}([a, b])$. *Hint: Find a counterexample.*

Exercise 8.3 Consider the function

$$f(x) = \begin{cases} \sin\frac{1}{x} & \text{when } x \neq 0, \\ 0 & \text{when } x = 0. \end{cases}$$

Prove the existence of $\int_{-1}^1 dx f(x)$. Note that this function is neither continuous in $x = 0$ nor monotonic in any of its neighbourhood. Are you able to compute the integral?

Exercise 8.4 Compute $\int_0^2 dx[x]$, $\int_0^2 dx[x]^2$, and $\int_0^2 dx[x^2]$, where $[x]$ denotes the integer part of x. Prove that for any positive integer n it is $\int_0^n dx[x] = n(n-1)/2$. Compute the integral $\int_{-2}^{10} dx[x]$.

Exercise 8.5 Consider the function $f : \mathbb{R}^2 \to \mathbb{R}$ defined as $f(x, y) = x^2 \int_0^y dt\, g(t)$, where $g(t) \in C^0(\mathbb{R})$. Find the domain of definition of $f(x, y)$. Is the function $f(x, y)$ continuous? Is it differentiable? Prove that $(0, 0)$ is a critical point of $f(x, y)$. What assumptions on g are necessary for $(0, 0)$ to be a local minimum?

Exercise 8.6 Compute the length of the arc of the curve $y = ae^x/4 + e^{-x}/a$ in the interval $[0, 1]$ as a function of $a > 0$.

Exercise 8.7 With reference to Example 8.11, compute the integral of the scalar field $F(x, y) = x + \sqrt{y}$ on the arc of the parabola $y = x^2$ in the interval $[0, a]$, with $a > 0$.

Exercise 8.8 Using Example 8.9 and Theorem 8.10, compute the following indefinite integrals *(Hint: In the second integral, try to use the tangent and hyperbolic tangent functions)*:

$$\int dx \frac{x}{(x^2 + a^2)^n}, \quad \int dx \frac{1}{(a^2 \pm x^2)^{3/2}}, \quad \int dx (e^x + e^{-x})^3.$$

Exercise 8.9 Using Example 8.9 and Theorem 8.10, compute the following indefinite integrals *(Hint: In the second integral, try to rewrite the numerator in terms of the derivative of the denominator)*:

$$\int dx \frac{1}{ax^2 + bx + c}, \quad \int dx \frac{x}{ax^2 + bx + c}.$$

Exercise 8.10 Using Example 8.9 and Theorem 8.11, compute the following indefinite integrals:

$$\int dx\, x^\alpha \log x, \quad \int dx \sqrt{ax^2 + bx + c}, \quad \int dx \arcsin x.$$

Exercise 8.11 Find the quadratic polynomial $P(x)$ such that $P(0) = P(1) = 0$ and $\int_0^1 dx\, P(x) = 1$.

Exercise 8.12 Find a reduction formula for the computation of the following integrals:

$$\int dx \sin^n x, \quad \int dx \cos^n x, \quad \text{and} \quad \int dx \cos^n x, \sin^m x.$$

Exercise 8.13 Find a measure α such that for any function f continuous in $[-2, 2]$ it is $\int_{-2}^{2} d\alpha f = f(-1)/2 + f(1)/2$.

Exercise 8.14 If I know that $f^2 \in \mathcal{R}(\alpha, [a, b])$ can I conclude that $f \in \mathcal{R}(\alpha, [a, b])$?

Exercise 8.15 If I know that $f^3 \in \mathcal{R}(\alpha, [a, b])$ can I conclude that $f \in \mathcal{R}(\alpha, [a, b])$?

Exercise 8.16 Consider the following measures:

$$\alpha_1(x) = \begin{cases} 0 & x \le 0, \\ 1 & x > 0, \end{cases} \quad \alpha_2(x) = \begin{cases} 0 & x < 0, \\ 1 & x \ge 0, \end{cases} \quad \alpha_3(x) = \begin{cases} 0 & x < 0, \\ 1/2 & x = 0, \\ 1 & x > 0. \end{cases}$$

Prove that

1. $f \in \mathcal{R}(\alpha_1, [-1, 1])$ if and only if $\lim_{t \to 0^+} f(x) = f(0)$;
2. $f \in \mathcal{R}(\alpha_2, [-1, 1])$ if and only if $\lim_{t \to 0^-} f(x) = f(0)$;
3. $f \in \mathcal{R}(\alpha_3, [-1, 1])$ if and only if $f(x)$ is continuous in $x = 0$.

Exercise 8.17 Consider the measure

$$\alpha(x) = \begin{cases} x & x \le 1 \\ x^2 & x > 1. \end{cases}$$

Find the quadratic polynomial $P(x)$ such that $P(0) = P(2) = 0$ and $\int_0^2 d\alpha(x) P(x) = 1$.

Exercise 8.18 Find the continuous measure $\alpha(x)$ such that

$$\int_a^b d\alpha(x) f(x) = \int_0^1 dx f(a + (b - a)x)$$

for any integrable function f.

Exercise 8.19 Consider a random variable X distributed according to $F(x)$ with finite mean $E[X] = \int dF(x) x$ and variance $V[X] = \int dF(x) (x - E[x])^2$. Define $v_2(a) = \int dF(x) (x - a)^2$, with $a \in \mathbb{R}$. Find $a^* = \arg \min_a v_2(a)$ and $v_2(a^*)$.

Exercise 8.20 With $F(x)$ as in the previous exercise and a a real number, define the function $v_1(a) = \int dF(x) |x - a|$. Find the value of a which minimises $v_1(a)$. *Hint: Prove that $v_1(a)$ can be derived.*

Exercise 8.21 Let $0 < a < b < 1$ Find the measure $\alpha(x)$ such that for any function $f \in C^0([0, 1])$,

$$\int_0^1 d\alpha(x)f(x) = f(a) + f(b) + \int_a^b dx f(x).$$

Is the measure $\alpha(x)$ unique? Can the property above be extended to all Riemann-integrable functions in $[0, 1]$?

Exercise 8.22 Let $\alpha \in C^1([a, b])$, positive and increasing. Compute $\int_a^b d\alpha\alpha^2$, $\int_a^b d\alpha^2\alpha$ and $\int_a^b d\sqrt{\alpha}\sqrt{\alpha}$.

Measure Theory

9

9.1 Algebras, Measurable Spaces, and Measures

Consider a set X and the set of its subsets, its power set, 2^X.

Definition 9.1 *(Algebra and σ-algebra)* A subset $\mathcal{A} \subseteq 2^X$ is an *algebra* if

1. $\emptyset \in \mathcal{A}$, $X \in \mathcal{A}$;
2. $\forall A \in \mathcal{A}$, $A^c \in \mathcal{A}$;
3. $\forall A_1, A_2, ..., A_n \in \mathcal{A}$, $\bigcup_{i=1}^{n} A_n \in \mathcal{A}$.

$\mathcal{A}$ is a *σ-algebra* if, in addition,

(4) for any sequence (A_n) of elements of $\mathcal{A}$, $\cup_{i=1}^{\infty} A_i \in \mathcal{A}$.

Property (3) together with (2) implies that $\cap_{i=1}^{n} A_n \in \mathcal{A}$. Property (4) extends property (3) from finite to countable unions. This is different from the definition of a topology, which is required to be closed under the union of any number of open sets (Definition 2.1). The second difference is represented by property (2). In a topology, the complement of an open set is in general not open, the only exceptions being the whole space itself and the parts in which a not connected topology might be split. Properties (2) and (4) imply that a σ-algebra is closed with respect to countable intersections, $\cap_{i=1}^{\infty} A_i = (\cup_{i=1}^{\infty} A_i^c)^c$. For any set X, the power set 2^X and the smallest algebra $\{\emptyset, X\}$ are σ-algebras and are called the *discrete* and *trivial* σ-algebra, respectively.

▶ **Example 9.1** *(Toy σ-algebra)* Consider the set $X = \{a, b, c\}$. Verify that $\mathcal{A}_1 = \{\{a\}, \{b, c\}, X, \emptyset\}$ and $\mathcal{A}_2 = \{\{b\}, \{a, c\}, X, \emptyset\}$ are algebras and σ-algebras. Note that $\mathcal{A}_1 \cap \mathcal{A}_2 = \{\emptyset, X\}$ is again a σ-algebra.

© The Author(s), under exclusive license to Springer Nature Switzerland AG 2023
G. Bottazzi, *Advanced Calculus for Economics and Finance*, Classroom
Companion: Economics, https://doi.org/10.1007/978-3-031-30316-6_9

The result of the previous example is general: given two σ-algebras $\mathcal{A}_1, \mathcal{A}_2 \subseteq 2^X$, it is easy to verify that $\mathcal{A}_1 \cap \mathcal{A}_2$ is a σ-algebra. This observation leads to the following.

Definition 9.2 *(Generated σ-algebra)* Given any collection of sets $C \subseteq 2^X$, the σ-algebra $\sigma(C)$ *generated* by C is defined as the intersection of all the σ-algebras that contains C. Formally

$$\sigma(C) = \cap_\alpha \{\mathcal{A}_\alpha \mid C \subseteq \mathcal{A}_\alpha \text{ and } \mathcal{A}_\alpha \text{ is a } \sigma\text{-algebra}\}.$$

The set of all σ-algebras that contain C has at least one element, 2^X; therefore, Definition 9.2 is always meaningful. If $C \subseteq C'$, then $\sigma(C) \subseteq \sigma(C')$. The σ-algebra generated by a σ-algebra is obviously the original σ-algebra. In particular, $\sigma(\sigma(C)) = \sigma(C)$. If $C' \subseteq \sigma(C)$, then $\sigma(C') \subseteq \sigma(C)$. If, in addition, $C \subseteq \sigma(C')$, then they generate the same σ-algebra, $\sigma(C) = \sigma(C')$.

▶ **Example 9.2** *(Toy generated σ-algebras)* Consider a set X and $A \subset X \neq \emptyset$. Then $\sigma(\{A\}) = \sigma(\{A^c\}) = \sigma(\{A, A^c\}) = \{\emptyset, X, A, A^c\}$. If $A \cup B \subset X$, $A, B \neq \emptyset$, and $A \cap B = \emptyset$, then $\sigma(\{A, B\}) = \{\emptyset, X, A, B, A^c, B^c, A \cup B, A^c \cap B^c\}$.

Definition 9.3 *(Measurable space)* A *measurable space* $(X, \mathcal{A})$ is a set X together with a σ-algebra $\mathcal{A} \subseteq 2^X$.

The elements of $\mathcal{A}$ are *measurable sets*. On measurable spaces, we can define a measure.

Definition 9.4 *(Measure space)* Let $(X, \mathcal{A})$ be a measurable space. A function $\mu : \mathcal{A} \to \mathring{\mathbb{R}}_{\geq 0}$ is a *measure* if

1. $\mu(\emptyset) = 0$;
2. given a sequence (A_n) of elements of $\mathcal{A}$ such that $\forall i \neq j, A_i \cap A_j = \emptyset$, $\mu(\cup_{n=1}^\infty A_n) = \sum_{n=1}^\infty \mu(A_n)$.

The triple $(X, \mathcal{A}, \mu)$ is a *measure space*.

The measure of a measurable set can be a nonnegative element of the set of extended real numbers, that is a nonnegative real number or $+\infty$. Property (2) is called *countable additivity*. A measure is countably additive on the union of disjoint measurable sets. Definition 9.4 conforms to our intuition of how a measure should behave. If $A, B \in \mathcal{A}$, then $A \setminus B = A \cap B^c \in \mathcal{A}$. Since $A \cap B^c$ and $A \cap B$ are disjoint, $\mu(A) = \mu(A \cap B^c) + \mu(A \cap B)$, which implies $\mu(A \setminus B) = \mu(A) - \mu(A \cap B)$. In other terms, if the elements of a measurable set are removed from the elements of another measurable set, the measure of the latter is proportionally reduced, depending on how much the two sets overlap.

▶ **Example 9.3** *(Counting measure)* Consider a measurable space $(X, \mathcal{A})$ and $\forall A \in \mathcal{A}$ define the function $\mu^{\#}(A)$ as number of elements in A. This function is a measure and is called *counting measure*.

▶ **Example 9.4** *(Point mass measure)* Consider a measurable space $(X, \mathcal{A})$ and let $x \in X$. The *point mass* measure δ_x is defined $\forall A \in \mathcal{A}$ as $\delta_x(A) = 1$ if $x \in A$ and $\delta_x(A) = 0$ if $x \notin A$. Given a collection of points $(x_1, x_2, \ldots, x_n) \subset X$, and a set of positive weights $(w_1, w_2, \ldots, w_n) \subset \mathbb{R}_{>0}$, one can easily prove that $\mu = \sum_{i=1}^{n} w_i \delta_{x_i}$ is a measure. This is a special case of the functions we introduce later in Definition 9.16.

Two useful properties of the measure follow directly from its definition. First, subsets cannot have a greater measure than the set that contains them.

Theorem 9.1 *Let $(X, \mathcal{A}, \mu)$ be a measure space. If $A, B \in \mathcal{A}$ and $A \subseteq B$, then $\mu(A) \leq \mu(B)$.*

Proof Let $C = B \setminus A$. As $A \cap C = \emptyset$, $\mu(B) = \mu(A \cup C) = \mu(A) + \mu(C) \geq \mu(A)$. $\square$

Second, any measure is *countably subadditive* on any sequence of sets.

Theorem 9.2 *Let $(X, \mathcal{A}, \mu)$ be a measure space. If $A_1, A_2, \ldots \in \mathcal{A}$ and $A = \cup_{i=1}^{\infty} A_i$ then $\mu(A) \leq \sum_{i=1}^{\infty} \mu(A_i)$.*

Proof Let $A_1' = A_1$, $A_2' = A_2 \setminus A_1$, $A_3' = A_3 \setminus (A_1 \cup A_2)$, ..., in general, $A_i' = A_i \setminus \cup_{j=1}^{i-1} A_j$. First, notice that $\cup_{i=1}^{\infty} A_i = \cup_{i=1}^{\infty} A_i'$ but $A_i' \cap A_j' = \emptyset$. Then $\mu(\cup_{i=1}^{\infty} A_i) = \mu(\cup_{i=1}^{\infty} A_i') = \sum_{i=1}^{\infty} \mu(A_i') \leq \sum_{i=1}^{\infty} \mu(A_i)$, the last inequality due to the fact that, $\forall i$, $A_i' \subseteq A_i$ and Theorem 9.1 applies. $\square$

The measure presented in Examples 9.3 and 9.4 can be defined in any measurable space. On the contrary, the next example discusses a measure that is impossible to build.

▶ **Example 9.5** *(Vitali set)* In applications, we often want measures that satisfy some specific properties. When dealing with real numbers, we would like to have a measure that conforms to our intuitive notion of "length", so that, for instance, the interval $[a, b]$ has measure $b - a$. Intuition also suggests that the measure should be translation-invariant. For example, the "length" of the set $[a + c, b + c]$ should be the same as the set $[a, b]$. Formally, given a real set A and a real number x, the translated set is defined as $T_x(A) = \{z \mid \exists y \in A, z = y + x\}$. One would hope to find a measure such that $\mu(T_x(A)) = \mu(A)$ for any x. We will show that such a measure can never measure all sets.

Take $[0, 1]$ and partition it according to the equivalence relation $x \sim y$ if there exists a $z \in \mathbb{Q}$ such that $x = y + z$. Now, build a set $E \subset [0, 1]$ by picking a single number from each equivalence class of the relation. The number of sets we can build is infinite, but they all share common properties. An element of E will come from $[0]$, the equivalent class to which 0 belongs, which contains all rational numbers between 0 and 1, included. All other elements of E are irrational numbers whose difference is not rational. Sets like E are called *Vitali sets* from the Italian mathematician Giuseppe Vitali who first studied them in 1905. Let us try to measure E. For any $y \in [0, 1]$, $\exists! z \in E$ such that $y - z = w \in [-1, 1] \cup \mathbb{Q}$. This means that $y \in T_w(E)$. At the same time, it is clear that for any rational $w \in [-1, 1]$, $T_w(E) \subseteq [-1, 2]$. Thus, $[0, 1] \subseteq \cup_{w \in [-1,1] \cap \mathbb{Q}} T_w(E) \subseteq [-1, 2]$ so that

$$\mu([0, 1]) \leq \mu \left(\cup_{w \in [-1,1] \cap \mathbb{Q}} T_w(E) \right) \leq \mu([-1, 2]).$$

According to our intuition, it must be $\mu([0, 1]) = 1$ and $\mu([-1, 2]) = 3$. For any $x \neq y$, $x, y \in \mathbb{Q}$, $T_x(E) \cap T_y(E) = \emptyset$, so that

$$\mu(\cup_{w \in [-1,1] \cap \mathbb{Q}} T_w(E)) = \sum_{w \in [-1,1] \cap \mathbb{Q}} \mu(T_w(E)).$$

Since we are dealing with translated sets, it must be $\mu(T_w(E)) = \mu(E)$. But this is impossible, because the summation in the previous equation is on a countably infinite set, and to fulfil both inequalities, the measure $\mu(E)$ should be both positive and zero. We can conclude that the set E cannot be measured by any measure that is, at the same time, invariant by translation and reduces to the intuitive notion of length when intervals are considered.

A measure can be a bounded or unbounded function on $\mathcal{A}$. In both cases, it is clear that it reaches its maximum on the whole space X.

Definition 9.5 *(Finite and σ-finite measures)* A measure μ is *finite* if $\mu(X) < +\infty$. It is *σ-finite* if there exists a sequence (A_n) of elements of $\mathcal{A}$ such that $X = \cup_{i=1}^{\infty} A_i$ and $\mu(A_i) < +\infty, \forall i$.

If μ is finite, $(X, \mathcal{A}, \mu)$ is a *finite measure space*. If μ is σ-finite, then $(X, \mathcal{A}, \mu)$ is a *σ-finite measure space*.

▶ **Example 9.6** *(Finite measure space)* Consider the set of natural numbers $\mathbb{N}$ and let $\mathbb{O}$ and $\mathbb{E}$ be the set of odd and even numbers, respectively. Define $\mathcal{A} = \{\emptyset, \mathbb{N}, \mathbb{O}, \mathbb{E}\}$. The counting measure of Example 9.3 is neither finite nor σ-finite on the measurable space $(\mathbb{N}, \mathcal{A})$, as all measurable sets apart $\emptyset$ have an infinite measure. Conversely, on the measurable space $(\mathbb{N}, 2^{\mathbb{N}})$, the counting measure is σ-finite as we have $\mathbb{N} = \cup_{n=1}^{\infty} \{n\}$ and $\mu^{\#}(\{n\}) = 1$. The point mass measure in Example 9.4 is finite for any algebra.

9.1.1   Complete Measure Space

Subsets of sets having measure zero play an important role in measure theory and have a special name.

Definition 9.6 *(Null set)* Given the measure space $(X, \mathcal{A}, \mu)$, a set $N \subseteq X$ is a *null set* if there exists an $A \in \mathcal{A}$, such that $N \subseteq A$ and $\mu(A) = 0$.

Note that in the above definition, the set N is not required to be in $\mathcal{A}$. We will say that some property or relation is valid *almost everywhere*, abbreviated with a.e., when all elements of X that violate that property or relation belong to a null set. We will write μ-a.e. when it is important to specify which measure is used for the definition of null sets. In applications, especially in integration theory, it is important that the algebra contains all null sets.

Definition 9.7 *(Complete measure space)* If $\mathcal{A}$ contains all null sets, then $(X, \mathcal{A}, \mu)$ is a *complete* measure space.

The property of being complete depends not only on the σ-algebra but also on the measure.

▶ **Example 9.7** *(Incomplete measure space)* Consider the set $X = \{a, b, c\}$ and the σ-algebra $\mathcal{A} = \{\emptyset, X, \{a\}, \{b, c\}\}$. Assign the measure $\mu(\{a\}) = 1$ and $\mu(\{b, c\}) = 0$. Then $\{b\}$ and $\{c\}$ are both null sets that do not belong to the algebra. Thus, $(X, \mathcal{A}, \mu)$ is not complete. One can make the space complete by adding both singlets to the algebra, defining a new algebra $\bar{\mathcal{A}} = \{\emptyset, X, \{a\}, \{b\}, \{c\}, \{b, c\}\}$, and setting $\mu(\{b\}) = \mu(\{c\}) = 0$. Now $(X, \bar{\mathcal{A}}, \mu)$ is a complete space. If we had posed $\mu(\{b, c\}) = 1$, the original measure space would have been complete.

Example 9.7 serves as an inspiration for a more general consideration. If a measure space is not complete, one can consider the σ-algebra generated by the elements of the original σ-algebra and all null sets and extend the original measure on this new σ-algebra by imposing that all null sets have measure zero. The next theorem details the procedure.

Theorem 9.3 (Measure space completion) *Consider a measure space $(X, \mathcal{A}, \mu)$ and let N be the collection of all its null sets. Define $\bar{\mathcal{A}} = \{A \cup N \mid A \in \mathcal{A}, N \in \mathcal{N}\}$ and, $\forall A \in \mathcal{A}$ and $N \in \mathcal{N}$, define $\bar{\mu}(A \cup N) = \mu(A)$. Then $(X, \bar{\mathcal{A}}, \bar{\mu})$ is a complete measure space.*

Proof We start by proving that $\bar{\mathcal{A}}$ is a σ-algebra. First, $\emptyset, X \in \bar{\mathcal{A}}$, because $\emptyset$ is a null set. Second, because $\mathcal{A}$ is closed under countable unions, then also $\mathcal{N}$ is closed under countable unions. In fact, $\bigcup_{i=1}^{\infty} N_i \subseteq \bigcup_{i=1}^{\infty} M_i$, where $M_i \in \mathcal{A}$ is the set with measure zero that contains N_i. Since $\mu(\bigcup_{i=1}^{\infty} M_i) \leq \sum_{i=1}^{\infty} \mu(M_i) = 0$, the countable union of null sets is a null set. This implies that $\bar{\mathcal{A}}$ is closed under countable unions, as

$\cup_{i=1}^{\infty} A_i \cup N_i = (\cup_{i=1}^{\infty} A_i) \cup (\cup_{i=1}^{\infty} N_i)$ and, in the last expression, the first union of sets belongs to $\mathcal{A}$ and the second to $\mathcal{N}$. Third, note that $N^c = (M \setminus N) \cup M^c$, where M is the set with measure zero that contains N. Then $(A \cup N)^c = A^c \cap ((M \setminus N) \cup M^c) = (A^c \cap M^c) \cup (A \cap (M \setminus N))$. The first part of the union is an element of $\mathcal{A}$. For the second part, notice that $A \cap (M \setminus N) \subseteq M$, so it is a null set.

Now consider the function $\bar{\mu} : \bar{\mathcal{A}} \to \mathbb{R}_{\geq 0}, \bar{\mu}(A \cup N) = \mu(A)$. It is trivially $\bar{\mu}(\emptyset) = 0$ and we can also show that it is additive on disjoint sets. Consider $\cup_{i=1}^{\infty} A_i \cup N_i$ where the sets $A_i \cup N_i$ are all disjoint. Then

$$\bar{\mu}\left(\cup_{i=1}^{\infty} A_i \cup N_i\right) = \bar{\mu}\left((\cup_{i=1}^{\infty} A_i) \cup (\cup_{i=1}^{\infty} N_i)\right) =$$

$$\mu(\cup_{i=1}^{\infty} A_i) = \sum_{i=1}^{\infty} \mu(A_i) = \sum_{i=1}^{\infty} \bar{\mu}(A_i \cup N_i).$$

$\bar{\mathcal{A}}$ contains all its null sets by construction.   □

The σ-algebra of the completed space in Theorem 9.3 is the algebra generated by the elements of the original algebra and the null sets, $\bar{\mathcal{A}} = \sigma(\mathcal{A} \cup \mathcal{N})$. In fact, $\mathcal{A} \cup \mathcal{N} \subseteq \bar{\mathcal{A}}$, which implies that $\sigma(\mathcal{A} \cup \mathcal{N}) \subseteq \bar{\mathcal{A}}$, but, at the same time, $\bar{\mathcal{A}} \subseteq \sigma(\mathcal{A} \cup \mathcal{N})$. Any element of $\sigma(\mathcal{A} \cup \mathcal{N})$ can be written as the union of an element of $\mathcal{A}$ and a null set.

9.1.2 Borel σ-Algebra

To make a set X measurable, it suffices to define a σ-algebra on it. There are several ways in which this can be done, but if the set considered is a topological space (X, T), then the use of the topology seems natural.

Definition 9.8 *(Borel σ-algebra)* Consider the topological space (X, T). The Borel σ-algebra $\mathcal{B}$ is the σ-algebra generated by the collection of open sets T.

These special topologies take their name from the French mathematician Félix Borel (1871–1956). The elements of the Borel σ-algebra are often called *Borel sets*. Using the Borel σ-algebra, any topological space (X, T) becomes a measurable space $(X, \mathcal{B})$.

▶ **Example 9.8** *(Nested algebra)* Consider the nested topology as in Example 2.4 with $T = \{\emptyset, X, A_1, A_2\}$ and $\emptyset \subset A_1 \subset A_2 \subset X$. The associated Borel σ-algebra is $\mathcal{B} = \{\emptyset, X, A_1, A_2, A_1^c, A_2^c, A_1^c \cap A_2, A_1 \cup A_2^c\}$. We want to define a finite measure on the measurable space $(X, \mathcal{B})$ such that $\mu(A_1) = 1$, $\mu(A_2) = 2$, and $\mu(X) = 3$. From the property of the measure, $\mu(A_1^c) = \mu(X) - \mu(A_1) = 2$ and $\mu(A_2^c) = \mu(X) - \mu(A_2) = 1$. Because $A_1 \cup A_2^c = \emptyset$, $\mu(A_1 \cup A_2^c) = \mu(A_1) + \mu(A_2^c) = 2$, and because $A_1^c \cap A_2 = (A_1 \cup A_2^c)^c$, $\mu(A_1^c \cap A_2) = \mu(X) - \mu(A_1 \cup A_2^c) = 1$.

If instead we had required $\mu(A_1) = 2$ and $\mu(A_2) = 1$, the properties of the measure would have led to a contradiction. In other terms, it is not possible to define such a measure on the considered algebra.

We will use the notion of Borel σ-algebra mainly in conjunction with finite-dimensional real normed spaces. In this case, the Borel σ-algebra can be generated by the natural base of the topology.

Lemma 9.1 *Consider the normed space $(\mathbb{R}^n, \rho)$ and let $\mathbb{B}$ be the base of open balls. Then the Borel σ-algebra is the σ-algebra generated by $\mathbb{B}$, $\mathcal{B} = \sigma(\mathbb{B})$.*

Proof Because $\mathbb{B} \subset T$, $\sigma(\mathbb{B}) \subseteq \sigma(T) = \mathcal{B}$. At the same time, as discussed in Theorem 4.8, any open set is a countable union of disjoint open balls, thus $T \subseteq \sigma(\mathbb{B})$, which implies $\mathcal{B} \subseteq \sigma(\mathbb{B})$.   □

The previous theorem cannot be extended to any Borel σ-algebra. Its proof uses a result that has been proved only for finite-dimensional normed spaces. Lemma 9.1 does not specify which norm is used to define the balls. According to Theorem 4.11, this is not relevant, as all norms generate the same topology (see Exercise 4.20) and, as such, the same Borel σ-algebra. For example, the open balls in Lemma 9.1 can be "open spheres" $B(\mathbf{x}, r) = \{\mathbf{z} \in \mathbb{R}^n \mid \|\mathbf{z} - \mathbf{x}\| < r\}$ or open n-cells (see Exercise 4.13).

In fact, the Borel σ-algebra of the normed space topology can be generated starting from different collections of sets, not only from a base of the topology. Let $C_1 = \{\times_{i=1}^n (a_i, b_i) \mid a_i < b_i, \forall i\}$ be the collection of open n-cells, $C_2 = \{\times_{i=1}^n (a_i, b_i] \mid a_i < b_i, \forall i\}$ the collection of semi-open (or semi-closed) n-cells, and $C_3 = \{\times_{i=1}^n [a_i, b_i] \mid a_i < b_i, \forall i\}$ the collection of closed n-cells. According to Lemma 9.1, $\sigma(C_1) = \mathcal{B}$. Now note that along each dimension i, $(a_i, b_i] = \cap_{h=1}^\infty (a_i, b_i + 1/h)$ and $(a_i, b_i) = \cup_{h=1}^\infty (a_i, b_i - (b_i - a_i)/(2h)]$. Hence, $C_2 \subset \sigma(C_1)$ and $C_1 \subset \sigma(C_2)$, which implies $\sigma(C_2) = \sigma(C_1) = \mathcal{B}$. A similar argument can be used to prove that $\sigma(C_3) = \mathcal{B}$. Despite the fact that C_2 generates a different topology than C_1 (see Exercise 2.11) and C_3 cannot be a base of a topology (see Exercise 2.10), the three sets generate the same Borel σ-algebra.

In Example 9.5, we have seen that the idea of a translation-invariant measure that satisfies our geometric intuition of length, area, or volume cannot be extended to all sets. We can define it on Borel sets, but this would lead to an incomplete measure space. A better approach is illustrated in the next section.

9.1.3 Lebesgue Measure

Because the Borel measure does not provide a translation-invariant complete measure on normed real spaces, we have to look further in search of an appropriate algebra and an appropriate measure. Our search is organised into two steps. First, we will define a very reasonable approximation of the measure, called an "outer measure", which conforms to our intuition of what a measure should be. Next, we will prove that when this approximation is restricted to a suitable algebra of sets, it becomes a fully fledged complete measure.

Definition 9.9 *(Outer Measure)* Consider a set X. An *outer measure* $\mu^* : 2^X \to \bar{\mathbb{R}}_{\geq 0}$ is a function such that

1. $\mu^*(\emptyset) = 0$;
2. if $A \subseteq B$ then $\mu^*(A) \leq \mu^*(B)$;
3. if $A = \cup_{i=1}^{\infty} A_i$ then $\mu^*(A) \leq \sum_{i=1}^{\infty} \mu^*(A_i)$, that is the outer measure is subadditive.

While a measure is defined on a σ-algebra of subsets, an outer measure is defined on all subsets. However, an outer measure is not required to be additive on disjoint sets.

▶ **Example 9.9** *(Toy outer measure)* Let us define an outer measure on the set $X = \{a, b, c, d\}$. Specify $\mu^*(\emptyset) = 0$ and for any nonempty $A \in 2^X$, $\mu^*(A) = l_1 > 0$ if $A \subseteq \{a, b\}$, $\mu^*(A) = l_2 > 0$ if $A \subseteq \{c, d\}$, and $\mu^*(A) = \mu^*(A \cap \{a, b\}) + \mu^*(A \cap \{c, d\})$ otherwise. So that, for example, $\mu^*(\{d\}) = l_2$ and $\mu^*(\{a, b, c\}) = \mu^*(\{a, b\}) + \mu^*(\{c\}) = l_1 + l_2$. It is easy to verify that μ^* satisfies all properties of Definition 9.9.

The following proposition suggests a general way of building an outer measure: select a set of "boxes" of given sizes and define the outer measure of any set as the infimum of the size of all possible combinations of boxes that contain it.

Theorem 9.4 *Let $C \subseteq 2^X$ be a collection of sets such that*

1. *$\emptyset \in C$;*
2. *there exists a countable cover of X in C, that is $X = \cup_{i=1}^{\infty} C_i$ with $C_i \in C$.*

Define a function $l : C \to \mathbb{R}_{\geq 0}$, with $l(\emptyset) = 0$. Then the function

$$\mu^*(A) = \inf \left\{ \sum_i l(C_i) \,\middle|\, A \subseteq \cup_i C_i, \, C_i \in C \right\}$$

is an outer measure.

Proof Given $A \subseteq X$, let $C(A) = \{\cup_i C_i \mid A \subseteq \cup_i C_i\}$ be the collection of countable covers of the set A with elements of C. Since $\emptyset \in C(\emptyset)$, $\mu^*(\emptyset) = 0$. If $A \subseteq B$, then $C(B) \subseteq C(A)$, thus $\mu^*(A) \leq \mu^*(B)$.

Now consider $A = \cup_{j=1}^{\infty} A_j$. Set $\epsilon > 0$. For any A_j, there exist an integer n_j and a collection of sets $\{C_1^j, \ldots, C_{n_j}^j\}$ such that $A_j \subseteq \cup_{i=1}^{n_j} C_i^j$ and $\sum_{i=1}^{n_j} l(C_i^j) \leq$

$\mu^*(A_j) + \epsilon/2^j$. In fact, if this were not the case, then $\mu^*(A_j) \geq \mu^*(A_j) + \epsilon/2^j$, which is absurd. Clearly $A \subseteq \cup_{j=1}^{\infty} \cup_{i=1}^{n_j} C_i^j$ and

$$\mu^*(A) \leq \sum_{j=1}^{\infty} \sum_{i=1}^{n_j} l(C_i^j) \leq \sum_{j=1}^{\infty} \left(\mu^*(A_j) + \frac{\epsilon}{2^j} \right) = \sum_j \mu^*(A_j) + \epsilon.$$

Since this is true for any $\epsilon > 0$, $\mu^*(A) \leq \sum_{j=1}^{\infty} \mu^*(A_j)$. □

We use the procedure illustrated in Theorem 9.4 to define important outer measures on $\mathbb{R}$ and $\mathbb{R}^n$.

Definition 9.10 (*Lebesgue outer measure in* $\mathbb{R}$) In $\mathbb{R}$, consider the set $C = \{(a, b] \mid a \leq b \in \mathbb{R}\}$ of all semi-open intervals, and define $l((a, b]) = b - a$.

The hypotheses of Theorem 9.4 are verified as $\mathbb{R} \subseteq \cup_{h=1}^{\infty} (-h, h]$. Using sequences of semi-open intervals, it is easy to show that with the outer measure μ^* obtained from Definition 9.10, $\mu^*([a, b]) = \mu^*((a, b)) = \mu^*([a, b)) = b - a$ and $\mu^*(\{x\}) = 0$, $\forall x \in \mathbb{R}$.

Definition 9.11 (*Lebesgue–Stieltjes outer measure in* $\mathbb{R}$) Let α be a nondecreasing right continuous real-valued function on $\mathbb{R}$. In $\mathbb{R}$, consider the set $C = \{(a, b] \mid a \leq b \in \mathbb{R}\}$ of all semi-open intervals and let $l_\alpha((a, b]) = \alpha(b) - \alpha(a)$.

The outer measure μ_α^* derived from Definition 9.11 has properties similar to μ^* from Definition 9.10. In particular, $\mu_\alpha^*([a, b]) = \mu_\alpha^*((a, b)) = \mu_\alpha^*([a, b)) = \alpha(b) - \alpha(a)$ and $\mu_\alpha^*(\{x\}) = 0$, $\forall x \in \mathbb{R}$. The requirement for α to be right continuous is essential, otherwise for some points $a < b < c$, it can be $l_\alpha((a, b] \cup (b, c]) \neq l_\alpha((a, b]) + l_\alpha((b, c])$.

Definition 9.12 (*Lebesgue outer measure in* $\mathbb{R}^n$) On $\mathbb{R}^n$ consider the set $C = \{\times_{j=1}^n (a_j, b_j] \mid a_j \leq b_j, \forall j\}$ of all semi-open rectangles and let $l_n \left(\times_{j=1}^n (a_j, b_j] \right) = \prod_{j=1}^n (b_j - a_j)$.

In this case, $\mathbb{R}^n \subseteq \sum_{h=1}^{\infty} \times_{j=1}^n (-h, h]$. In the associated outer measure,

$$\mu_n^*(\times_{j=1}^n [a_j, b_j]) = \mu_n^*(\times_{j=1}^n (a_j, b_j)) = \prod_{j=1}^n (b_j - a_j),$$

and $\forall \mathbf{x} \in \mathbb{R}^n$, $\mu_n^*(\{\mathbf{x}\}) = 0$.

Since in $\mathbb{R}^n$ any open set is the countable union of disjoint open intervals, we can easily compute the outer measure of these sets. The same is true for the union of closed and open sets and for their intersection. Using the definition above, one can actually compute the outer measure of all sets belonging to the Borel σ-algebra and prove that those outer measures restricted to their respective Borel σ-algebras are measures. This would be cumbersome, as one should consider countable unions and intersections of countable unions and intersections, which are uncountable. Moreover, the resulting spaces will not be complete, significantly reducing their importance for integration theory (see Sect. 9.3 and, in particular, the analysis of the Riemann integral in Sect. 9.3.1).[1] We follow a different approach. We start by identifying a specific collection of sets.

Definition 9.13 *(Outer measurable set)* Let μ^* be an outer measure on X. The set $A \subset X$ is μ^*-*measurable* if $\forall E \subseteq X$, $\mu^*(E) = \mu^*(E \cap A) + \mu^*(E \cap A^c)$. The collection of all μ^*-measurable subsets of X is denoted by $\mathcal{A}^*$.

Note that $E \cap A$ and $E \cap A^c$ are disjoint sets, and $E = (E \cap A) \cup (E \cap A^c)$. So, for any set E, $\mu^*(E) \le \mu^*(E \cap A) + \mu^*(E \cap A^c)$. Definition 9.13 restricts the weak inequality to be an equality. Clearly, $\emptyset$ belongs to $\mathcal{A}^*$ and, for symmetry, if A is μ^*-measurable, A^c is μ^*-measurable.

▶ **Example 9.10** *(Toy outer measure)* Consider the outer measure in Example 9.9. By construction, $\{a, b\}$ and $\{c, d\}$ are μ^*-measurable. Note that $\mu^*(X) = l_1 + l_2$ and for any singlet $\{x\} \subset X$, $\mu^*(X) < \mu^*(X \cap \{x\}) + \mu^*(X \cap \{x\}^c)$, so singlets are not μ^*-measurable. This implies that sets with three elements are also not μ^*-measurable. Considering $\{b, c\}$, $\mu^*(\{a, b\}) = l_1 < \mu^*(\{a, b\} \cap \{b, c\}) + \mu^*(\{a, b\} \cap \{b, c\}^c) = 2l_1$. In summary, $\mathcal{A}^* = \{\emptyset, X, \{a, b\}, \{c, d\}\}$. Note that $\mathcal{A}^*$ is a σ-algebra.

Now, we can use outer measurable sets to define a measure space.

Theorem 9.5 (Lebesgue measure space) $\mathcal{A}^*$ *is a σ-algebra on X, the restriction of μ^* on $\mathcal{A}^*$ is a measure, and $(X, \mathcal{A}^*, \mu^*)$ is a complete measure space.*

Proof First, we prove that $\mathcal{A}^*$ is an algebra by proving that it is closed with respect to the union of sets. Let $A, B \in \mathcal{A}^*$. Applying the definition of μ^*-measurable set subsequently for A and B, $\forall E \subseteq X$,

$$\mu^*(E) = \mu^*(E \cap A) + \mu^*(E \cap A^c) = \mu^*(E \cap A \cap B) +$$
$$\mu^*(E \cap A \cap B^c) + \mu^*(E \cap A^c \cap B) + \mu^*(E \cap A^c \cap B^c).$$

Note that $(A \cap B) \cup (A \cap B^c) \cup (A^c \cap B) = A \cup B$. So, from the property of the outer measure,

$$\mu^*(E \cap A \cap B) + \mu^*(E \cap A \cap B^c) + \mu^*(E \cap A^c \cap B) \ge \mu^*(E \cap (A \cup B))$$

---

[1] I do not provide an explicit example of a null set which is not Borel. Its construction, based on the so-called *Cantor function*, requires some substantial work and would not add much to the present discussion.

and, substituting in the previous equation, $\mu^*(E) \geq \mu^*(E \cap (A \cup B)) + \mu^*(E \cap (A \cup B)^c)$. Because μ^* is an outer measure, this inequality is actually an equality. This proves the point and also that

$$\mu^*(E \cap A \cap B) + \mu^*(E \cap A \cap B^c) + \mu^*(E \cap A^c \cap B) = \mu^*(E \cap (A \cup B)).$$

Then, if A and B are disjoint elements of $\mathcal{A}^*$, $\mu^*(E \cap (A \cup B)) = \mu^*(E \cap A) + \mu^*(E \cap B)$. In particular, by setting $E = X$, $\mu^*(A \cup B) = \mu^*(A) + \mu^*(B)$. We have proved that the outer measure μ^* is additive on disjoint outer measurable sets and, consequently, it is a measure on the algebra $\mathcal{A}^*$.

To prove that $\mathcal{A}^*$ is a σ-algebra, consider $A = \cup_{i=1}^{\infty} A_i$, $A_i \in \mathcal{A}^*$. Define $\tilde{A}_1 = A_1$, $\tilde{A}_2 = A_2 \setminus A_1$, $\tilde{A}_3 = A_3 \setminus (A_1 \cup A_2)$, ... such that $\tilde{A}_i \cap \tilde{A}_j = \emptyset$ if $i \neq j$ and $A = \cup_{i=1}^{\infty} \tilde{A}_i$. Let $B_n = \cup_{i=1}^{n} \tilde{A}_i$. Because $\mathcal{A}^*$ is an algebra, $\tilde{A}_n, B_n \in \mathcal{A}^*, \forall n$. For any $E \subseteq X$,

$$\mu^*(E) = \mu^*(E \cap B_n) + \mu^*(E \cap B_n^c) = \sum_{i=1}^{n} \mu^*(E \cap \tilde{A}_i) + \mu^*(E \cap B_n^c).$$

When $n \to \infty$, $B_n^c \to A^c$ and $\sum_{i=1}^{\infty} \mu^*(E \cap \tilde{A}_i) \geq \mu^*(E \cap A)$, so that $\mu^*(E) \geq \mu^*(E \cap A) + \mu^*(E \cap A^c)$. But because μ^* is an outer measure, this is actually an equality.

Finally, consider a set A such that $\mu^*(A) = 0$. For the property of the outer measure, for any $E \in X$, $\mu^*(E \cap A) = 0$ and $\mu^*(E) \leq \mu^*(E \cap A) + \mu^*(E \cap A^c) = \mu^*(E \cap A^c) \leq \mu^*(E)$, so that the relation is actually an equality. We have proved that any set A whose outer measure is equal to zero belongs to $\mathcal{A}^*$. By the definition of the outer measure, if A has outer measure equal to zero, any subset of A has an outer measure equal to zero. Therefore, they all belong to $\mathcal{A}^*$, which implies that $\mathcal{A}^*$ is complete.   □

By applying the previous theorem, Definitions 9.10, 9.11, and 9.12 lead respectively to the Lebesgue space $(\mathbb{R}, \mathcal{L}, l)$, the Lebesgue–Stieltjes space $(\mathbb{R}, \mathcal{L}, l_\alpha)$, and the Lebesgue space $(\mathbb{R}^n, \mathcal{L}_n, l_n)$ in more than one dimension. Definitions 9.10 and 9.12 introduce translation-invariant outer measures. The first two spaces share the same σ-algebra as they are based on the same set of semi-open intervals. The first space is actually a special case of the other two, with $\alpha(x) = x$ and $n = 1$. These are the spaces that we most often encounter in applications. We conclude the analysis of the Lebesgue measure by investigating the relationship between the Lebesgue and Borel σ-algebras.

Theorem 9.6 *All sets in the Borel σ-algebra of $\mathbb{R}^n$ are l_n-measurable.*

Proof We start by proving the theorem in $\mathbb{R}$. With reference to Definition 9.10, consider a semi-open interval $(a, b] \in C$ and any subset $E \subseteq \mathbb{R}$. According to the definition of the Lebesgue outer measure, $\forall \epsilon > 0$ there exists a collection of intervals $\{(a_i, b_i]\}$ such that $E \subseteq \cup_i (a_i, b_i]$, $\sum_i l((a_i, b_i]) \leq \mu^*(E) + \epsilon$. The intersection of two semi-open intervals is a semi-open interval, so that

$$\mu^*((a_i, b_i] \cap (a, b]) + \mu^*((a_i, b_i] \cap (a, b]^c) = l((a_i, b_i]).$$

Hence,

$$\mu^*(E \cap (a, b]) + \mu^*(E \cap (a, b]^c) \leq \sum_i \mu^*((a_i, b_i] \cap (a, b])$$

$$+ \sum_i \mu^*((a_i, b_i] \cap (a, b]^c) = \sum_i l((a_i, b_i]) \leq \mu^*(E) + \epsilon.$$

Since ϵ can be chosen as small as desired and because of the property of the outer measure, the inequality above implies $\mu^*(E \cap (a, b]) + \mu^*(E \cap (a, b]^c) = \mu^*(E)$, that is $(a, b] \in \mathcal{L}$. We have shown that $C \subset \mathcal{L}$. But because $\mathcal{B} = \sigma(C)$, the statement follows.

The proof can be replicated similarly for $\mathbb{R}^n$ using the semi-open rectangles of Definition 9.12. The key property is that the intersection of two semi-open rectangles is a semi-open rectangle.   □

Let $\mathcal{N}_n$ be the collection of null sets in $(\mathbb{R}^n, \mathcal{B}_n)$ with respect to the Lebesgue measure l_n. Consider the completion of the Borel σ-algebra, $\sigma(\mathcal{B}_n \cup \mathcal{N}_n)$. Since $\mathcal{B}_n \subseteq \mathcal{L}_n$ and $\mathcal{N}_n \subseteq \mathcal{L}_n$, we have $\sigma(\mathcal{B}_n \cup \mathcal{N}_n) \subseteq \mathcal{L}_n$. But does the Lebesgue σ-algebra contain sets that are not in the completion of the Borel σ-algebra? The answer is no as shown by the following corollary.

Corollary 9.1 $\mathcal{L}_n$ *is the completion of the Borel σ-algebra in $\mathbb{R}^n$ with respect to the Lebesgue measure, $\mathcal{L}_n = \sigma(\mathcal{B}_n \cup \mathcal{N}_n)$.*

Proof Since we know that $\sigma(\mathcal{B}_n \cup \mathcal{N}_n) \subseteq \mathcal{L}_n$, we have only to prove that $\mathcal{L}_n \subseteq \sigma(\mathcal{B}_n \cup \mathcal{N}_n)$. By Theorem 9.4, $\forall A \in \mathcal{L}_n$ there should be a sequence of elements (C_i) of C such that $A \subseteq B = \cup_i C_i$ and $l_n(A) = l_n(B)$. $B \in \mathcal{B}_n$ because the elements of C belong to the Borel σ-algebra. Because A is outer measurable, $l_n(B^c) = l_n(A \cap B^c) + l_n(A^c \cap B^c)$. But $A^c \cap B^c = (A \cup B)^c = B^c$, so that $l_n(A \cap B^c) = 0$. Since $A = B \cup (A \cap B^c)$, the set A is the union of a set of the Borel σ-algebra, B, and a null set, $A \cap B^c$. Thus, $\mathcal{L}_n \subseteq \mathcal{B}_n \cup \mathcal{N}_n$, and the statement follows.   □

9.2 Measurable Functions

Having defined measurable spaces, it is time to see how we can connect them. The fundamental tool we will use is the notion of a measurable function.

Definition 9.14 *(Measurable function)* Let $(X, \mathcal{A}_X)$ and $(Y, \mathcal{A}_Y)$ be two measurable spaces. A function $f : X \rightarrow Y$ is measurable if $\forall A_Y \in \mathcal{A}_Y$, $f^{-1}(A_Y) \in \mathcal{A}_X$.

In other words, a function is measurable if the preimage of any measurable set is measurable. When necessary, we will specify that a function is measurable with respect to a specific σ-algebra by saying that it is $\mathcal{A}_X$-measurable or $\mathcal{A}_Y$-measurable, when the emphasis is, respectively, on the domain or image space of the map f. In Definition 9.14, there is no reference to the actual measure possibly implemented on the σ-algebras of the two spaces that the function connects. Note that a measurable function is not required to be bounded (see Example 9.14).

Lemma 9.2 (Composition of measurable functions) *The composition of measurable functions is measurable.*

Proof Consider three measurable spaces $(X, \mathcal{A}_X)$, $(Y, \mathcal{A}_Y)$, and $(Z, \mathcal{A}_Z)$ and two measurable functions $f : X \to Y$ and $g : Y \to Z$. Consider a measurable set $A_Z \in \mathcal{A}_Z$. Because g is measurable, $g^{-1}(A_z) \in \mathcal{A}_Y$, and because f is measurable, $f^{-1}(g^{-1}(A_z)) \in \mathcal{A}_X$. □

Obviously, not all functions are measurable with respect to all σ-algebras, but there is one notable exception.

▶ **Example 9.11** *(Constant function)* Let $c \in Y$ be a constant. If $f(x) = c$ for all $x \in X$, then $f^{-1}(A_Y)$ is either X (if $c \in A_Y$) or the empty set (if $c \notin A_Y$). Thus, f is measurable with respect to any algebra $\mathcal{A}$ on 2^X.

Complete metric spaces possess an important property.

Theorem 9.7 *Consider two functions f and g from a complete metric space $(X, \mathcal{A}_X)$ to a metric space $(Y, \mathcal{A}_Y)$. If f is measurable and $g = f$ a.e., then g is measurable.*

Proof Define $N = \{x \in X \mid g(x) \neq f(x)\}$. By assumption N is a null set and since $\mathcal{A}_X$ is complete, $N \in \mathcal{A}_X$. For any $A_Y \in \mathcal{A}_Y$, $g^{-1}(A_Y) = (g^{-1}(A_Y) \cap N) \cup (g^{-1}(A_Y) \cap N^c)$. Note that $g^{-1}(A_Y) \cap N$ is a null set, thus it belongs to $\mathcal{A}_X$, and $g^{-1}(A_Y) \cap N^c = f^{-1}(A_Y) \cap N^c$ is the intersection of two elements of $\mathcal{A}_X$, thus it belongs to $\mathcal{A}_X$ as well. □

Note that for the previous result, we did not make any completeness assumption about the image space. In the domain space, completeness is essential. Otherwise, the set of points on which the two functions differ may not be measurable. Theorem 9.7 and the discussion in Example 9.11 make clear that any function defined over a complete measurable space which is a.e. constant is measurable. In general, any function defined over a measurable space can be used to define an algebra in the image space.

Lemma 9.3 *Consider the function* $f : X \to Y$ *and let* $\mathcal{A}_X$ *be a* σ*-algebra on* X. *Then the set* $\sigma(f) = \{Y \in 2^Y \mid f^{-1}(Y) \in \mathcal{A}_X\}$ *is a* σ*-algebra on* Y.

Proof The proof is trivial once one realises that $f^{-1}(A_Y^c) = (f^{-1}(A_Y))^c$ and $f^{-1}(A_Y \cup A_Y') = f^{-1}(A_Y) \cup f^{-1}(A_Y')$. $\square$

In other words, given any σ-algebra on X, the collection of sets with measurable preimages forms a σ-algebra on Y. In Example 9.11, for example, the σ-algebra induced by the constant is the entire power set, 2^Y. From the definition of a measurable function, it follows that a function f between two measurable spaces $(X, \mathcal{A}_X)$ and $(Y, \mathcal{A}_Y)$ is measurable if and only if $\mathcal{A}_Y \subseteq \sigma(f)$. In particular, a function $f : X \to Y$ defined over a measurable space $(X, \mathcal{A}_X)$ is always measurable if the σ-algebra on Y is $\sigma(f)$. A similar mechanism is in place for the measure: any measurable function defines a measure on the image space based on any given measure on the domain space.

Lemma 9.4 (Pushforward measure) *Let* $(X, \mathcal{A}_X)$ *and* $(Y, \mathcal{A}_Y)$ *be two measurable spaces and* $f : X \to Y$ *be a measurable function between them. Consider the measure* μ_X *on* $(X, \mathcal{A}_X)$. *The function* $\mu_f : \mathcal{A}_Y \to \mathbb{R}_{\geq 0}$ *defined by* $\mu_f(A_Y) = \mu_X(f^{-1}(A_Y)), \forall A_Y \in \mathcal{A}_Y$ *is a measure on* $(Y, \mathcal{A}_Y)$. *It is called the* pushforward *measure or* image measure *defined by the function* f.

Proof The proof is immediate as clearly $f^{-1}(\emptyset) = \emptyset$, and if A_Y and A_Y' are disjoint elements of $\mathcal{A}_Y$, then also $f^{-1}(A_Y)$ and $f^{-1}(A_Y')$ are disjoint elements of $\mathcal{A}_X$. $\square$

The pushforward measure is often denoted by $\mu_X \circ f^{-1}$. It will be used in Sect. 9.3 to discuss the change of variables in Lebesgue integrals and in Sect. 9.5 to study random variables.

▶ **Example 9.12** *(Measurable functions and* σ*-algebras)* Consider the measure space of Example 9.7 with $\mu(\{a\}) = \mu(\{b, c\}) = 1$ and the real-valued function $f(a) = 1$, $f(b) = 2$, and $f(c) = 4$. The σ-algebra induced in $\mathbb{R}$ consists of all sets that contain both 2 and 4 or none of the two. This collection of sets is clearly closed under union and intersection.

If we consider the measure space $(\mathbb{R}, \mathcal{B})$ as the arrival space, then the function f is not measurable. Indeed, the preimage of the measurable set $(3, 5)$ is the set $\{c\}$, which is not in the σ-algebra of X.

If we change the definition of the function to $f(b) = f(c) = 2$, then it becomes measurable, and the pushforward measure induced on $(\mathbb{R}, \mathcal{B})$ is the point mass measure (see Example 9.4) $\mu_X \circ f^{-1} = \delta_1 + \delta_2$.

When the algebra $\mathcal{A}_X$ in the domain of the function is a Borel σ-algebra, the function is said to be *Borel measurable*. In the case $X = \mathbb{R}^n$, if not specified otherwise, this is meant to refer to the measure generated by the Euclidean topology.

Analogously, the function is said to be *Lebesgue measurable* if $\mathcal{A}_X = \mathcal{L}_n$. Since $\mathcal{B}_n \subseteq \mathcal{L}_n$, a Borel measurable function is also Lebesgue measurable, while the opposite is generally not true.

Assume that the algebra $\mathcal{A}_Y$ in the image space is generated by a collection of sets C, $\mathcal{A}_Y = \sigma(C)$. Then if $f^{-1}(C) \in \mathcal{A}_X$ for any $C \in C$, $C \subseteq \sigma(f)$. Because $\mathcal{A}_Y$ is the smallest σ-algebra that contains C, $\mathcal{A}_Y \subseteq \sigma(f)$, which means that the function f is measurable.

Corollary 9.2 *Consider two topological spaces (X, T_X) and (Y, T_Y). Any continuous function $f : X \rightarrow Y$ is Borel measurable.*

Proof By the definition of Borel σ-algebra, $\mathcal{B}_X = \sigma(T_X)$ and $\mathcal{B}_Y = \sigma(T_Y)$. Because the function is continuous, the preimage of any elements of T_y belongs to T_x, and hence to $\mathcal{B}_X$. Thus, $T_y \subseteq \sigma(f)$ and the statement follows. $\square$

9.2.1 Measurable Real-Valued Functions

Let us focus on functions $f : X \rightarrow \mathbb{R}$ defined over a generic measurable space $(X, \mathcal{A})$ and with images in the measurable space $(\mathbb{R}, \mathcal{B})$, where $\mathcal{B}$ is the Borel σ-algebra generated by the Euclidean topology. In this specific case, Definition 9.14 is usually replaced by the following.

Definition 9.15 *(Measurable real-valued function)* A function $f : X \rightarrow \mathbb{R}$ is measurable on the space $(X, \mathcal{A})$ if $\forall a \in \mathbb{R}$, $D_a(f) = \{x \in X \mid f(x) > a\} \in \mathcal{A}$.

The equivalence of the definition above with Definition 9.14 follows from the fact that the set of open intervals $C = \{(a, +\infty)\}$ generates $\mathcal{B}$.

▶ **Example 9.13** *(Alternative definitions)* One can easily devise other definitions equivalent to Definition 9.15 using different collections of sets that generate $\mathcal{B}$, for instance those discussed in Sect. 9.1.2. The function $f : X \rightarrow \mathbb{R}$ is Borel measurable if one of the following statements is true:

1. $\forall a \in \mathbb{R}$, $\{x \in X \mid f(x) \leq a\} \in \mathcal{A}$;
2. $\forall a \in \mathbb{R}$, $\{x \in X \mid f(x) \geq a\} \in \mathcal{A}$;
3. $\forall a \in \mathbb{R}$, $\{x \in X \mid f(x) < a\} \in \mathcal{A}$.

▶ **Example 9.14** *(Unbounded measurable function)* Consider the function $f : \mathbb{R} \rightarrow \mathbb{R}$ defined by

$$f(x) = \begin{cases} x & \text{if } x \leq 0; \\ 1/x & \text{if } x > 0. \end{cases}$$

Note that $D_a(f) = (a, +\infty)$ if $a \leq 0$ and $D_a(f) = (0, 1/a)$ if $a > 0$. Therefore, the function f is measurable with respect to the Borel σ-algebra.

Because continuous functions are Borel measurable, the property of measurability is preserved by composition with a continuous function. For this reason, the property of being measurable interacts quite fairly with the usual operations on real functions.

Theorem 9.8 *Let $(X, \mathcal{A})$ be a measurable space, f and g be measurable real-valued functions, and c a real constant, then $f + g$, cf, $|f|$, $f g$, $\max\{f, g\}$, and $\min\{f, g\}$ are measurable.*

Proof Note that $\forall a \in \mathbb{R}$, $D_a(f+g) = \cup_{r \in \mathbb{Q}} D_r(f) \cap D_{a-r}(g)$. Since $\mathbb{Q}$ is countable, and the algebra is closed under countable unions and intersections, $D_a(f + g)$ is measurable. If $c > 0$, $D_a(cf) = D_{a/c}(f) \in \mathcal{A}$. If $c < 0$, $D_a(cf) = \{f < a/c\} = f^{-1}([a/c, +\infty)) \in \mathcal{A}$ because $[a/c, +\infty) \in \mathcal{B}$. Since $f g = (f+g)^2/2 - f^2/2 - g^2/2$ and $|f| = \sqrt{f f}$, the two functions are measurable because they are the sum of measurable functions and composition with continuous functions. Finally, $D_a(\max\{f, g\}) = D_a(f) \cup D_a(g)$ while $D_a(\min\{f, g\}) = D_a(f) \cap D_a(g)$. Hence, both sets belong to $\mathcal{A}$. $\qquad\qquad\square$

The infimum and supremum of countable sequences of measurable functions are measurable as well.

Theorem 9.9 *Let $(X, \mathcal{A})$ be a measurable space and (f_n) be a sequence of real-valued measurable functions $f_n : X \to \mathbb{R}$. Then $\sup_n f_n$, $\inf_n f_n$, $\limsup_{n \to \infty} f_n$, and $\liminf_{n \to \infty} f_n$ are measurable.*

Proof For any $a \in \mathbb{R}$, $D_a(\sup_n f) = \cup_n D_a(f_n)$ and $D_a(\min_n f) = \cap_n D_a(f_n)$. Both sets belong to $\mathcal{A}$. Remember that $\limsup_{n \to \infty} f_n = \lim_{n \to \infty} \sup_{i \geq n} f_i$. Since $\sup_{i \geq n} f_i$ is nonincreasing in n, we have that $D_a(\limsup_{n \to \infty} f_n) = \cap_n \cup_{i \geq n} D_a(f_i)$ and, for the properties of σ-algebras, this set belongs to $\mathcal{A}$, $\forall a \in \mathbb{R}$. Analogously, we have that $D_a(\liminf_{n \to \infty} f_n) = \cup_n \cap_{i \geq n} D_a(f_i) \in \mathcal{A}$. $\qquad\square$

In particular, if a sequence of measurable functions is pointwise convergent $f_n \to f$, (see Definition 5.18), then the limit function f is a measurable function. This is also true if the limit function is unbounded on measurable sets. In addition, we have a more powerful result for complete metric spaces.

Corollary 9.3 (Almost everywhere convergence) *Let (f_n) be a sequence of measurable real-valued functions defined over a complete measure space $(X, \mathcal{A})$. If $\lim_{n \to \infty} f_n = f$ a.e., then f is measurable.*

Proof Consider $g = \lim_{n\to\infty} f_n$. For Theorem 9.9, this is a measurable function. By hypothesis, g differs from f on a null set, so that the statement follows from Theorem 9.7. □

In some applications, most notably applications that have to do with probability theory, one usually wants to work with sequences that converge almost everywhere. In these applications, the completeness of the measure space is usually assumed, even if the theory we are developing does not forcibly require it.

▶ **Example 9.15** *(Infimum and supremum of sequences of Riemann integrable functions)* The result of Theorem 9.9 about the superior and inferior limits of sequences of measurable functions do not hold for sequences of Riemann integrable functions. Consider the sequence of real functions (f_n) defined in $[0, 1]$ as

$$f_n(x) = \begin{cases} 1 & \text{if } x = h/n, \text{ for } h = 1, \ldots, n, \\ 0 & \text{otherwise.} \end{cases}$$

The function f_n is Riemann integrable for any n, $\int_0^1 dx f_n(x) = 0$. However, $f(x) = \limsup_{n\to\infty} f_n(x)$ takes the value 1 for all rational numbers and 0 for all irrational numbers in $[0, 1]$. Therefore, it is not Riemann integrable (see Example 8.2).

9.3 Lebesgue Integral

This section extends the theory of integration to all real-valued measurable functions defined on a generic measure space $(X, \mathcal{A}, \mu)$. If not explicitly stated otherwise, we consider functions taking values in $(\mathbb{R}, \mathcal{B})$, where $\mathcal{B}$ is the Borel σ-algebra generated by the Euclidean topology. We will proceed in steps. First, we define the integral of a very specific subset of measurable functions. We then extend the definition to all nonnegative measurable functions, proving several fundamental theorems. Finally, using these theorems, we extend the definition further to a large class of measurable functions in an easy and straightforward way.

Definition 9.16 *(Indicator function)* Consider a set X and a subset $A \subseteq X$. The function $I_A : X \to \mathbb{R}_{\geq 0}$

$$I_A(x) = \begin{cases} 1 & \text{if } x \in A \\ 0 & \text{otherwise} \end{cases}$$

is said to be the *characteristic* or *indicator* function of the set A.

In a measurable space $(X, \mathcal{A})$, the function I_A is measurable if and only if $A \in \mathcal{A}$. Starting from the indicator function, we can build a special class of measurable functions.

Definition 9.17 *(Simple function)* In a measurable space $(X, \mathcal{A})$, a *simple* function $s(x)$ is built from a finite collection $(A_1, ..., A_n)$ of elements of $\mathcal{A}$ and of nonnegative constants $(a_1, ..., a_n)$, such that $s(x) = \sum_{i=1}^{n} a_i I_{A_i}(x)$.

The domain of the simple function is the measurable set $\cup_i A_i$. Simple functions are measurable by construction. By adding $a_{n+1} = 0$ and $A_{n+1} = (\cup_{i=1}^{n} A_i)^c$, to the set (a_i) and (A_i), any simple function can be considered defined on the whole space X. It is immediately apparent that if $s_1(x)$ and $s_2(x)$ are simple functions defined on the same space, then $s_1(x) + s_2(x)$ and $s_1(x) s_2(x)$ are simple functions. Similarly, $c \, s_1(x)$ is a simple function as long as $c \geq 0$. More generally, given $f : \mathbb{R}_{\geq 0} \to \mathbb{R}_{\geq 0}$, $f(s(x))$ is simple if $s(x)$ is simple. The elements (A_i) can always be chosen to be mutually disjoint. For any measurable set $A \in \mathcal{A}$, if $s(x)$ is a simple function, $s(x) I_A(x)$ is a simple function. Simple functions are the building blocks of the notion of integration based on measure theory.

Definition 9.18 *(Integral of simple functions)* The Lebesgue integral of a simple function $s(x) = \sum_{i=1}^{n} a_i I_{A_i}(x)$ is defined on the measure space $(X, \mathcal{A}, \mu)$ as $\int d\mu \, s = \sum_{i=1}^{n} a_i \, \mu(A_i)$.

If $\forall x, s_1(x) \leq s_2(x)$, then $\int d\mu \, s_1 \leq \int d\mu \, s_2$. To see it, simply redefine them on a finer common partition of X. The previous definition can be seen as an integral on the domain of the simple function. In general, given a measurable set E, we define $\int_E d\mu \, s = \int d\mu \, s I_E = \sum_{i=1}^{n} a_i \, \mu(A_i \cap E)$. For the properties of simple functions, if $E_1 \cap E_2 = \emptyset$, then $\int_{E_1 \cup E_2} d\mu \, s = \int_{E_1} d\mu \, s + \int_{E_2} d\mu \, s$.

▶ **Example 9.16** *(Integral of Dirichlet's function)* Consider the measure space $(\mathbb{R}, \mathcal{B}, l)$. Any countable union of singlets belongs to $\mathcal{B}$ and has Lebesgue measure zero. Consider the set of rational numbers $[0, 1] \cap \mathbb{Q}$. This is a countable union of singlets, so that $l([0, 1] \cap \mathbb{Q}) = 0$. Consequently, $l([0, 1] \setminus \mathbb{Q}) = l([0, 1]) - l([0, 1] \cap \mathbb{Q}) = 1$. Dirichlet's function in Example 8.2 is the simple function $f = I_{[0,1] \setminus \mathbb{Q}}$. It is Borel and, consequently, Lebesgue measurable. According to Definition 9.18, $\int d\mu \, f = l([0, 1] \setminus \mathbb{Q}) = 1$. In Example 8.2, we have seen that this function cannot be integrated in the Riemann sense.

Using the simple functions of Definition 9.18, we can extend the notion of integral to any nonnegative measurable function.

Definition 9.19 *(Integral of nonnegative measurable functions)* The Lebesgue integral of a nonnegative measurable function $f : X \to R$ on a measure space $(X, \mathcal{A}, \mu)$ is

$$\int d\mu \, f = \sup \left\{ \int d\mu \, s \mid s \text{ is simple and } s(x) \leq f(x), \forall x \right\}.$$

The supremum is computed over all simple functions that are equal to or lower than the function f at any point. It can be a nonnegative real number or $+\infty$. This is the central definition of this section. It is worth spending some time on it. First of all, note that Definition 9.19 is restricted to measurable functions. In principle, one can define it on bounded functions, but then the resulting integral would lack desirable properties.

▶ **Example 9.17** *(Integral and generic bounded functions)* Take the measure space $(\mathbb{R}, \mathcal{A}, \mu)$ with $\mathcal{A} = \{\emptyset, \mathbb{R}\}$, $\mu(\emptyset) = 0$, and $\mu(\mathbb{R}) = 1$. Consider the functions $f = I_{[0,1]}$ and $g = 1 - I_{[0,1]}$. Note that, on this measure space, the only simple function lower than or equal to f and g is the constant $s(x) = 0$. Conversely, the greater simple function lower than or equal to $f + g$ is the constant $s(x) = 1$. Then, using Definition 9.19, one would get $\int d\mu \, f = \int d\mu \, g = 0$ but $\int d\mu \, (f + g) = 1$.

If $f \leq M$, and the measure space is finite, the simple constant function $s(x) = M$ is greater than all the simple functions considered in Definition 9.19. Thus, the integral of f is finite, $\int d\mu \, f \leq M\mu(X)$. If the function is unbounded or the measure space is not finite, then the integral can be infinite. In any case, from the property of the supremum, it follows that if $g(x) \leq f(x)$, then $\int d\mu \, g \leq \int d\mu \, f$.

A second consideration has to do with the domain of integration. If f is a measurable function and E is a measurable set, then $f I_E$ is a measurable function and we define $\int_E d\mu \, f = \int d\mu \, f I_E$. Definition 9.19 guarantees the following.

Corollary 9.4 *Let f be a nonnegative measurable function and E_1, E_2 two disjoints measurable sets, then $\int_{E_1 \cup E_2} d\mu \, f = \int_{E_1} d\mu \, f + \int_{E_2} d\mu \, f$.*

Proof For any measurable set E, define

$$L_f(E) = \left\{ \int_E d\mu \, s \mid s \text{ is simple and } s(x) \leq f(x), \forall x \in E \right\}.$$

Note that $\int_E d\mu f = \sup L_f(E)$. For the property of simple functions, if $l_1 \in L_f(E_1)$ and $l_2 \in L_f(E_2)$, then $l_1 + l_2 \in L_f(E_1 \cup E_2)$. Thus, $\sup L_f(E_1 \cup E_2) \geq \sup L_f(E_1) + \sup L_f(E_2)$. At the same time, $\forall l \in L_f(E_1 \cup E_2)$, there exist $l_1 \in L_f(E_1)$ and $l_2 \in L_f(E_2)$ such that $l = l_1 + l_2$. Indeed, if $l = \int_{E_1 \cup E_2} d\mu \, s$, consider the integrals of $s I_{E_1}$ and $s I_{E_2}$. Thus, $\sup L_f(E_1 \cup E_2) \leq \sup L_f(E_1) + \sup L_f(E_2)$. The statement follows from the two inequalities. □

By the definition of integral, if there exists $a, b \in \mathbb{R}$ such that $\forall x \in A, a \leq f(x) \leq b$, then $a\,\mu(A) \leq \int_A d\mu \, f(x) \leq b\,\mu(A)$.

Theorem 9.10 *Consider a nonnegative measurable function f. If $\int d\mu \, f = 0$ then $f = 0$ almost everywhere.*

Proof Define $E_n = \{x \mid f(x) > 1/n\}$. Because f is measurable, $E_n \in \mathcal{A}$. Observe that $\int d\mu \, f \geq \int_{E_n} d\mu \, f \geq \mu(E_n)/n$, which implies $\mu(E_n) = 0$, and that

$\{x \mid f(x) > 0\} = \cup_n E_n \in \mathcal{A}$. From the property of the measure, $\mu(\cup_n E_n) \leq \sum_n \mu(E_n) = 0$, which proves the statement. $\square$

▶ **Example 9.18** *(Almost everywhere zero function)* Consider a nonnegative real-valued function f defined over a measure space and assume that almost everywhere $f(x) = 0$. If f is measurable, then $N = \{x \in X \mid f(x) \neq 0\}$ is a measurable set and $\int d\mu \, f = \int_N d\mu \, f + \int_{N^c} d\mu \, f$. The first term of the sum is zero because $\mu(N) = 0$, the second because $f(x) = 0$ for $x \in N^c$. However, in general, if we know that the function f is zero a.e., we cannot conclude that it is measurable. We know for sure only if the metric space is complete.

Another consequence of Definition 9.19 is that, for the property of the supremum, given a nonnegative measurable function f, there are sequences of simple functions (s_n), with $s_n \leq f$, such that their integral converges from below to the integral of f. As the next example shows, it is also possible to build nondecreasing sequences of simple functions pointwise converging to f.

▶ **Example 9.19** *(Simple function approximation)* Consider a nonnegative measurable function f. For any integer n, define the sets

$$E_i^n = \left\{ x \, \middle| \, \frac{i-1}{2^n} \leq f(x) < \frac{i}{2^n} \right\}, i = 1, 2, ..., n2^n,$$

and $\bar{E}^n = \{x \in \mathbb{R} \mid f(x) \geq n\}$. Given n, all E_i^n and $\bar{E}^n$ are mutually disjoint and form a partition of X. Since the function f is measurable, $E_i^n, \bar{E}^n \in \mathcal{A}$. We define the simple function

$$\underline{s}_n(x) = \sum_{i=1}^{n2^n} \frac{i-1}{2^n} I_{E_i^n}(x) + n I_{\bar{E}^n}(x).$$

Note that, $\forall x$, $\underline{s}_n(x) \leq \underline{s}_{n+1}(x) \leq f(x)$ and $\lim_{n \to \infty} \underline{s}_n(x) = f(x)$. In fact, if $n > f(x)$, then $f(x) - \underline{s}_n(x) < 1/2^n$. If the function f is bounded, the convergence is uniform. In this case, for n sufficiently large, $f(x) - \underline{s}_n(x) < 1/2^n$ for any x.

Monotone converging sequences of functions are the subject of a fundamental result of Lebesgue integration theory: for nondecreasing sequences of functions, we can pass the limit inside the integral.

Theorem 9.11 (Monotone convergence) *Let (f_n) be a nondecreasing sequence of nonnegative measurable functions such that $\forall x \in X$, $f_n(x) \leq f_{n+1}(x)$ and $\lim_{n \to \infty} f_n(x) = f(x)$. Then $\lim_{n \to \infty} \int d\mu \, f_n(x) = \int d\mu \, f(x)$.*

Proof Define $\lim_{n\to\infty} \int d\mu \, f_n(x) = L$. Since $f_n(x) \le f(x)$, $L \le \int d\mu \, f$. Consider a simple function $s = \sum_i a_i I_{A_i}$ such that $s(x) \le f(x)$. Take $c \in (0, 1)$ and let $E_n = \{x \mid f_n(x) \ge cs(x)\}$. By the definition of integral,

$$\int d\mu \, f_n \ge \int_E d\mu \, f_n \ge c \int_E d\mu \, s = c \sum_i a_i \mu(A_i \cap E_n).$$

As $f(x) \ge c\, s(x)$, $\lim_{n\to\infty} E_n = X$, so that

$$\lim_{n\to\infty} \sum_i a_i \mu(A_i \cap E_n) = \sum_i a_i \mu(A_i) = \int d\mu \, s.$$

This implies that $\lim_{n\to\infty} \int d\mu \, f_n \ge c \int d\mu \, s$ for any simple function $s \le f$. Thus $L = \lim_{n\to\infty} \int d\mu f_n \ge c \int d\mu \, f$. Since this is true $\forall c \in (0, 1)$ we must have $L \ge \int d\mu \, f$. Because $L \le \int d\mu \, f$, this proves the assertion. $\square$

The pointwise convergence of a sequence of functions, which proved to be rather weak in other cases (for instance, it does not preserve continuity, see Example 5.28), is sufficient, in the Lebesgue theory, to ensure that the integral of the limit is the limit of the integral. In particular, we now know that the integral of a nondecreasing sequence of simple functions pointwise converging to a nonnegative measurable function converges to the integral of that function.

▶ **Example 9.20** (*Lebesgue integral of the linear function*) Consider the function $f(x) = x$ defined over $(\mathbb{R}, \mathcal{B}, l)$. The function is continuous and consequently measurable. We want to compute its Lebesgue integral in the interval $[a, b] \subseteq \mathbb{R}_{\ge 0}$. For $n \in \mathbb{N}$, consider the equispaced partition $\{x_i = a + (b - a)i/n \mid i = 0, \ldots, n\}$ and the simple function $s_n(x) = \sum_{i=1}^{n} x_{i-1} I_{[x_{i-1}, x_i]}$. Notice that $f(x) \ge s_{n+1}(x) \ge s_n(x)$ and $\lim_{n\to\infty} s_n(x) = x$, $\forall x \in [a, b]$. Using the definition of the integral for simple functions, by the same computation as Example 8.3,

$$\int_{[0,1]} dl \, s_n = \sum_{i=1}^{n}(x_i - x_{i-1})x_{i-1} = a(b - a) + \frac{(b - a)^2}{n^2} \frac{n(n + 1)}{2},$$

and, by the monotone convergence theorem,

$$\int_{[0,1]} dl \, x = \lim_{n\to\infty} \int_{[0,1]} dl \, s_n = \frac{1}{2}b^2 - \frac{1}{2}a^2.$$

In this case, the Lebesgue and Riemann integrals yield the same result.

Another important consideration concerns the integral of the limit inferior of sequences of functions. Consider a sequence of nonnegative measurable functions (f_n) and define $\underline{f}(x) = \inf_n f_n(x)$. According to Theorem 9.9, this is a measurable function. For any n, $\underline{f}(x) \leq f_n(x)$, $\int d\mu\, \underline{f}(x) \leq \int d\mu\, f_n(x)$ such that $\int d\mu\, \inf_n f_n(x) \leq \inf_n \int d\mu\, f_n(x)$. This result is not surprising as, in general, different points of the domain correspond to different functions in the sequence that take the smallest values. The previous consideration can be extended to the limit.

Theorem 9.12 (Fatou's lemma) *Consider a sequence of nonnegative and measurable functions* (f_n), *then*

$$\int d\mu \liminf_{n\to\infty} f_n(x) \leq \liminf_{n\to\infty} \int d\mu\, f_n(x).$$

Proof Define $\underline{f}_n(x) = \inf_{i \geq n} f_i(x)$ so that $\liminf_{n\to\infty} f_n(x) = \lim_{n\to\infty} \underline{f}_n(x) = f(x)$. For any $i \geq n$, $\int d\mu\, \underline{f}_n(x) \leq \int d\mu\, f_i(x)$, and consequently $\int d\mu\, \underline{f}_n(x) \leq \inf_{i \geq n} \int d\mu\, f_i(x)$. Note that $\underline{f}_n(x)$ is a nondecreasing sequence. Using the result of Theorem 9.11 and taking the limit for $n \to \infty$ prove the assertion. □

▶ **Example 9.21** (*Limit inferior and Lebesgue integral*) The reason for the presence of the inequality in Theorem 9.12 can be better illustrated with an example. Let I_n be the indicator function relative to the interval $A_n = [-1/n, 1/n]$ and consider the sequence of simple functions $f_n(x) = n I_n$. Using the Lebesgue measure, $\int d\mu\, f_n = n\mu(A_n) = 2$ for any n, so that $\liminf_{n\to+\infty} \int d\mu\, f_n = 2$. On the other hand, $\liminf_{n\to+\infty} f_n(x) = 0$ if $x \neq 0$, so that $\int d\mu \liminf_{n\to+\infty} f_n(x) = 0$, as $\{0\}$ is a measure zero set.

To extend the notion of integral to functions that can take negative values, we need a preliminary definition.

Definition 9.20 (*Integrable function*) A measurable function f is *integrable* on the set $A \in \mathcal{A}$ if $\int_A d\mu\,|f| < +\infty$.

A function is integrable if it is integrable on the entire space.

Definition 9.21 (*Integral of integrable functions*) Consider an integrable function f and define $E_f^{\pm}(x) = \{x \mid f(x) \lessgtr 0\}$. Then its Lebesgue integral is defined as $\int d\mu\, f = \int_{E_f^+} d\mu\, f - \int_{E_f^-} d\mu\,(-f)$.

Alternatively, we can define the integral of an integrable function f as $\int d\mu\, f = \int d\mu\,(|f|+f)/2 - \int d\mu\,(|f|-f)/2$. The functions to be integrated in the two terms of the sum are measurable and nonnegative. Definition 9.21 is restricted to integrable functions because if both $\int_{E_f^+} d\mu\, f$ and $\int_{E_f^-} d\mu\,(-f)$ are infinite, it reduces to an indeterminate expression. From Definition 9.21, it follows that $\int d\mu\,|f| \geq |\int d\mu\, f|$.

▶ **Example 9.22** *(Chebyshev inequality)* Let f be an integrable function and $\forall a > 0$, consider the measurable set $E(a) = \{x \in X \mid |f(x)| \geq a\}$. Note that $\forall p \geq 1$, $|f(x)|^p/a^p \geq 1$ if $x \in E(a)$. Thus,

$$\frac{1}{a^p} \int d\mu \, |f|^p \geq \int_{E(a)} d\mu \, \frac{|f|^p}{a^p} \geq \int_{E(a)} d\mu = \mu(E(a)).$$

This inequality can be used to find an upper bound of the measure of the set $E(a)$ or a lower bound of the integral of $|f|^p$.

The Lebsegue integral is linear on integrable functions. For any $c \in \mathbb{R}$, $\int d\mu \, cf = c \int d\mu \, f$. Concerning the integral of the sum of functions, we have the following.[2]

Theorem 9.13 *If f and g are integrable, then $f + g$ is integrable and $\int d\mu \, (f+g) = \int d\mu \, f + \int d\mu \, g$.*

Proof The proof is carried out in three steps. First, we will prove it for simple functions, then for nonnegative measurable functions, and finally for all measurable functions.

Let $s_1 = \sum_i a_i I_{A_i}$ and $s_2 = \sum_i b_i I_{A_i}$ be simple functions that, without loss of generality, we can assume are defined using the same collection of measurable sets (A_i). Then $\int d\mu \, (s_1 + s_2) = \sum_i (a_i + b_j)\mu(A_i) = \sum_i a_i \mu(A_i) + \sum_i b_i \mu(A_i) = \int d\mu \, s_1 + \int d\mu \, s_2$.

Next, let f and g be nonnegative integrable functions. According to Theorem 9.8, $f + g$ is a nonnegative measurable function. Consider two nondecreasing sequences of simple functions that converge from below to the two functions, $\lim_{n \to \infty} s_{f,n} = f$ and $\lim_{n \to \infty} s_{g,n} = g$ (see Example 9.19). Then $(s_{f,n} + s_{g,n})$ is a sequence of simple functions converging from below to $f + g$. By Theorem 9.11,

$$\int d\mu \, (f + g) = \lim_{n \to \infty} \int d\mu \, (s_{f,n} + s_{g,n}) =$$

$$\lim_{n \to \infty} \int d\mu \, s_{f,n} + \lim_{n \to \infty} \int d\mu \, s_{g,n} = \int d\mu \, f + \int d\mu \, g.$$

For the last step of the proof, note that given two integrable functions f and g, $f + g$ is integrable. In fact, $|f + g| \leq |f| + |g|$ and using the previous step of the proof, $\int d\mu \, |f + g| \leq \int d\mu \, (|f| + |g|) = \int d\mu \, |f| + \int d\mu \, |g|$. Consider the following algebraic equality:

$$(|f + g| + f + g) + (|f| - f) + (|g| - g) =$$
$$(|f + g| - f - g) + (|f| + f) + (|g| + g).$$

---

[2] The reason why this theorem is not stated for generic measurable functions is because the integral of measurable functions may be infinite and arithmetic operations between infinities are not properly defined.

For the previous consideration, all the terms in parentheses are positive integrable functions, and, when taking the integral, we can apply it separately to each term of both sides. Rearranging the integrals, we finally get

$$\int d\mu \; (|f + g| + f + g) - \int d\mu \; (|f + g| - f - g) =$$

$$\int d\mu \; (|f| + f) - \int d\mu \; (|f| - f) + \int d\mu \; (|g| + g) - \int d\mu \; (|g| - g).$$

The left-hand side is $2 \int d\mu \; (f + g)$ and the right-hand side $2 \int d\mu \; f + 2 \int d\mu \; g$. $\square$

We conclude this section by reviewing several results that directly come from the definition of a Lebesgue integral and are often used in applications. We start by showing that if we can find an integrable function as an upper bound, we can apply Theorem 9.12 and extend Theorem 9.11 to any pointwise converging sequence of functions.

Theorem 9.14 (Dominated convergence) *Let (f_n) be a converging sequence of measurable functions, $\lim_{n \to \infty} f_n(x) = f(x)$. If there exists an integrable function $g(x)$ such that $\forall n$, $|f_n(x)| \leq g(x)$, then $\lim_{n \to \infty} \int d\mu \; f_n = \int d\mu \; f$.*

Proof Note that the functions f_n are integrable by hypothesis, as $\int d\mu \; |f_n(x)| \leq \int d\mu \; |g(x)| < +\infty$. Apply Theorem 9.12 to the sequences of nonnegative functions $(g + f_n)$ and $(g - f_n)$ to obtain

$$\liminf_{n \to \infty} \int d\mu \; (g + f_n) \geq \int d\mu \; \liminf_{n \to \infty} (g + f_n),$$

$$\liminf_{n \to \infty} \int d\mu \; (g - f_n) \geq \int d\mu \; \liminf_{n \to \infty} (g - f_n).$$

From the linearity of the integral, remembering that when the elements of a sequence are multiplied by -1, the inferior limit of the new sequence is the opposite of the superior limit of the original sequence, and simplifying the finite integral of g,

$$\liminf_{n \to \infty} \int d\mu \; f_n \geq \int d\mu \; \liminf_{n \to \infty} f_n = \int d\mu \; f =$$

$$= \int d\mu \; \limsup_{n \to \infty} f_n \geq \limsup_{n \to \infty} \int d\mu \; f_n,$$

which proves the assertion. $\square$

Theorems 9.11 and 9.14 are powerful results. The following example shows that they are not generally valid in the case of Riemann integrable functions.

▶ **Example 9.23** (*Failure of monotone and dominated convergence for Riemann integrals*) For any natural number n, consider the real function $f_n : [0, 1] \to \mathbb{R}$,

$$
f_n(x) = \begin{cases} 0 & \text{if } x \in \mathbb{R}\backslash\mathbb{Q}, \\ 1 & \text{if } x \in \mathbb{Q} \text{ and } x = p/q \text{ with } q \leq n, \\ \frac{1}{q} & \text{if } x \in \mathbb{Q} \text{ and } x = p/q \text{ with } q > n, \end{cases}
$$

where p, q are mutually prime natural numbers. The function $f_n(x)$ differs from Thomae's function in Example 8.4 only in a finite set of points; thus, we can conclude that its Riemann integral exists and is equal to zero for any n. Moreover, note that for any n, $f_n(x) \geq f_{n-1}(x)$ and $f_n(x) \leq 1$. On the other hand, $f(x) = \lim_{n\to\infty} f_n(x)$ is Dirichlet's function of Example 8.2, which is not Riemann integrable. In summary, the sequence of Riemann integrable functions (f_n) is nondecreasing and bounded, but its limit is not Riemann integrable. We showed that, in general, neither the theorem of monotone convergence nor the theorem of dominated convergence apply to Riemann integrals.

The second application is a useful generalisation of the Jensen inequality discussed in Corollary 1.1 from finite summations to integrals.

Corollary 9.5 (Jensen's inequality, integral form) *Let g be integrable on a finite measure set A and let $f : \mathbb{R} \to \mathbb{R}$ be a bounded function such that $f \circ g$ is also integrable. Then, if f is concave,*

$$
f\left(\frac{1}{\mu(A)} \int_A d\mu\, g\right) \geq \frac{1}{\mu(A)} \int_A d\mu\, f \circ g,
$$

while if f is convex,

$$
f\left(\frac{1}{\mu(A)} \int_A d\mu\, g\right) \leq \frac{1}{\mu(A)} \int_A d\mu\, f \circ g.
$$

Proof Assume that f is concave. Then $\forall x_0$ in its domain, there exists a $c(x_o)$ such that $f(x) \leq f(x_o) + c(x_0)(x - x_0)$ (see the discussion after Theorem 6.4) so that $f(g(x)) \leq f(x_0) + c(x_0)(g(x) - x_0)$. Integrating both sides

$$
\int_A d\mu\, f \circ g \leq f(x_0)\, \mu(A) + c(x_0)\left(\int_A d\mu\, g - \mu(A)\, x_0\right).
$$

Setting $x_0 = \int_A d\mu\, g/\mu(A)$ proves the first assertion. If instead f is convex, then $\forall x_0$ in its domain, there exists a $c(x_o)$ such that $f(x) \geq f(x_o) + c(x_0)(x - x_0)$, and repeating the reasoning above, the second assertion is proved.   □

Fig. 9.1 Relation between spaces and functions in Theorem 9.15

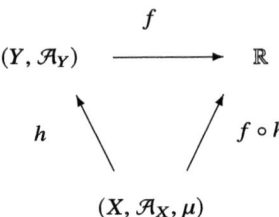

If the measure is finite, then the set A can be the entire space. Corollary 9.5 is often stated by assuming $\mu(A) = 1$.

Finally, two results are provided that expand on similar results obtained in Chap. 8 for the Riemann integral. The first result is the derivation of the general "change of variable" procedure for Lebesgue integrals. We have a real-valued function $f : Y \to \mathbb{R}$ defined over a space and a map $h : X \to Y$ between two spaces. Both f and g are measurable. We want to compute the integral of the composed function $f \circ h : X \to \mathbb{R}$ with respect to a measure μ on X (see Fig. 9.1).

Theorem 9.15 (Substitution Theorem) *Let $(X, \mathcal{A}_X)$ and $(Y, \mathcal{A}_Y)$ be two measurable spaces and $h : X \to Y$ be a measurable function between them. Consider the measure μ on $(X, \mathcal{A}_X)$ and a measurable function $f : Y \to \mathbb{R}$. The function $f \circ h$ is integrable on X with respect to μ if and only if the function f is integrable on Y with respect to the pushforward measure $\mu \circ h^{-1}$ and*

$$\int d\mu \, f \circ h = \int d(\mu \circ h^{-1}) \, f.$$

Proof We will directly prove the last equality. Let $s = \sum_{i=1}^{n} a_i I_{A_i}$ be a simple function on $(Y, \mathcal{A}_Y)$. Because h is measurable, the function $s \circ h = \sum_{i=1}^{n} a_i I_{h^{-1}(A_i)}$ is a simple function on $(X, \mathcal{A}_X)$. From the definition of pushforward measure,

$$\int d(\mu \circ h^{-1}) \, s = \sum_{i=1}^{n} a_i \mu(h^{-1}(A_i)) = \int d\mu \, s \circ h.$$

Therefore, the statement is true for simple functions. Let f be a nonnegative measurable function on $(Y, \mathcal{A}_Y)$ and (s_n) a nondecreasing sequence of simple functions that converges to it. Then the sequence $(s_n \circ h)$ is a nondecreasing sequence of simple functions that converges to $f \circ h$, and using Theorem 9.11, the statement is proved. Finally, for a generic integrable function f, just apply the previous result to the nonnegative integrable functions $f I_{\{f > 0\}}$ and $-f I_{\{f < 0\}}$. $\qquad\qquad\square$

If the integral is performed over a measurable set $A \in \mathcal{A}_X$, $\int_A d\mu \, f \circ h = \int_{h(A)} d(\mu \circ h^{-1}) \, f$.

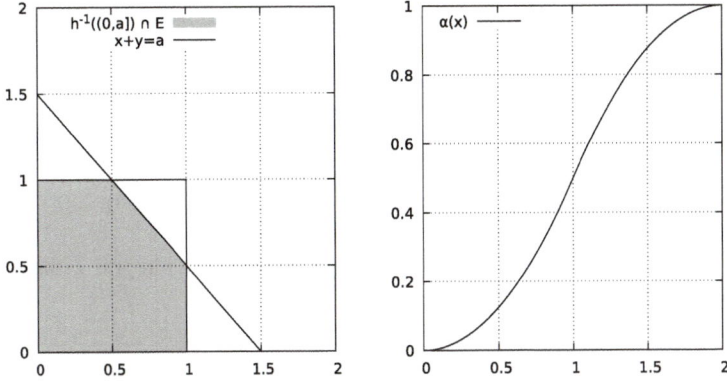

Fig. 9.2 The preimage of an interval (left) and the Stieltjes pushforward measure induced by the function h (right) of Example 9.24

▶ **Example 9.24** *(Simple double integral)* Consider a Borel measurable function $f : \mathbb{R} \to \mathbb{R}$ and the rectangle $E = [0, 1] \times [0, 1]$. We want to compute $\int_E dl_2 \, f(x + y)$. The function $f(x + y)$ is the composition of a Borel measurable function on $\mathbb{R}$ with a continuous function from $\mathbb{R}^2$ to $\mathbb{R}$, $h(x, y) = x + y$, so it is Borel measurable. Using Theorem 9.15 with $h(x, y) = x + y$, $\int_E dl_2 \, f(x + y) = \int_{h(E)} d(l_2 \circ h^{-1}) f$. Note that $h(E) = [0, 2]$. To compute the pushforward measure, define

$$\alpha(a) = l_2 \circ h^{-1}((0, a] \cap h(E)) = l_2(h^{-1}((0, a]) \cap E).$$

Because $h^{-1}((0, a]) = \{(x, y) \mid 0 < x + y \le a\}$, if $a < 0$ the argument of l_2 is an empty set. If $a > 0$, it is the intersection of the rectangle E with an isosceles right angled triangle in the first quadrant with catheti of length a along the axis. A simple geometric computation (see Fig. 9.2 for an example) delivers the area of this surface,

$$\alpha(a) = \begin{cases} 0 & a \le 0, \\ a^2/2 & 0 \le a \le 1, \\ 2a - a^2/2 - 1 & 1 \le a \le 2, \\ 1 & 2 \le a. \end{cases}$$

Since for any semi-open interval $I = (a, b]$, $l_2 \circ h^{-1}(I) = \alpha(b) - \alpha(a)$, we conclude that the pushforward measure is the Lebesgue–Stieltjes measure l_α, so that $\int_E dl_2 \, f(x + y) = \int_{[0,2]} dl_\alpha f$.

▶ **Example 9.25** *(Continuous and strictly increasing Stieltjes measure)* Consider a Borel measurable function $f : A \subseteq \mathbb{R} \to \mathbb{R}$ and let $\alpha : \mathbb{R} \to \mathbb{R}$ be a continuous strictly increasing function. Apply Theorem 9.15 with $\mu = l_\alpha$, $h = \alpha$, and $E = \alpha^{-1}(A)$ to get $\int_{\alpha^{-1}(A)} dl_\alpha \, f \circ \alpha = \int_A d(l_\alpha \circ \alpha^{-1}) \, f$. Note that because α is continuous with a continuous inverse, $l_\alpha \circ \alpha^{-1}((a, b]) = l_\alpha((\alpha^{-1}(a), \alpha^{-1}(b)]) = (a, b]$. Hence,

$l_\alpha \circ \alpha^{-1}$ is the Lebesgue measure l and we recover the usual change of variable formula:

$$\int_{\alpha^{-1}(A)} dl_\alpha \, f \circ \alpha = \int_A dl \, f.$$

If α is not continuous with a continuous inverse, then, in general, $l_\alpha \circ \alpha^{-1} \neq l$. For example, let $\alpha = I_{\{x \geq 0\}}$. Then $l_\alpha \circ \alpha^{-1}((0, 2]) = l_\alpha([0, +\infty)) = 0$.

The second result is about the derivative of the Lebesgue integral with respect to a parameter of the integrated function.

Corollary 9.6 (Derivative of the Lebesgue integral) *Let $f(x, t) : \mathbb{R} \times \mathbb{R} \to \mathbb{R}$ be a Lebesgue measurable function in x for all t. Assume that the function can be derived with respect to t and that there exists an integrable function g such that $|\partial_t f| < g$. Then the function $I(t) = \int dl(x) f(x, t)$ can be derived and*

$$\frac{d}{dt} I(t) = \int dl(x) \, \partial_t f(x, t).$$

Proof By assumption, $\lim_{n \to \infty} n \, (f(x, t + 1/n) - f(x, t)) = \partial_t f$. For Theorem 9.9, given the hypothesis, this guarantees that $\partial_t f$ is a measurable function. Note that for sufficiently large n, $n|f(x, t + 1/n) - f(x, t)| < g(x)$. Then, by dominated convergence (Theorem 9.14),

$$\frac{d}{dt} I(t) = \lim_{n \to \infty} n \, (I(t + 1/n) - I(t)) = \lim_{n \to \infty} \int dl(x) n \, (f(x, t + 1/n) - f(x, t)) =$$

$$\int dl(x) \lim_{n \to \infty} n \, (f(x, t + 1/n) - f(x, t)) = \int dl(x) \, \partial_t f(x, t),$$

proving the assertion. $\square$

There is nothing special about the Lebesgue measure in the previous result. You can replace it with any other measure defined on $\mathbb{R}$, for example, with a Lebesgue–Stieltjes measure. Note that the requirements of Corollary 9.6 are weaker than those of Theorem 8.12. In particular, the derivative $\partial_t f$ is not required to be continuous.

▶ **Example 9.26** *(Mollification) Mollification* is the procedure with which a smooth approximation, in general infinitely derivable, is built to a given continuous function. We will see how it can be done via *convolution* with appropriately smooth functions.

Let ϕ be a nonnegative measurable function such that $\phi(x) = 0$ if $|x| > 1$ and $\int dl \, \phi = \int_{-1}^{1} dl \, \phi = 1$. Then $\forall \eta > 0$, the function $\phi_\eta(x) = \eta \, \phi(x/\eta)$ is such that $\phi_\eta(x) = 0$ if $|x| > \eta$ and $\int dl \, \phi_\eta = \int_{-\eta}^{\eta} dl \, \phi_\eta = 1$ (use the change of variable). The *convolution* of the function ϕ_η with a measurable function f is defined as

$$\tilde{f}_\eta(x) = \int dl(y) \, f(y) \phi_\eta(x - y).$$

If $\phi \in C^m$, trivially $\phi_\eta \in C^m$ but also $\tilde{f}_\eta \in C^m$. In fact, using Corollary 9.6, $\forall j \leq m$,

$$\frac{d^j \tilde{f}_\eta}{dx^j} = \int dl(y) \, f(y) \, \phi_\eta^{(j)}(x - y).$$

Finding an integrable function that dominates the integrand to apply Corollary 9.6 is easy and the search is left to the reader. A suitable C^∞ function that can serve the purpose of "smoothing" any function f is (see Exercise 6.5)

$$\phi(x) = \begin{cases} 0 & |x| \geq 1 \\ c \, e^{-1/(1-x^2)} & |x| < 1, \end{cases}$$

where $c > 0$ is selected to normalise the integral of ϕ to 1.

Consider now a continuous function f and let the interval $[a, b]$ be internal to its domain. Then, f is uniformly continuous in any closed interval $[a - \delta', b + \delta']$ for a sufficiently small $\delta' > 0$. This implies that $\forall \epsilon > 0$ there exists a $\delta > 0$ such that $|f(x) - f(x')| < \epsilon$ if $x, x' \in [a - \delta', b + \delta']$ and $|x - x'| < 2\delta$. Take $\eta < \min\{\delta, \delta'\}$, then $\forall x \in [a, b]$,

$$|\tilde{f}_\eta(x) - f(x)| = \left| \int_{x-\eta}^{x+\eta} dl(y) f(y) \phi_\eta(x - y) - f(x) \right| \leq$$

$$\int_{x-\eta}^{x+\eta} dl(y) \, |f(y) - f(x)| \, \phi_\eta(x - y) < \epsilon,$$

where we have used the translation invariance of the Lebesgue measure,

$$\int_{x-\eta}^{x+\eta} dl(y) \phi_\eta(x - y) = \int_{-\eta}^{\eta} dl(y) \phi_\eta(y) = 1, \forall x.$$

The same procedure can be applied to a continuous function $f : E \subseteq \mathbb{R}^n \to \mathbb{R}$. Consider a compact set $K \subset \text{int } E$, and let $K' \supset K$ be a compact set such that $K' \subseteq E$ and $\delta' = \inf\{\|\mathbf{x} - \mathbf{y}\|, \mathbf{x} \in K, \mathbf{y} \in \partial K'\} > 0$. Then, for a sufficiently small $\eta < \delta'$, the function

$$\tilde{f}_\eta(\mathbf{x}) = \int dl_n(\mathbf{y}) \, f(\mathbf{y}) \, \phi_\eta(\|\mathbf{x} - \mathbf{y}\|)$$

is such that $\forall \epsilon > 0$, $|\tilde{f}_\eta - f| < \epsilon$ on the whole K. To prove it, exploit the fact that f is uniformly continuous in K'. Note that the coefficient c in the definition of ϕ should be modified to take into account the fact that the integral is now performed over a n-dimensional ball (see Exercise 9.22). If K is made of a finite union of disjoint pieces, one can apply the theorem on each piece separately, which means that the function f needs not to be everywhere continuous. Due to Theorem 4.11, one can also use, with little modification, any other norm instead of the Euclidean one.

We have found that any continuous function can be uniformly approximated on any compact subset of its domain by a smooth function, with infinitesimal precision. This is an important result in applications because it means that, under many circumstances, we can actually assume that the continuous functions we are dealing with are in fact smooth. Once we assume continuity, we get smoothness almost for free.

The results in this section, most notably monotone convergence (Theorem 9.11) and dominated convergence (Theorem 9.14), are not modified if the sequence of functions considered fails to converge on a set of measure zero. If N is that set, simply replace any function f with the function $f I_{N^c}$. This substitution will not modify the result of any integral. However, in applications, it might be difficult to prove that the set on which convergence fails is a set of measure zero. It is often simpler to prove that it is a null set. In this case, complete metric spaces turn out to be really useful. Now every null set is measurable by construction, and the previous trick can be adopted without any problems. In particular, as we shall see below, the completeness of the Lebesgue measure is essential for the proper extension of the Riemann integration theory.

9.3.1 Lebesgue and Riemann Integral on $\mathbb{R}$

When we need to actually compute an integral, the best way to proceed is often the Riemann way, possibly exploiting the powerful result of Theorem 8.9. This section compares the Lebesgue and Riemann integrals on $\mathbb{R}$. Through this comparison, we will learn how to use the Riemann theory to compute Lebesgue integrals. We will also learn the necessary and sufficient conditions for a function to be Riemann integrable.

Consider the nonnegative bounded function $f : [a, b] \subset \mathbb{R} \to \mathbb{R}_{\geq 0}$. For any partition $P = \{x_0 = a, < x_1 < \ldots < x_n = b\}$ of $[a, b]$ consider the lower sum $L(P)$ and the upper sum $U(P)$ of Definition 8.2. Note that

$$s_P^L(x) = \sum_{i=0}^{n-1} I_{(x_i, x_{i+1}]}(x) \inf_{[x_i, x_{i+1}]} f$$

and

$$s_P^U(x) = \sum_{i=0}^{n-1} I_{(x_i, x_{i+1}]}(x) \sup_{[x_i, x_{i+1}]} f$$

are simple functions such that $\int dl\, s_P^L(x) = L(P)$ and $\int dl\, s_P^U(x) = U(P)$. In this way, we have built a correspondence between the partitions of the interval $[a, b]$ and a collection of special simple functions defined on it. This correspondence will be useful to prove the next two theorems.

Theorem 9.16 *If $f \geq 0$ and $f \in \mathcal{R}([a, b])$, then f is continuous l-a.e. in $[a, b]$ and its Lebesgue integral is equal to its Riemann integral,*

$$\int_{[a,b]} dl\, f = \int_a^b dx f(x).$$

Proof Because f is Riemann integrable in $[a, b]$, according to Definition 8.3, there exists a sequence of partitions (P_n) with $P_n \subset P_{n+1}$, such that

$$\lim_{n\to\infty} (U(P_n) - L(P_n)) = \lim_{n\to\infty} \int dl \left(s_{P_n}^U(x) - s_{P_n}^L(x)\right) = 0.$$

The functions $s_{P_n}^U - s_{P_n}^L$ are all bounded above by $\sup_{[a,b]} f - \inf_{[a,b]} f$. Thus, by dominated convergence (Theorem 9.14), we can pass the limit inside the integral and, remembering Theorem 9.10, conclude that l-a.e. $\lim_{n\to\infty} s_{P_n}^U(x) - s_{P_n}^L(x) = 0$. Let x be a point such that the last limit is zero. We will prove that the function $f(x)$ is continuous in x. First, we can assume that x does not belong to the partitions P_n. In fact, if the partition P_n is such that $x_i < x = x_{i+1} < x_{i+2}$, one can take a new partition P_n' by removing the point x_{i+1} and adding the points x_{i+1}' and x_{i+1}'' such that $x_i < x_{i+1}' < x < x_{i+1}'' < x_{i+2}$. From the definition of supremum and infimum, it is immediate to verify that $s_{P_n'}^U \leq s_{P_n}^U$ and $s_{P_n'}^L \geq s_{P_n}^L$. Thus, the limit of the sequence is still zero if the new sequence of partitions (P_n') is considered. For any $\epsilon > 0$, take n sufficiently large such that $s_{P_n'}^U(x) - s_{P_n'}^L(x) < \epsilon$. Then there exists an interval $I_n = [x_{i_n}, x_{i_n+1}]$, with $x_{i_n}, x_{i_n+1} \in P_n'$, such that $x \in I_n$ and $\sup_{I_n} f - \inf_{I_n} f \leq \epsilon$. This implies that in any open neighbourhood $x \in N(x) \subset I_n$, $|f(y) - f(x)| < \epsilon, \forall y \in N(x)$, which proves that the function is continuous in x. Then let $A = \{x \mid \lim_{n\to\infty} s_{P_n}^U(x) - s_{P_n}^L(x) = 0\}$ and notice that

$$\int_{[a,b]} dl\, s_{P_n}^L(x) = \int_A dl\, s_{P_n}^L(x)$$

because $l([a, b] \cap A^c) = 0$. For any $x \in A$, $\lim_{n\to\infty} s_{P_n}^L(x) = f(x)$ and the simple functions $s_{P_n}^L$ are nondecreasing in n. Thus, by monotone convergence (Theorem 9.11),

$$\int_a^b dx\, f(x) = \lim_{n\to\infty} \int_{[a,b]} dl\, s_{P_n}^L = \lim_{n\to\infty} \int_A dl\, s_{P_n}^L =$$

$$\int_A dl \lim_{n\to\infty} s_{P_n}^L = \int_A dl\, f = \int_{[a,b]} dl\, f,$$

proving the assertion. $\qquad\square$

The previous theorem establishes a necessary condition for a nonnegative bounded function f to be integrable on an interval $[a, b]$: the set of its discontinuities must have Lebesgue measure zero. The following theorem clarifies that this condition is also sufficient.

Theorem 9.17 *If $f \geq 0$ is bounded and continuous l-a.e. on $[a, b]$, then the Riemann integral of f on $[a, b]$ exists and is equal to the Lebesgue integral.*

Proof Let A be the set of points in $[a, b]$ where the function f is continuous. Let P_n be the equispaced partition that divides $[a, b]$ into n equal parts. If $x \in A$, then $\lim_{n \to \infty} s_{P_n}^L(x) = f(x) = \lim_{n \to \infty} s_{P_n}^U(x)$. Thus, by dominated convergence (Theorem 9.14),

$$\lim_{n \to \infty} L(P_n) = \lim_{n \to \infty} \int_A dl \, s_{P_n}^L = \int_A dl \, \lim_{n \to \infty} s_{P_n}^L =$$
$$\int_A dl \, f(x) = \lim_{n \to \infty} \int_A dl \, s_{P_n}^U = \lim_{n \to \infty} U(P_n).$$

But since $l([a, b] \cup A^c) = 0$, $\int_A dl \, f = \int_{[a,b]} dl \, f$.   □

In summary, a nonnegative bounded function is Riemann integrable in an interval $[a, b]$ if and only if the Lebesgue measure of the set of points in which it is discontinuous is zero. In previous theorems, we have made use of the fact that the Lebesgue measure is complete. In fact, we have assumed that if some property is valid almost everywhere, the set of points which violate it is measurable. Thus, set-wise, we can remove those points from a measurable domain and remain with a measurable set. A similar analysis would not have been possible using an incomplete measure space, such as the Borel measure.

The extension of the previous results to generic bounded functions is made by separately considering the set in which the function is positive and that in which it is negative.

▶ **Example 9.27** *(Change of variable in Lebesgue integrals)* Let α be a strictly increasing differentiable function with a Riemann integrable derivative α'. On $(\mathbb{R}, \mathcal{L})$ consider the measure $\mu(A) = \int_A dl \, \alpha'$. For any semi-open interval

$$\mu((a, b]) = \int_a^b dl \, \alpha' = \alpha(b) - \alpha(a) = l_\alpha((a, b]),$$

where we have used the equivalence between the Lebesgue and the Riemann integral. Thus, we can conclude that $\mu = l_\alpha$. From Example 9.25, we know that for any Lebesgue measurable set E,

$$\int_E dl \, f = \int_{\alpha^{-1}(E)} dl_\alpha \, f \circ \alpha,$$

and substituting the alternative definition of the Lebesgue–Stieltjes measure with differentiable and invertible function, we have that

$$\int_E dl\, f = \int_{\alpha^{-1}(E)} dl\, \alpha'\, f \circ \alpha.$$

This is the Riemann integral change of variable discussed in Theorem 8.10.

The analysis of the Lebesgue–Stieltjes measure, described in Definition 9.11, is analogous to the Lebesgue case. The relation between the upper and lower sums and the Lebesgue integral, computed according to the Lebesgue–Stieltjes measure, of suitably defined simple functions is still valid. To prove Theorems 9.16 and 9.17, we have only used this relation and general results from measure theory. The requirement of the integrated function to have a zero measure set of discontinuities remains. In this case, however, consider that the zero measure sets can also be entire intervals, where the function α is constant. In order to ensure the existence of the Riemann–Stieltjes integral, the function f must be continuous at the points where α is discontinuous. Nonetheless, as long as the set of these points has Lebesgue measure zero, the Lebesgue–Stieltjes integral exists.

▶ **Example 9.28** *(Lebesgue integral of a Stieltjes measure)* Consider the function $\alpha = I_{x \geq 0}$. We want to integrate α over $A = [-1, 1]$ using α itself as a measure. For the Lebesgue–Stieltjes integral, α is a simple function and $\int_A dl_\alpha\, \alpha = l_\alpha(A \cap [0, \infty)) = l_\alpha([0, 1]) = \alpha(1) - \alpha(0) = 0$. Conversely, the Riemann–Stieltjes integral $\int_{-1}^{1} d\alpha\, \alpha$ does not exist.

9.4 Product Measure Space

In this section, we will see how to use the algebras and measures defined on two σ-finite measure spaces $(X, \mathcal{A}_X, \mu_X)$ and $(Y, \mathcal{A}_Y, \mu_Y)$ to define an algebra and a measure on their Cartesian product $X \times Y$. We start with generic set-theoretic definitions. Given any set $E \subseteq X \times Y$, and any point $x \in X$ and $y \in Y$, let $S_x(E) = \{y \mid (x, y) \in E\} \subseteq Y$ be the set of points of Y that appears in at least one ordered couple with x, and $S_y(E) = \{x \mid (x, y) \in E\} \subseteq X$ the set of points of X that appears in at least one ordered couple with y. These are "slices" of the subset in the product space, associated with elements of the original spaces; see Fig. 9.3. Depending on the points x and y and on the set E, $S_x(E)$ and $S_y(E)$ can be a proper subset, the whole space, or the empty set.

Fig. 9.3 The set $E \subset X \times Y$
sliced along $x \in X$ and
$y \in Y$

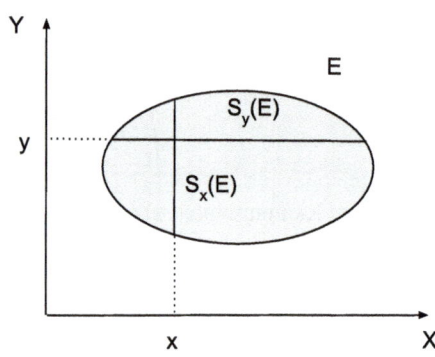

9.4.1   Product σ-Algebra

Given two generic subsets $A_X \subseteq X$ and $A_Y \subseteq Y$, define the *rectangle* $A_X \times A_Y = \{(x, y) \mid x \in A_x, y \in A_Y\}$. Then consider the collection of *measurable rectangles*

$$\mathcal{R}_{X \times Y} = \{A_X \times A_Y \mid A_X \in \mathcal{A}_X, A_Y \in \mathcal{A}_Y\},$$

obtained considering all measurable subsets of the original spaces. If $E \in \mathcal{R}_{X \times Y}$, then $\forall x \in X, S_x(E) \in \mathcal{A}_Y$, and $\forall y \in Y, S_y(E) \in \mathcal{A}_x$. This is easy to see, as $S_x(A_X \times A_Y)$ is A_Y if $x \in A_X$ and $\emptyset$ otherwise. The same applies to S_y.

Definition 9.22 *(Product σ-algebra)* The *product σ-algebra* $\mathcal{A}_{X \times Y}$ on the set $X \times Y$ is the σ-algebra generated by the measurable rectangles $\mathcal{A}_{X \times Y} = \sigma(\mathcal{R}_{X \times Y})$. The measurable space $(X \times Y, \mathcal{A}_{X \times Y})$ is the *product measurable space*.

▶ **Example 9.29** *(Discrete and trivial product σ-algebra)* If $\mathcal{A}_X = 2^X$ and $\mathcal{A}_X = 2^Y$, then according to Definition 9.22, $\mathcal{A}_{X \times Y} = 2^{X \times Y}$. If $\mathcal{A}_X = \{\emptyset, X\}$ and $\mathcal{A}_Y = \{\emptyset, Y\}$, then $\mathcal{A}_{X \times Y} = \{\emptyset, X \times Y\}$, as $\emptyset \times Y = X \times \emptyset = \emptyset$.

▶ **Example 9.30** *(Product measure and Borel measure in $\mathbb{R}^n$)* Consider the Borel measurable spaces $(\mathbb{R}, \mathcal{B})$ and $(\mathbb{R}^2, \mathcal{B}_2)$, where $\mathcal{B}$ is the σ-algebra generated by the Euclidean topology T in $\mathbb{R}$ and $\mathcal{B}_2$ is the σ-algebra generated by the Euclidean topology T_2 in $\mathbb{R}^2$. Let $\mathcal{R}$ be the set of all rectangles $(a, b) \times (c, d)$, where (a, b) and (c, d) are open intervals of $\mathbb{R}$.

Since $\mathcal{R}$ generates T_2 (see Exercise 4.7), $\mathcal{R} \subseteq T_2$ and $\mathcal{B} \times \mathcal{B} = \sigma(\mathcal{R}) \subseteq \sigma(T_2) = \mathcal{B}_2$. Likewise, since the elements of T_2 can be written as countable unions of the elements of $\mathcal{R}$ (see Theorem 4.8 and Exercise 4.13), $T_2 \subseteq \sigma(\mathcal{R})$, that is $\mathcal{B}_2 \subseteq \mathcal{B} \times \mathcal{B}$. Thus, we can conclude that $\mathcal{B}_2 = \mathcal{B} \times \mathcal{B}$. That is, the Borel σ-algebra on $\mathbb{R}^2$ is the product of the Borel σ-algebra on $\mathbb{R}$ with itself. The same reasoning applies to the product σ-algebra $\mathcal{B}_n$ of any finite-dimensional normed space $\mathbb{R}^n$, $\mathcal{B}_n = \times_{i=1}^n \mathcal{B} = \mathcal{B}^n$.

The next theorem shows that all sets of the product σ-algebra have measurable slices.

Theorem 9.18 *For all $(x, y) \in X \times Y$ and $E \in \mathcal{A}_{X \times Y}$, $S_x(E) \in \mathcal{A}_Y$ and $S_y(E) \in \mathcal{A}_X$.*

Proof Fix $x \in X$ and define $C_x = \{E \subseteq \mathcal{A}_{X \times Y} \mid S_x(E) \in \mathcal{A}_Y\}$. This is the set of all elements of the product σ-algebra whose x-slices are measurable with respect to the measure defined on Y. Notice that $\mathcal{R}_{X \times Y} \subseteq C_x$. Thus, since $\mathcal{A}_{X \times Y}$ is the smallest σ-algebra that contains $\mathcal{R}_{X \times Y}$, if we prove that C_x is a σ-algebra, we have proved that $C_x = \mathcal{A}_{X \times Y}$, because, by definition, $C_x \subseteq \mathcal{A}_{X \times Y}$.

First, notice that both $\emptyset$ and $X \times Y$ trivially belong to C_x. In fact, they are rectangles. Moreover, $\forall E \in C_x$, $S_x(E^c) = \{y \mid (x, y) \notin E\} - \{y \mid y \notin S_x(E)\} - S_x(E)^c \in \mathcal{A}_y$. Thus, $E^c \in C_x$ and we have proved that C_x contains the complement of all its sets. Finally, $S_x(\bigcup_{i=1}^{\infty} E_i) = \bigcup_{i=1}^{\infty} S_x(E_i)$, so that C_x is closed under countable union.

The analysis is identical for y-slices.   □

The previous result serves to prove that the restrictions of a measurable function on a product measurable space are measurable in the original spaces.

Theorem 9.19 *Let $f : X \times Y \to Z$ be a measurable function from the product space $(X \times Y, \mathcal{A}_{X \times Y})$ to $(Z, \mathcal{A}_Z)$. Then $\forall x \in X$ and $\forall y \in Y$, the restrictions $f_x : Y \to Z$ with $f_x(y) = f(x, y)$ and $f_y : X \to Z$ with $f_y(x) = f(x, y)$ are measurable functions in the respective domain spaces.*

Proof It suffices to observe that $f_x^{-1}(A_Z) = S_x(f^{-1}(A_Z))$ and $f_y^{-1}(A_Z) = S_y(f^{-1}(A_Z))$, $\forall A_Z \in \mathcal{A}_Z$.   □

Often, measurable functions on the product σ-algebra are built starting from measurable functions on the original σ-algebras.

▶ **Example 9.31** *(Real-valued functions on product space)* If the real-valued function $f : X \to \mathbb{R}$ is measurable in $(X, \mathcal{A}_X)$, then the function $h(x, y) = f(x)$ is measurable in $(X \times Y, \mathcal{A}_{X \times Y})$, indeed $D_a(h) = D_a(f) \times Y$. Moreover, if $g : Y \to \mathbb{R}$ is measurable in $(Y, \mathcal{A}_Y)$, then $h(x, y) = f(x)g(y)$ and $h(x, y) = f(x) + g(y)$ are measurable in $(X \times Y, \mathcal{A}_{X \times Y})$ (see Theorem 9.8).

9.4.2   Product Measure

The next step is to define a measure on the measurable space $(X \times Y, \mathcal{A}_{X \times Y})$. We will start by discussing a handy way of building a σ-algebra starting from an algebra.

Definition 9.23 *(Monotone class)* Given a set X, a *monotone class* $M \subseteq 2^X$ on X is a collection of subsets such that

- given a monotone increasing sequence of elements of M, $E_1 \subseteq E_2 \subseteq \ldots$, then $\cup_{i=1}^{\infty} E_i \in M$;
- given a monotone decreasing sequence of elements of M, $E_1 \supseteq E_2 \supseteq \ldots$, then $\cap_{i=1}^{\infty} E_i \in M$.

Essentially, M is a collection of subsets that is closed with respect to the intersection and union of monotone sequences of sets. Note that a σ-algebra is a monotone class, but a monotone class is generally not a σ-algebra. For example, a singlet $\{x\}$ is itself a monotone class, but not a σ-algebra. It is immediate to see that, as with algebras, the intersection of monotone classes is a monotone class. Thus, given a collection of subsets $C \subseteq 2^X$, it is possible to define the monotone class generated by the collection C as the intersection of all the monotone classes that contain it. When the starting collection of subsets is an algebra $\mathcal{A}^0$, something special happens: the monotone class generated by it is precisely the σ-algebra generated by it.

Theorem 9.20 *Consider a set X, an algebra $\mathcal{A}^0 \subseteq 2^X$, and the monotone class M generated by $\mathcal{A}^0$. Then $M = \sigma(\mathcal{A}^0)$.*

Proof Clearly $\mathcal{A}^0 \subseteq M \subseteq \sigma(\mathcal{A}^0)$. If we show that M is a σ-algebra, then we know that $M = \sigma(\mathcal{A}^0)$, because there are no algebras smaller than $\sigma(\mathcal{A}^0)$ that contain $\mathcal{A}^0$.

We start by proving that M is closed with respect to the union of sets. Define the set $M_1 = \{E \in M \mid E \cup A \in M, \forall A \in \mathcal{A}^0\}$ and consider an increasing sequence of elements in M_1, $E_1 \subseteq E_2 \subseteq \ldots$. By definition, $E = \cup_{i=1}^{\infty} E_i \in M$. Take any $A \in \mathcal{A}^0$, then $E \cup A = \cup_{i=1}^{\infty} E_i \cup A$, but $\{E_i \cup A\}$ is an increasing sequence, and therefore, $E \cup A \in M$, which implies that $E \in M_1$. You can repeat the same analysis for the intersection of nested sequences, thus proving that M_1 is a monotone class. Since clearly $\mathcal{A}^0 \subseteq M_1$ (because it is an algebra), $M_1 = M$.

Now consider $M_2 = \{E \in M \mid E \cup E' \in M, \forall E' \in M\}$. For the previous result, we know that $\mathcal{A}^0 \subseteq M_2$ and following the same procedure we did before, we can easily prove that M_2 is a monotone class. Thus, $M_2 = M$, which implies, by definition, that M is closed with respect to the union.

To prove that M is closed with respect to taking the complement, define $M_3 = \{E \in M \mid E^c \in M\}$. Given an increasing sequence $E_1 \subseteq E_2 \subseteq \ldots$ of elements in M_3, and setting $E = \cup_{i=1}^{\infty} E_i$, $E^c = \cap_{i=1}^{\infty} E_i^c$. Since $\{E_i^c\}$ is a decreasing sequence of elements, $E^c \in M$ and $E \in M_3$. You can repeat the same analysis for the intersection of nested sequences. This shows that M_3 is a monotone class. Since $\mathcal{A}^0$ is an algebra, $\mathcal{A}^0 \subseteq M_3$, that is, $M_3 = M$.

We have proved that M is an algebra. However, notice that any countable union of elements of M can be written as the union of an increasing sequence $E_1 \cup E_2 \cup E_3 \ldots = E_1 \cup (E_1 \cup E_2) \cup (E_1 \cup E_2 \cup E_3) \ldots$. This means that M is also closed under countable unions, and hence, it is a σ-algebra. $\qquad\square$

Using the previous result, we can prove the existence and clarify the computation of the product measure. Let us start with the case of finite measure spaces.

Theorem 9.21 (Product measure) *Given two finite measure spaces* $(X, \mathcal{A}_X, \mu_X)$ *and* $(Y, \mathcal{A}_Y, \mu_Y)$, *for all* $E \in \mathcal{A}_{X \times Y}$, *the function* $h_X^E : X \to \mathbb{R}_{\geq 0}$, $h_X^E(x) = \mu_Y(S_x(E))$ *is measurable in* $\mathcal{A}_X$, *the function* $h_Y^E : Y \to \mathbb{R}_{\geq 0}$, $h_Y^E(y) = \mu_X(S_y(E))$ *is measurable in* $\mathcal{A}_Y$, *and* $\int d\mu_X \, h_X^E = \int d\mu_Y \, h_Y^E$ *is the* product measure $\mu_{X \times Y}(E)$ *of the set* E.

Proof We need to prove that the two integrals in the definition of the product measure are actually equal. Consider the collection of elements of the product σ-algebra for which the relation in the statement is true

$$C = \left\{ E \in \mathcal{A}_{X \times Y} \middle| \int d\mu_X \, h_X^E = \int d\mu_Y \, h_Y^E \right\}.$$

This set C is not empty, as it obviously contains $\emptyset$ and $X \times Y$. We will prove that, in fact, $C = \mathcal{A}_{X \times Y}$.

First, we show that C is a monotone class. Consider a monotone increasing sequence of elements of C, $E_1 \subseteq E_2 \subseteq \ldots$, and let $E = \cup_n E_n$. The sequences of measurable functions $(h_X^{E_n})$ and $(h_Y^{E_n})$ are such that $\lim_{n \to \infty} h_X^{E_n} = h_X^E$ and $\lim_{n \to \infty} h_Y^{E_n} = h_Y^E$. These functions are dominated by $\mu_X(X) \times \mu_X(Y)$ and applying Theorem 9.14,

$$\int d\mu_X \, h_X^E = \lim_{n \to \infty} \int d\mu_X \, h_X^{E_n} = \lim_{n \to \infty} \int d\mu_Y \, h_Y^{E_n} = \int d\mu_Y \, h_Y^E,$$

so that $E \in C$. Analogously, let $E_1 \supseteq E_2 \supseteq \ldots$ be a monotone nondecreasing sequence of elements of C and $E = \cap_n E_n$. Then $(h_X^{E_n})$ and $(h_Y^{E_n})$ are sequences of measurable functions such that $\lim_{n \to \infty} h_X^{E_n} = h_X^E$ and $\lim_{n \to \infty} h_Y^{E_n} = h_Y^E$. Again, using Theorem 9.14 twice, we can prove that $\int d\mu_X \, h_X^E = \int d\mu_Y \, h_Y^E$, that is $E \in C$.

Second, we consider the set containing finite unions of all measurable rectangles, $\bar{\mathcal{R}}_{X \times Y}$. The intersection of rectangles is a rectangle, and the complement of a rectangle can be written as union of rectangles (see Fig. 9.4). Thus, $\bar{\mathcal{R}}_{X \times Y}$ is an algebra. This implies that all the elements of $\bar{\mathcal{R}}_{X \times Y}$ can be written as the finite union of disjoint rectangles.

Third, we show that $\bar{\mathcal{R}}_{X \times Y} \subseteq C$. For any measurable rectangle $E = A_X \times A_y$, $h_X^E = I_{A_X} \mu_Y(A_y)$ and $h_Y^E = I_{A_Y} \mu_X(A_X)$ are measurable. Consider two disjoint measurable rectangles $E = (A_X \times A_y) \cup (B_X \times B_y)$. We have $h_X^E = I_{A_X} \mu_Y(A_y) + I_{B_X} \mu_Y(B_y)$ and $h_Y^E = I_{B_Y} \mu_Y(A_X) + I_{A_Y} \mu_Y(B_X)$, so that the statement of the theorem is clearly satisfied. The same reduction to simple functions is obtained also when the union of more than two disjoint rectangles is considered, thus $\bar{\mathcal{R}}_{X \times Y} \subseteq C$.

In conclusion, $\bar{\mathcal{R}}_{X \times Y}$ is an algebra and C a monotone class that contains it. But according to Theorem 9.20, the smallest monotone class that contains $\bar{\mathcal{R}}_{X \times Y}$ is precisely $\mathcal{A}_{X \times Y}$. Thus, $\mathcal{A}_{X \times Y} \subseteq C$ and the statement follows. $\qquad\square$

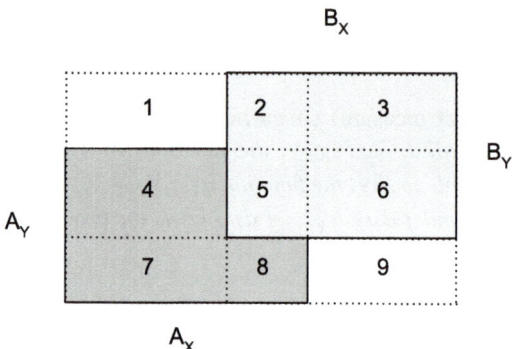

Fig. 9.4 The intersection of $(A_X \times A_Y) \cup (B_X \times B_Y)$ is the rectangle number 5. The union of $(A_X \times A_Y) \cup (B_X \times B_Y)$ can be written as the union of the seven disjoint rectangles identified by a number from 2 to 8. The complement of $(A_X \times A_Y)$ are the rectangles number 1, 2, 3, 6, and 9

The previous definition satisfies all the properties of a measure. Obviously, $\mu_{X \times Y}(\emptyset) = 0$. Given two disjoint measurable sets $E_1, E_2 \in \mathcal{A}_{X \times Y}, \forall x \in X, S_x(E_1)$ and $S_x(E_2)$ are disjoint measurable sets in $\mathcal{A}_Y$ and $\forall y \in Y, S_y(E_1)$ and $S_y(E_2)$ are disjoint measurable sets in $\mathcal{A}_X$. Thus, $h_X^{E_1 \cup E_2} = h_X^{E_1} + h_X^{E_2}$ and $h_Y^{E_1 \cup E_2} = h_Y^{E_1} + h_Y^{E_2}$. In Theorem 9.21, the requirement for the measures to be finite is important. If the original measures are not finite, then the product measure might not be well defined.

▶ **Example 9.32** *(Uncertain measure)* To have an intuition of the kind of problem one finds when dealing with infinite measures, consider the product of the counting measure and of the Lebesgue measure on $\mathbb{R}^2$. Consider the set $E_n = \bigcup_{h=1}^{n} \{h\} \times (h - \delta_n, h + \delta_n) \subseteq \mathbb{R}^2$. This is the union of disjoint rectangles and one has $\mu(E_n) = 2n\delta_n$. If one sets $\delta_n = 1/n$, then $\lim_{n \to \infty} \mu(E_n) = 2$, while if one sets $\delta_n = 1/n^2$, $\lim_{n \to \infty} \mu(E_n) = 0$, despite the fact that in both cases $\lim_{n \to \infty} E_n = \mathbb{N} \times \mathbb{N}$.

Conveniently, however, the definition can be extended from finite to σ-finite measures, like the Borel measure space on $\mathbb{R}$.

Theorem 9.22 (Product of σ-finite measures) *The result in Theorem 9.21 is valid also when σ-finite measure spaces are considered.*

Proof Let $(X, \mathcal{A}_X, \mu_X)$ and $(Y, \mathcal{A}_Y, \mu_Y)$ be two σ-finite measure spaces. Thus, there exist two nested sequences $X_1 \subseteq X_2 \subset \ldots$ and $Y_1 \subseteq Y_2 \subset \ldots$ of $\mathcal{A}_X$-measurable and $\mathcal{A}_Y$-measurable sets, respectively, such that $\bigcup_{i=1}^{\infty} X_i = X$, $\bigcup_{i=1}^{\infty} Y_1 = Y$, and $\forall i$, $\mu_X(X_i), \mu_Y(Y_i) < +\infty$.

On $(X, \mathcal{A}_X), \forall E \in \mathcal{A}_X$ define the measure $\mu_X^i(E) = \mu_X(E \cap X_i)$, and on $(Y, \mathcal{A}_Y), \forall E \in \mathcal{A}_Y$ define the measure $\mu_Y^i(E) = \mu_Y(E \cap Y_i)$ (see Exercise 9.12). These measures are finite. Note that $\forall E \in \mathcal{A}_X$, if $i > j$, $\mu_X^i(E) \geq \mu_X^j(E)$, and $\lim_{i \to \infty} \mu_X^i(E) = \mu_X(E)$. Moreover for any $\mathcal{A}_X$-measurable function f, it follows that $\int d\mu_X^i \, f = \int d\mu_X \, f I_{X_i}$ (see Exercise 9.12). The measures μ_Y^i have the same properties on $(Y, \mathcal{A}_Y)$.

According to Theorem 9.21, $\forall E \in \mathcal{A}_{X \times Y}$, $\int d\mu_X^i \, h_{X_i}^E = \int d\mu_Y^i \, h_{Y_i}^E$, where $h_{X_i}^E(x) = \mu_Y^i(S_x(E))$ and $h_{Y_i}^E(y) = \mu_X^i(S_y(E))$. From the previous discussion, $\int d\mu_X^i \, h_i^E = \int d\mu_X \, h_{X_i}^E I_{X_i}$ and $\int d\mu_Y^i \, h_i^E = \int d\mu_Y \, h_{Y_i}^E I_{Y_i}$. But $(h_{X_i}^E I_{X_i})$ and $(h_{Y_i}^E I_{Y_i})$ are nondecreasing sequences of measurable functions such that $\lim_{i \to \infty} h_{X_i}^E(x) I_{X_i} = \mu_Y(S_x(E))$ and $\lim_{i \to \infty} h_{Y_i}^E(y) I_{Y_i} = \mu_X(S_y(E))$ so that, by monotone convergence (Theorem 9.11), the statement follows. $\square$

▶ **Example 9.33** *(Lebesgue measure and product measure)* The Lebesgue measure on $\mathbb{R}^2$ has been built on the basis of the outer measure introduced in Definition 9.12. The measure space $(\mathbb{R}^2, \mathcal{L}_2, l_2)$ is not the product measure of $(\mathbb{R}, \mathcal{L}, l)$ with itself. This can be easily understood by looking at the completeness of the measure. Consider the rectangle $I = [0, 1] \times \{0\}$. This is a zero measure set in $l \times l$. Now, take the Vitali set $E \notin \mathcal{L}$ defined in Example 9.5. The rectangle $E \times \{0\} \subset I$ is a null set, but it is not measurable. Thus, $l \times l$ is not a complete measure, while we know from Theorem 9.5 that l_2 is complete.

9.4.3   Multiple Integrals

Having defined a measure on the product space $X \times Y$, we can proceed to integrate real-valued measurable functions on it. The actual computation can be facilitated by a powerful theorem, which is named Fubini's (or "reduction") theorem after the Italian mathematician Guido Fubini (1879 – 1943). This theorem reduces the computation of a multivariate integral to a sequence of integrals in a single variable. We will start by proving a preliminary result.

Lemma 9.5 *Let* $s(x, y) = \sum_{i=1}^n a_i I_{A_i}$ *be a real-valued simple function on the product measure space* $(X \times Y, \mathcal{A}_{X \times Y}, \mu)$ *and denote with* s_x *and* s_y *its restrictions on* Y *and* X, *respectively, associated with generic points* $x \in X$ *and* $y \in Y$. *Then*

$$\int d\mu \, s = \int d\mu_X \left(\int d\mu_Y \, s_x \right) = \int d\mu_Y \left(\int d\mu_X \, s_y \right).$$

Proof Without loss of generality, we can assume that the sets A_i are mutually disjoint. For the properties of the product measure and from Definition 9.18,

$$\int d\mu \, s = \sum_{i=1}^n a_i \mu(A_i) = \int d\mu_X \sum_{i=1}^n a_i \mu_Y(S_x(A_i)).$$

At the same time $s_x = \sum_{i=1}^{n} a_i I_{S_x(A_i)}$ and from the linearity property of integrals,

$$\int d\mu_Y \, s_x = \sum_{i=1}^{n} a_i \int d\mu_Y \, I_{S_x(A_i)} = \sum_{i=1}^{n} a_i \, \mu_Y(S_x(A_i)),$$

thus proving the first equality. The second equality is proved similarly.   □

This result can be generalised to nonnegative measurable functions and to integrable functions.

Theorem 9.23 (Fubini) *Let f be a nonnegative measurable or integrable function on the product measure space in Lemma 9.5 and f_x and f_y its restrictions on Y and X, respectively, associated with generic points $x \in X$ and $y \in Y$. Then*

$$\int d\mu \, f = \int d\mu_X \left(\int d\mu_Y \, f_x \right) = \int d\mu_Y \left(\int d\mu_X \, f_y \right).$$

Proof If the function f is nonnegative, there is a nondecreasing sequence of simple functions (s_n) that converges to f. It is immediate to see that the sequences of their restrictions, $(s_{n,x})$ and $(s_{n,y})$, are nondecreasing and converging to f_x and f_y, respectively. Then, by monotone convergence, using Lemma 9.5,

$$\int d\mu \, f = \lim_{n \to \infty} \int d\mu \, s_n = \lim_{n \to \infty} \int d\mu_X \left(\int d\mu_Y \, s_{n,x} \right) =$$
$$\int d\mu_X \left(\int d\mu_Y \lim_{n \to \infty} s_{n,x} \right) = \int d\mu_X \left(\int d\mu_Y \, f_x \right).$$

We have used the fact that $(\int d\mu_Y \, s_{n,x})$ is a pointwise increasing sequence of functions for each x because $(s_{n,x})$ is a pointwise increasing sequence of functions for each y. The other equality is proved analogously. If the function is integrable, then the result can be applied separately for $f I_{f>0}$ and $-f I_{f<0}$.   □

From Example 9.33, we know that the measure induced by the outer measure in Definition 9.12 is a genuine new measure on $\mathbb{R}^2$ and is not $l \times l$. However, it remains true that $(\mathbb{R}^2, \mathcal{L}_2)$ is the completion, with respect to l_2, of $(\mathbb{R}^2, \mathcal{B}_2)$, which is itself a product measurable space (see Example 9.30). We have $\mathcal{B}_2 = \mathcal{B} \times \mathcal{B} \subset \mathcal{L} \times \mathcal{L} \subset \mathcal{L}_2$, and the only complete space when considering the measure l_2 is the latter (and largest). When applying Theorem 9.23 to Riemann integrals, one has to proceed with caution. A sufficient condition is that all the restrictions of the function satisfy the requirements of Theorem 9.17. In practice, one often integrates piecewise continuous functions on compact domains, so that no issues arise.

▶ **Example 9.34** *(The order of integration might matter)* In $(\mathbb{R}^2, \mathcal{B}_2, l_2)$ consider the function

$$f(x, y) = \begin{cases} 1 & \text{if } x \geq 0 \text{ and } x \leq y < x + 1, \\ -1 & \text{if } x \geq 0 \text{ and } x + 1 \leq y < x + 2, \\ 0 & \text{otherwise.} \end{cases}$$

The function and its restriction are piecewise continuous; therefore, we should be able to compute the one-dimensional integral of the restrictions using the Riemann approach. Let l_X and l_Y denote the Lebesgue measures on the two component spaces. Then $\forall x$, $\int dl_Y\, f_x(y) = \int_x^{x+2} dy\, f(x, y) = 0$, so that $\int dl_X \int dl_Y\, f_x(y) = 0$. At the same time,

$$\int dl_X\, f_y(x) = \int_{\max\{0, y-2\}}^{\max\{0, y\}} dx\, f(x, y) = \begin{cases} y & \text{if } 0 \leq y \leq 1, \\ 2 - y & \text{if } 1 \leq y \leq 2, \\ 0 & \text{otherwise.} \end{cases}$$

Therefore, $\int dl_Y \int dl_X\, f_y(x) = 1$. The reason for the discrepancy is that the function f is neither positive nor integrable. In fact, $\int_{f>0} dl\, f = \int_{f<0} dl\, (-f) = +\infty$.

When performing multiple integrals on $\mathbb{R}^n$ using Fubini's theorem, the Lebesgue measure dl_n is typically expressed using individual variables $dx_1 dx_2 \ldots dx_n$. The following examples present the computation of multiple integrals in different situations. The reader is advised to study them carefully.

▶ **Example 9.35** *(Integral of a two-variable function)* Consider the function $f(x, y) = x^2 + xy$ on $\mathbb{R}^2$ and the set $A = \{(x, y) : x \geq 0, y \geq 0, x + y \leq 1\}$. The function f is continuous and consequently Riemann integrable. We want to compute $\int_A dl_2\, f$. Specifically, we will compute $\int dl_X \int_{S_x(A)} dl_Y\, f_x(y)$. If $x < 0$ or $x > 1$, $S_x(A) = \emptyset$. If $x \in [0, 1]$, $S_x(A) = [0, 1 - x]$. Then, $\int_A dl_2\, f = \int_0^1 dx \left(\int_0^{1-x} dy\, x^2 + xy \right)$. The integral in parentheses is performed keeping x constant. Since $\forall x$ the function is Riemann integrable, the integral can be computed from the primitive function $F_x(y) = x^2 y + xy^2/2$,

$$\int_0^{1-x} dy\, x^2 + xy = F_x(1 - x) - F_x(0) = \frac{1}{2}x - \frac{1}{2}x^3.$$

This is in turn a Riemann integrable function in x, and using its primitive $F(x) = x^2/4 - x^4/8$,

$$\int_A dl\, f = \int_0^1 dx\, \frac{1}{2}x - \frac{1}{2}x^3 = F(1) - F(0) = \frac{1}{8}.$$

The reader can check that reversing the order of integration, namely computing $\int_0^1 dy \left(\int_0^{1-y} dx\, x^2 + xy \right)$, leads to the same result.

▶ **Example 9.36** *(Change of variables in multiple integrals)* Consider a measurable real-valued function $f : \mathbb{R}^n \to \mathbb{R}$ and an open domain $E \in \mathcal{B}$ so that the integral $\int_E dl_n f$ exists and is finite. Let $\mathbf{H} : \mathbb{R}^n \to E$ be a differentiable function with a differentiable inverse. This function expresses any element of the space $\mathbf{x} \in E$ (the old coordinates) as a function of an element $\mathbf{y} \in \mathbf{H}^{-1}(E)$ (the new coordinates). According to Theorem 9.15 (replace h with $\mathbf{H}^{-1}$ and f with $f \circ \mathbf{H}$), $\int_E dl_n f = \int_{\mathbf{H}^{-1}(E)} d(l_n \circ \mathbf{H}) f \circ \mathbf{H}$.

We want to derive a simple expression for the pushforward measure $l_n \circ \mathbf{H}$. Let us start with the case $n = 2$. Consider an infinitesimal rectangle around a point $\mathbf{x} = (x_1, x_2)$, $\delta R(\mathbf{x}) = (x_1, x_1 + dx_1] \times (x_2, x_2 + dx_2]$. Ignoring higher order infinitesimals, $\mathbf{H}(\delta R(\mathbf{x}))$ is a parallelogram whose vertices are $\mathbf{H}(\mathbf{x})$, $\mathbf{H}(x_1 + dx_1, x_2) = \mathbf{H}(\mathbf{x}) + \mathbf{J}_1(\mathbf{x})dx_1$, $\mathbf{H}(x_1, x_2 + dx_2) = \mathbf{H}(\mathbf{x}) + \mathbf{J}_2(\mathbf{x})dx_2$, and $\mathbf{H}(x_1 + dx_1, x_2 + dx_2) = \mathbf{H}(\mathbf{x}) + \mathbf{J}_1(\mathbf{x})dx_1 + \mathbf{J}_2(\mathbf{x})dx_2$ where $\mathbf{J}_i$ is i^{th} column of the Jacobian matrix. A basic result in linear algebra is that the area of a parallelogram is the absolute value of the determinant of the matrix built from the vectors defining the couples of opposite hedges. Thus, $l_2(\mathbf{H}(\delta R)) = |\det J(\mathbf{x})|dx_1 dx_2 + o(dx_1\, dx_2)$.

Consider now a finite semi-open rectangle $R = (x, x + a] \times (y, y + b]$. Using two equispaced partitions on the two coordinates, divide it into n^2 disjoint small semi-open rectangles $R = \cup_{i=1}^n \cup_{j=1}^n R_{i,j}^n$, with opposing sides having length a/n and b/n. When n is large, $l_2(\mathbf{H}(R_{i,j}^n)) = |\det J(\mathbf{x}_{i,j})|ab/n^2 + o(1/n^2)$ where $\mathbf{x}_{i,j}$ is the lower left corner of the sub-rectangle $R_{i,j}^n$. Then

$$l_2(\mathbf{H}(R)) = \sum_{i=1}^n \sum_{j=1}^n l_2(\mathbf{H}(R_{i,j}^n)) =$$

$$\frac{ab}{n^2} \sum_{i=1}^n \sum_{j=1}^n |\det J(\mathbf{x}_{i,j})| + n^2 o(\frac{1}{n^2}) = \int_R dl_2\, s_n + n^2 o(\frac{1}{n^2}),$$

with $s_n = \sum_{i=1}^n \sum_{j=1}^n |\det J(\mathbf{x}_{i,j})| I_{R_{i,j}^n}$. The sequence of simple functions (s_n) is dominated by the constant function $\sup_R |\det J|$ and, because of the continuity of the elements of the Jacobian, pointwise converges to $|\det J|$. Thus, by Theorem 9.14, when $n \to \infty$, $l_2(\mathbf{H}(R)) = \int_R dl_2 |\det J|$. Because any open set is the countable union of disjoint open rectangles, the previous equality is valid for any open set. The same procedure is valid for any dimension n, so we can conclude that

$$\int_E dl_n f = \int_{\mathbf{H}^{-1}(E)} d(l_n \circ \mathbf{H}) f \circ \mathbf{H} = \int_{\mathbf{H}^{-1}(E)} dl_n |\det J| f \circ \mathbf{H}.$$

▶ **Example 9.37** *(Polar coordinates)* Consider the sector of the unit circle $A = \{(x, y) \mid x \geq 0, y \geq 0, x^2 + y^2 \leq 1\}$ and the function $f(x, y) = x\, y$. We want to compute $I = \int_A dl\, f$. Using Fubini's Theorem 9.23,

$$I = \int_0^1 dx \int_0^{\sqrt{1-x^2}} dy\, xy = \int_0^1 dx\, \frac{1}{2}x(1 - x^2) = \frac{1}{4} - \frac{1}{8} = \frac{1}{8}.$$

Consider the change of variables $x = \rho \cos\phi$, $y = \rho \sin\phi$ described in Example 7.1. The associated Jacobian (see Example 9.36) reads

$$J = \begin{pmatrix} \cos\phi & -\rho \sin\phi \\ \sin\phi & \rho \cos\phi \end{pmatrix}$$

so that $|\det J| = \rho$. This change of variables violates the requirement of Example 9.36 in $(0, 0)$. This is not an issue. Being a zero measure set, we can remove this point from the integration domain without affecting the computation of the integral. With the new variables, the integral is on a rectangle

$$I = \int_0^1 d\rho \int_0^{\pi/2} d\phi\, \rho^3 \cos\phi \sin\phi = \frac{1}{8} \int_0^{\pi/2} d\phi \sin 2\phi = \frac{1}{8}.$$

▶ **Example 9.38** (*Gaussian integral*) Consider the integral $I = \int_{-\infty}^{+\infty} dx\, e^{-x^2}$. The integrated function is continuous and nonnegative; hence, the Lebesgue integral exists. Note that

$$I^2 = \left(\int_{-\infty}^{+\infty} dx\, e^{-x^2} \right) \left(\int_{-\infty}^{+\infty} dy\, e^{-y^2} \right) = \int_{\mathbb{R}^2} dxdy\, e^{-x^2-y^2}.$$

Using the polar coordinates in Example 9.37, the integral can be rewritten as $I^2 = \int_{\mathbb{R}^2} d\phi\, d\rho\, \rho e^{-\rho^2}$. Applying Fubini's theorem, first performing the integral in ϕ and then expressing the remaining improper Riemann integral as a limit,

$$I^2 = 2\pi \int_0^{+\infty} d\rho\, \rho e^{-\rho^2} = \pi \lim_{L \to +\infty} (1 - e^{-L^2}) = \pi,$$

so that $I = \sqrt{\pi}$.

9.5 Probability Measure

There is a special instance of a finite measure space that deserves a specific definition.

Definition 9.24 (*Probability measure*) A probability space is a measure space $(X, \mathcal{A}, \mu)$ such that $\mu(X) = 1$. The measure μ is a *probability measure*.

When probability spaces are concerned, the notation usually changes to something like $(\Omega, \mathcal{F}, P)$, where Ω is a generic set of possible outcomes, $\mathcal{F} \subseteq 2^{\Omega}$ is the σ-algebra (or σ-field), whose elements are called realisations or events, and P the probability measure defined on those events. The *joint probability* of two events $A_1, A_2 \in \mathcal{F}$, that is, the probability that they are both realised, is $P(A_1 \cap A_2)$. The probability that at least one of the two events is realised is $P(A_1 \cup A_2)$.

Definition 9.25 (*Independent events*) Let $(\Omega, \mathcal{F}, P)$ be a probability space. Two events $A, B \in \mathcal{F}$ are called *independent* if $P(A \cap B) = P(A) P(B)$.

It is easy to see that if A and B are independent, then A^c and B^c are independent. Given a sequence of events (A_n), the probability that at least one event of the sequence is realised is $P(\cup_{n=1}^{\infty} A_n)$, while the probability that they are all realised is $P(\cap_{n=1}^{\infty} A_n)$. The next result is named after Félix Borel and the Italian mathematician Francesco Cantelli (1875–1966).

Theorem 9.24 (First and second Borel–Cantelli lemmas) *Consider a sequence of events* (A_n). *If* $\sum_{n=1}^{\infty} P(A_n) < +\infty$, *then the probability that an infinite number of them are realised is zero. If instead* $\sum_{n=1}^{\infty} P(A_n) = +\infty$ *and the events* (A_n) *are independent, then the probability that an infinite number of them are realised is one.*

Proof An infinite number of events are realised if and only if, for any n, there is an event $k \geq n$ that is realised. The probability of the latter is the probability of $\cup_{k=n}^{\infty} A_k$. Thus, the probability that an infinite number of events are realised is $P(\cap_{n=1}^{\infty} \cup_{k=n}^{\infty} A_k)$.

To prove the first statement, notice that $(\tilde{A}_n = \cup_{k=n}^{\infty} A_k)$ is a monotonically decreasing sequence of sets. Thus $P(\cap_{n=1}^{\infty} \cup_{k=n}^{\infty} A_k) = \lim_{n \to \infty} P(\tilde{A}_n)$. For the subadditive property of the measure, $P(\tilde{A}_n) \leq \sum_{k=n}^{\infty} P(A_k)$. By assumption $\sum_n P(A_n)$ converges, thus $\lim_{n \to \infty} \sum_{k=n}^{\infty} P(A_k) = 0$ and the first statement follows.

To prove the second statement, notice that only a finite number of events are realised if and only if there is a n such that no event with $k \geq n$ is realised. The probability of the latter is $P((\cup_{k=n}^{\infty} A_k)^c) = P(\cap_{k=n}^{\infty} A_k^c)$. If we prove that this probability is zero, then we have proved the second statement. In fact, since the events are independent, $P(\cap_{k=n}^{\infty} A_k^c) = \sum_{k=n}^{\infty} P(A_k^c) = \sum_{k=n}^{\infty} (1 - P(A_k)) \leq \exp(-\sum_{k=n}^{\infty} P(A_k)) = 0$, because $\sum_{k=n}^{\infty} P(A_k)$ diverges by assumption. $\qquad \square$

When dealing with probability, we will say that some property or relation is valid *almost surely*, abbreviated with *a.s.*, when the property or relation is violated at most on a null set.

▶ **Example 9.39** *(Coin flip)* We want to describe in probabilistic terms the flipping of one coin. The coin has two faces: head (h) or tail (t), so we set $\Omega = \{h, t\}$. When flipping the coin, one of the two sides will turn up, so the algebra of events is set to $\mathcal{F} = \sigma(\{\{h\}, \{t\}\}) = \{\Omega, \emptyset, \{h\}, \{t\}\}$. If the coin is fair, we set $P(\{h\}) = P(\{t\}) = 1/2$. If the coin is unfair, we assign a different measure (probability) to the two singlets. Notice that $P(\{h\} \cap \{t\}) = 0 \neq P(\{h\}) P(\{t\}) = 1/4$, thus the head and tail events are not independent.

Different algebras on the same space Ω represent different possible sets of events that describe different situations.

▶ **Example 9.40** *(Two coin flip)* Consider the flipping of two fair coins in sequence. The outcome of each coin can be a head (h) or a tail (t). Then the probability space $(\Omega, \mathcal{F}, P)$ is build by setting $\Omega = \{tt, th, ht, hh\}$, $\mathcal{F} = 2^{\Omega}$ and using the counting measure of Example 9.3, $P = \mu^{\#}/4$.

Conversely, if the two coins are flipped at the same time, that is one cannot distinguish the realisation of one coin from that of the other, the algebra is generated by the observable events: two heads $\{hh\}$, two tails $\{tt\}$, and one head and one tail $\{ht, th\}$, that is $\mathcal{F} = \sigma(\{\{hh\}, \{ht, th\}, \{tt\}\})$. The measure P remains the same.

On probability spaces, we normally consider real-valued functions.

Definition 9.26 *(Random variable)* A *random variable* on a probability space $(\Omega, \mathcal{F}, P)$ is a $\mathcal{F}$-measurable function with values in $(\mathbb{R}, \mathcal{B})$.

By Lemma 9.2, if X is a random variable and g a Borel measurable function, then $g(X)$ is a random variable. Different σ-algebras on the same set Ω allow different random variables because some function might be measurable with respect to one algebra but not measurable with respect to the other.

▶ **Example 9.41** *(Random variables for two coin flip)* Consider the probability spaces in Example 9.40 and define a function $X_1 : \Omega \rightarrow \mathbb{R}$ that counts the number of obtained tails, that is $X_1(ht) = X_1(th) = 1$, $X_1(tt) = 2$, and $X_1 = 0$ otherwise. The set $D_a(X_1)$ of Definition 9.15 is $\emptyset$ if $a \geq 2$, $\{tt\}$ if $1 \leq a < 2$, $\Omega \setminus \{hh\}$ if $0 \leq a < 1$, and finally Ω if $a < 0$. Thus, the random variable X_1 is measurable on both probability spaces, the sequential flipping and simultaneous flipping.

Assume instead that we have a function X_2 that measures whether a head is the outcome of the first flip, that is, $X_2(hh) = X_2(ht) = 1$ and $X_2 = 0$ otherwise. In this case $D_0(X_2) = \{hh, ht\}$. This set belongs only to the σ-algebra of the sequential flipping case. Thus, f is an acceptable random variable only on that probability space.

Being a measurable function, a random variable X does two things. First, it generates a σ-algebra on Ω through the mechanism of Lemma 9.3: the collection of the preimages of Borel sets, $\sigma(X) = \{X^{-1}(B) \mid B \in \mathcal{B}\} \subseteq \mathcal{F}$, is a σ-algebra itself. This is called the σ-algebra of the random variable. It is the smallest σ-algebra on which X is measurable. Second, it defines a probability measure μ_X on the Borel measure space $(\mathbb{R}, \mathcal{B})$ through the pushforward mechanism described in Lemma 9.4. This leads to the following.

Definition 9.27 *(Probability distribution)* Consider a random variable $X : \Omega \to \mathbb{R}$. For any $B \in \mathcal{B}$ let $\mu_X(B) = P(X^{-1}(B))$. Since X is P-measurable, μ_X is a probability measure on $\mathcal{B}$ and is said to be the *distribution law* or *probability distribution* of the random variable X.

Equivalently, the behaviour of X can be described as follows.

Definition 9.28 *(Distribution function)* Consider a random variable $X : \Omega \to \mathbb{R}$. Its *distribution function* is defined as

$$F_X(x) = P(X^{-1}((-\infty, x])) = P(\{\omega \in \Omega \mid X(\omega) \le x\}).$$

From the definition, $F_X(x) = \int_{-\infty}^{x} d\mu_X$, $\lim_{x \to -\infty} F_X(x) = 0$, and $\lim_{x \to +\infty} F_X(x) = 1$. This is the same object that we met in Example 8.17.

Theorem 9.25 *Let F_X be the distribution function of a random variable $X : \Omega \to \mathbb{R}$. Then F_X is nondecreasing and right continuous.*

Proof The first statement follows immediately by noting that if $x > x'$, then $X^{-1}((-\infty, x']) \subseteq X^{-1}((-\infty, x])$. Concerning the second point, for any sufficiently small $\epsilon > 0$, there exists n such that $(-\infty, x] \subseteq (-\infty, x + \epsilon] \subseteq (-\infty, x + 1/n]$ and hence $F_X(x) \le F_X(x+\epsilon) \le F_X(x+1/n)$. But $\lim_{n \to \infty}(-\infty, x+1/n] = (-\infty, x]$, and the statement follows from the comparison theorem. $\qquad\square$

Note that $\lim_{h \to 0^+} F_X(x - h) = F_X(x) - \mu_X(\{x\})$. The jump of the distribution function in x is equal to the probability measure of the singlet $\{x\}$. If $\mu_X(\{x\}) = 0$, then the distribution function is continuous in x. As $\mu_X((a, b]) = F_X(b) - F_X(a)$, the probability distribution μ_X can be seen as the Lebesgue–Stieltjes measure associated with the increasing function F_X. The integral with respect to the probability measure of the random variable X is often denoted by dF_X.

▶ **Example 9.42** *(The measure of a random variable)* Consider the random variable X_1 in Example 9.41. It is $X_1^{-1}(0) = \{hh\}$, $X_1^{-1}(1) = \{ht, th\}$, and $X_1^{-1}(2) = \{tt\}$. Thus, the probability measure induced by the random variable X_1 on $\mathbb{R}$ is a point mass measure (see Example 9.4)

$$\mu_{X_1} = \frac{1}{4}I_{\{0\}} + \frac{1}{2}I_{\{1\}} + \frac{1}{4}I_{\{2\}}.$$

The associated distribution function $F_{X_1}(x)$ has a value 0 if $x < 0$, $1/4$ if $0 \leq x < 1$, $3/4$ if $1 \leq x < 2$, and 1 if $x \geq 2$.

The probability distribution μ_X, or the distribution function F_X, contains all the necessary information about the behaviour of the random variable. Any random variable on any probability space can ultimately be seen as a measure on $(\mathbb{R}, \mathcal{B})$. This is the reason why a large part of the random variable theory deals directly with the distribution function F_X, rather than with the original probability spaces from which it derives. While the probability space describes the model one has in mind, it is the distribution function that ultimately enters into any practical computation.

Definition 9.29 (*Mean and variance*) The *mean* or *expected value* of a random variable X is defined as $E[X] = \int dF_x \, x$ when the integral exists. Its *variance* is defined as $V[X] = \int dF_x \, (x - E[X])^2$ when the integral exists.

The two quantities above are not guaranteed to exist. The mean of the random variable X exists if the identity function is integrable with respect to the measure μ_X. In general, for any Borel measurable function g, the expected value of the random variable $g(X)$ is $\int_\Omega dP \, g \circ X = \int dF_X \, g$ and is denoted with $E[g(X)]$ or $E_X[g]$.

▶ **Example 9.43** (*Variance bounds*) If the variance exists, it can be used to obtain an upper bound to the probability of the occurrence of large values. Applying the Chebyshev inequality in Example 9.22 to $X - E[X]$, $\forall a > 0$, $P\{|X - E[X]| \geq a\} \leq V[X]/a^2$.

▶ **Example 9.44** (*Binomial distribution*) Consider n sequential flips of an unfair coin with probability p of showing head and $q = 1 - p$ of showing tail. The space of events is the sequence of length n of h and t, $\Omega = \{c = (c_1, \ldots, c_n) \mid c_i = h, t\}$. The σ-algebra is simply 2^Ω, and the measure is $P(c) = \prod_{i=1}^n (p I_{\{h\}}(c_i) + q I_{\{t\}}(c_i))$. This is a good probability measure, as one can easily check that $\sum_{c \in \Omega} P(c) = 1$. Consider the random variable that counts the number of heads in a sequence of length n, $X(c) = \sum_{i=1}^n I_{\{h\}}(c_i)$. The possible values (the support) of the random variable are the integer numbers between 0 and n. Let l be one of these numbers, then $X^{-1}(\{l\})$ are all the sequences that contain exactly l heads. Their number is given by the binomial coefficient and their probability is $p^l q^{n-l}$, so that

$$\mu_X(\{l\}) = P(X^{-1}(\{l\})) = \binom{n}{l} p^l q^{n-l}.$$

The resulting probability distribution is known as the *binomial distribution*. It is a point mass or *discrete* distribution.

If the distribution function F_X is differentiable, then its derivative $f_X = F_X'$ is called the *probability density* of the random variable X. Although the probability measure μ_X and the distribution function F_X can be defined for any random variable

X, the probability density may not exist. When the probability density exists, we have $\mathrm{E}[g(X)] = \int dl\, f_X g$.

▶ **Example 9.45** *(Moments, Central Moments, and Cumulants)* Let l be an integer number. The lth order moment $m_l(X)$ of a random variable X is defined as $m_l(X) = \int dF_X\, x^l = \mathrm{E}[X^l]$. This is the same definition we have seen in Example 8.17. In particular, the first moment of a random variable is its expected value, $m_1(X) = \mathrm{E}[X]$. If the variable has a mean, its lth order *central moment* $M_l(X)$ is analogously defined as $M_l = \int dF_X\, (x - m_1(X))^l = \mathrm{E}[(X - m_1(X))^l]$. The central moment is the moment of the random variable obtained by subtracting from the original random variable its expected value. Clearly, $M_1 = 0$, while M_2 is the variance of X. A moment or central moment of a given order might not exist or exist and be infinite.

The definition of a *cumulant* is a bit more complex. The *cumulant generating function* $K_X : \mathbb{R} \to \mathbb{R}$ of a random variable X is defined as

$$K_X(t) = \log \int dF_X\, e^{tx}.$$

Then the lth order cumulant $C_l(X)$ of X is

$$C_l(x) = \left. \frac{d^l}{dt^l} K_X(t) \right|_{t=0}.$$

In other terms, the cumulants are the coefficients of the Maclaurin expansion of the cumulant generating function $K_X(t) = C_1 t + C_2 t^2/2 + \dots$. Taking the derivative inside the integral, it is immediate to see that $C_1 = m_1$ and $C_2 = M_2$. The nice thing about cumulants is how they behave under the addition of independent random variables. Let X and Y be two random variables and consider their sum $X + Y$. If the random variables are independent,

$$\int dP\, e^{t(X+Y)} = \int dF_X\, e^{tX} \int dF_Y\, e^{tY},$$

which in turn implies that $K_{X+Y}(t) = K_X(t) + K_Y(t)$. Thus, $C_l(X + Y) = C_l(X) + C_l(Y)$: the cumulant of a given order of the sum of independent random variables is the sum of their cumulants of the same order. This is a generalisation of the well-known fact that the expected value or the variance of the sum of independent random variables is the sum of their expected values, or their variances, respectively.

9.5.1 Multiple Random Variables

The definition of probability measure and distribution function can be easily generalised to multiple random variables using the notion of product measure space introduced in Sect. 9.4. A vector function with n components, $\mathbf{X} : \Omega \to \mathbb{R}^n$ is a *multivariate* or vector random variable in the probability space $(\Omega, \mathcal{F}, P)$ if it is measurable, that is, $\forall B \in \mathcal{B}_n$, $X^{-1}(B) \in \mathcal{F}$. Since $\mathcal{B}_n$ is the n times product σ algebra of $\mathcal{B}$ with itself (see Example 9.30), we know that the n components $X_1, \ldots, X_n$ are measurable functions from Ω to $\mathbb{R}$ (see Example 9.31).

Definition 9.30 (Joint probability distribution and distribution function) Let $\mathbf{X} : \Omega \to \mathbb{R}^n$ be a vector random variable in $(\Omega, \mathcal{F}, P)$. The *joint probability distribution* $\mu_{\mathbf{X}}$ is defined on any Borel subset $B \subseteq \mathbb{R}^n$ as $\mu_{\mathbf{X}}(B) = P(\mathbf{X}^{-1}(B))$.

Analogously, the *joint distribution function* $F : \mathbb{R}^n \to \mathbb{R}_{\geq 0}$ reads

$$F(\mathbf{x}) = P\left(\mathbf{X}^{-1}((-\infty, x_1] \times \ldots \times (-\infty, x_n])\right) =$$

$$P\left(\bigcap_{i=1}^{N} X_i^{-1}((-\infty, x_i])\right) = P(\{\omega \in \Omega \mid X_i(\omega) \leq x_i, i = 1, \ldots n\}).$$

For any Borel measurable function $g : \mathbb{R}^n \to \mathbb{R}$, $\int_\Omega dP\, g \circ \mathbf{X} = \int d\mu_{\mathbf{X}}\, g$. The notion of independent events, Definition 9.25, is naturally extended to random variables.

Definition 9.31 (*Independent random variables*) The random variables $X_1, \ldots, X_n$ are *independent* if the elements of their σ-algebras are all independent. That is, if $\forall A_i \in \sigma(X_i)$, $i = 1, \ldots n$, then $P(\bigcap_{i=1}^{n} A_i) = \prod_{i=1}^{n} P(A_i)$.

▶ **Example 9.46** (*Dependent and independent random variables*) Consider the probability space of two coin sequential flip in Example 9.40. Let X, X_1, and X_2 be the random variables that count, respectively, the total number of heads, the number of heads in the first flip, and the number of heads in the second flip. The first variable takes value in $\{0, 1, 2\}$ and $\sigma(X) = \sigma(\{\{hh\}, \{ht, th\}, \{tt\}\})$. The other two variables take values in $\{0, 1\}$ and $\sigma(X_1) = \sigma(\{\{th, tt\}, \{hh, ht\}\})$, $\sigma(X_2) = \sigma(\{\{ht, tt\}, \{hh, th\}\})$.

Note that, for $\{hh\} \in \sigma(X)$ and $\{hh, ht\} \in \sigma(X_1)$, $P(\{hh\} \cap \{hh, ht\}) = 1/4$ while $P(\{hh\})\, P(\{hh, ht\}) = 1/8$. Thus, X and X_1 are not independent.

Conversely, if $\omega_1 \in \{\{th, tt\}, \{hh, ht\}\}$ and $\omega_2 \in \{\{ht, tt\}, \{hh, th\}\}$, $\omega_1 \cap \omega_2$ is a singlet, so that $P(\omega_1 \cap \omega_2) = 1/4 = P(\omega_1)P(\omega_2)$. We can conclude that X_1 and X_2 are independent.

Theorem 9.26 (Probability distribution of independent random variables) *Let* $\mathbf{X} : \Omega \to \mathbb{R}^n$ *be a vector random variable and* $X_1, \ldots, X_n$ *its components. If* $X_1, \ldots, X_n$ *are independent, then their joint probability distribution* $\mu_{\mathbf{X}}$ *on* $\mathcal{B}_n$ *is the product measure of the probability distribution of its components:* $\mu_{\mathbf{X}} = \mu_{X_1 \times \ldots \times X_n}$.

Proof Take a rectangle $R = B_1 \times \ldots \times B_n \subseteq \mathcal{B}_n$. Then $\mathbf{X}^{-1}(R) = \bigcap_{i=1}^{n} X_i^{-1}(B_i)$. If the components are independent,

$$\mu_{\mathbf{X}}(R) = P(\mathbf{X}^{-1}(R)) = \bigcap_{i=1}^{n} P(X_i^{-1}(B_i)) = \prod_{i=1}^{n} \mu_{X_i}(B_i) = \mu_{X_1 \times \ldots \times X_n}(R).$$

Thus, the statement is true on the rectangles. For Theorem 9.18, we know that it also true on the σ-algebra generated by the rectangles, which is precisely $\mathcal{B}_n$.   □

A similar factorisation is also in place for the distribution function. If the random variables considered are independent, $\forall \mathbf{x} = (x_1, \dots, x_n) \in \mathbb{R}^n$,

$$F(\mathbf{x}) = P(\bigcap_{i=1}^{N} X_i^{-1}((-\infty, x_i))) = \prod_{i=1}^{n} P(X_i^{-1}((-\infty, x_i))) = \prod_{i=1}^{n} F(x_i).$$

Theorem 9.27 *Let $X_1, \dots, X_n$ be independent random variables on the probability space $(\Omega, \mathcal{F}, P)$ and $g_i : \mathbb{R} \to \mathbb{R}$, $i = 1, \dots, n$, Borel measurable functions. Then the expected value of the function $g(\mathbf{x}) = \prod_{i=1}^{n} g_i(x_i)$ from $\mathbb{R}^n$ to $\mathbb{R}$ exists if and only if $E_{X_i}[g_i]$ exists for each $i = 1, \dots, n$ and $E[g] = \prod_{i=1}^{n} E_{X_i}[g_i]$.*

Proof First, note that since all g_i's are Borel measurable in $\mathbb{R}$, then g is Borel measurable in $\mathbb{R}^n$ (see Example 9.31). Because the random variables are independent, applying Fubini's Theorem 9.23

$$\int d\mu_{\mathbf{x}} g(\mathbf{x}) = \int d\mu_{X_1} \int d\mu_{X_2 \times \dots \times X_n} g_{x_1} =$$

$$\left(\int d\mu_{X_1} g_1(x_1) \right) \int d\mu_{X_2 \times \dots \times X_n} \prod_{i=2}^{n} g_i(x_i),$$

where I used the fact that for the restriction $g_{x_1} = g(x_1) \prod_{i=2}^{n} g(x_i)$ (see Theorem 9.19). The repeated application of this factorisation procedure proves the assertion.   □

▶ **Example 9.47** *(Joint moments)* When multiple random variables $X_1, X_2, \dots, X_n$ are considered, their joint moment is defined as $m_{l_1, \dots, l_n} = E[\prod_{i=1}^{n} X_i^{l_i}]$. If the variables are independent, then the joint moments are the product of the individual moments, $m_{l_1, \dots, l_n} = \prod_{i=1}^{n} m_{l_i}(X_i)$. The definition of joint central moments for multiple random variables is directly derived from the definition of a single variable. Details are left to the reader.

9.5.2 Banach Space of Square Summable Random Variables

Consider a probability space $(\Omega, \mathcal{F}, P)$. Let $X : \Omega \to \mathbb{R}$ be a random variable and define $\|X\| = \sqrt{\int_{\Omega} d\omega\, X(\omega)^2}$. It is immediate to see that $\|X\| \geq 0$, that for any constant c, $\|c X\| = |c|\, \|X\|$, and that for any two random variables X and Y, $\|X + Y\| \leq \|X\| + \|Y\|$. The function just defined is similar to the norm in Definition 4.1, with the only difference that $\|X\| = 0$ does not imply that X is zero for all $\omega \in \Omega$, but rather that a.s. $X = 0$. We can define an equivalence relation in the set of random variables by saying that two random variables X and Y are equivalent if they are a.s. equal, that is $P\{\omega \mid X(\omega) = Y(\omega)\} = 1$.

Definition 9.32 *(Square-integrable random variables)* $L_2(\Omega, \mathcal{F}, P)$ is the set of all (equivalence classes of) random variables $X : \Omega \to \mathbb{R}$ of the space $(\Omega, \mathcal{F}, P)$ such that $\|X\| < +\infty$.

From the triangle inequality, the linear combination of elements of $L_2(\Omega, \mathcal{F}, P)$ clearly belongs to $L_2(\Omega, \mathcal{F}, P)$, so the set is a linear space. The function $\|.\|$ is a norm on L_2. We can prove that this space is complete.

Theorem 9.28 $(L_2(\Omega, \mathcal{F}, P), \|.\|)$ *is a Banach space.*

Proof Consider a Cauchy sequence of random variables (X_k). We have to prove that it converges almost surely. From the definition of a Cauchy sequence, there is a subsequence (X_{n_k}) such that $\|X_{n_k} - X_{n_{k-1}}\| < 1/2^k$ for $k \geq 2$. Applying the Jensen inequality, Corollary 9.5, to each element of the sum,

$$\mathrm{E}\left[\sum_{h=2}^{k} |X_{n_h} - X_{n_{h-1}}|\right] = \sum_{h=2}^{k} \mathrm{E}\left[|X_{n_h} - X_{n_{h-1}}|\right] \leq \sum_{h=2}^{k} \|X_{n_h} - X_{n_{h-1}}\| \leq \sum_{h=2}^{k} \frac{1}{2^h}.$$

Taking the limit $k \to \infty$ we see that $\mathrm{E}[\sum_{h=2}^{\infty} |X_{n_h} - X_{n_{h-1}}|] < +\infty$. This implies that a.s. $\sum_{h=2}^{\infty} |X_{n_h}(\omega) - X_{n_{h-1}}(\omega)| < +\infty$, or the expected value would be infinite. Thus, apart from a measure zero set, $(X_{n_h}(\omega) - X_{n_{h-1}}(\omega))$ is absolutely convergent, that is, convergent. But since $X_{n_k} = \sum_{h=2}^{k}(X_{n_h} - X_{n_{h-1}}) + X_{n_1}$, this implies that there exists a random variable X such that a.s. $\lim_{k \to \infty} X_{n_k}(\omega) = X(\omega)$.

For any element of the original sequence, using Fatou's Theorem 9.12, the fact that the limit inferior is equal to the limit when the limit exists, and the continuity of the norm, one has

$$\|X_n - X\|^2 = \int_{\Omega} d\omega \liminf_{k \to \infty} |X_n - X_{n_k}|^2 \leq \liminf_{k \to \infty} \|X_n - X_{n_k}\|^2.$$

For the Cauchy property, the right-hand side can be made as small as desired by selecting sufficiently large values for n and k. Thus, $\lim_{n \to \infty} \|X_n - X\|^2 = 0$, which implies that a.s. $\lim_{n \to \infty} X_n = X$. □

The previous analysis can be easily extended to the case of vector random variables. Consider a probability space $(\Omega, \mathcal{F}, P)$, a vector random variable $\mathbf{X} : \Omega \to \mathbb{R}^n$ and define $\|\mathbf{X}\| = \sqrt{\sum_{i=1}^{n} \int_{\Omega} d\omega \, X_i(\omega)^2}$. The space $L_2(\Omega, \mathcal{F}, P)$ of all vector random variables $\mathbf{X}$ such that $\|\mathbf{X}\| < +\infty$ is a linear space. With the stipulation that two vector random variables are considered equal if they differ only on a null set, the function above is a norm on that space and one can easily prove that the vector version of $(L_2(\Omega, \mathcal{F}, P), \|.\|)$ is a Banach space by proving the convergence of Cauchy sequences of vector random variables component by component.

Exercises

Exercise 9.1 Consider the interval $[0, 1] \subset \mathbb{R}$. Prove that the set

$$\mathcal{A} = \{\emptyset, (0, 1/2], (1/2, 1], [0, 1]\}$$

is a σ-algebra on $[0, 1]$. Is this also true for the set $\mathcal{A}' = \{\emptyset, [0, 1/2], [1/2, 1], [0, 1]\}$?

Exercise 9.2 Let $X = \{a, b, c\}$. Compute the σ-algebra generated by $C_1 = \{\emptyset\}$, by $C_2 = \{\{a\}\}$, and by $C_3 = \{\{a\}, \{b\}\}$.

Exercise 9.3 Let $(X, \mathcal{A}, \mu)$ be a measure space. Prove that $\mu(A) + \mu(B) = \mu(A \cup B) + \mu(A \cap B)$ for any $A, B \in \mathcal{A}$.

Exercise 9.4 Let $(X, \mathcal{A}, \mu)$ be a measure space and $E \in \mathcal{A}$ a measurable subset. Prove that the function $\mu_E(A) = \mu(A \cap E)$ is a measure on $(X, \mathcal{A})$.

Exercise 9.5 Consider the Euclidean topology on $\mathbb{R}$. Prove that the sets $(0, 1), [1, 2], (3, 4], \{5\}$, and $(6, +\infty)$ belong to the Borel σ-algebra.

Exercise 9.6 What is the counting measure of $\{1, 2, 3\}$ and of $[0, 1]$ in $(\mathbb{R}, \mathcal{B})$?

Exercise 9.7 Consider a function $f : X \to Y$ and the smallest σ-algebra on X, $\mathcal{A} = \{\emptyset, X\}$. Assume that f takes two different values at two different points, $f(x_1) \neq f(x_2)$. Prove that if the σ-algebra on Y is the Borel σ-algebra of a Hausdorff topology, the function f is not measurable.

Exercise 9.8 Consider the outer measure μ^* defined starting from the sets in Definition 9.10. Prove that for any closed interval $[a, b], \mu^*([a, b]) = b - a$, while for any singlet $\{a\}, \mu^*(\{a\}) = 0$.

Exercise 9.9 In the set of measurable functions on a measure space $(X, \mathcal{A}_X, \mu)$ define the relation $f \sim g$ if a.e. $f = g$. Prove that $\sim$ is an equivalence relation.

Exercise 9.10 Given a sequence of nonnegative measurable functions (f_n), prove that $\int d\mu \sup_n f_n(x) \geq \sup_n \int d\mu \, f_n(x)$.

Exercise 9.11 Consider a sequence of nonnegative measurable functions (f_n) and assume that there exists an integrable function g such that $f_n(x) \leq g(x)$. Prove that

$$\int d\mu \limsup_{n \to \infty} f_n(x) \geq \limsup_{n \to \infty} \int d\mu \, f_n(x).$$

(Hint: Apply Theorem 9.12 to the sequence $(g - f_n)$.)

Exercise 9.12 With reference to Exercise 9.4, prove that for any measurable function f on $(X, \mathcal{A})$, $\int d\mu_E \, f = \int_E d\mu \, f$. *(Hint: Start by proving it on simple functions.)*

Exercise 9.13 Consider the function $\mu(A) = \int_A dl \, e^{-x}$ defined on any Lebesgue measurable set A. Prove that $\mu(A)$ is a finite measure on $\mathbb{R}_{\geq 0}$ and compute $\int_{\mathbb{R}_{\geq 0}} d\mu \, x$.

Exercise 9.14 Using the measure in the previous exercise, define the Lebesgue–Stieltjes measure on $\mathbb{R}^2$ with $l_\alpha((a, b] \times (c, d]) = \mu((a, b]) \, \mu((c, d])$ (see Definition 9.11). Let $A_1 = \mathbb{R}_{\geq 0} \times \mathbb{R}_{\geq 0}$ and $A_2 = [0, 1] \times [0, 1]$. Prove that they are measurable with respect to l_α. Compute $\int_{A_1} dl_\alpha \, xy$ and $\int_{A_2} d\mu_\alpha \, xy$.

Exercise 9.15 Let μ_α and μ_β be the Lebesgue–Stieltjes measures defined on $\mathbb{R}$ starting from the functions $\alpha(x) = \exp x$ and $\beta(y) = y^3$, respectively. Consider the completion of the product measure $\mu = \mu_\alpha \times \mu_\beta$ on $\mathbb{R}^2$ and the two subsets $A_1 = \{|x| + |y| \leq 1\}$ and $A_2 = \{|x| + |y| = 1\}$. Compute the integral of the functions $f(x, y) = x^2 y$ and $f(x, y) = xy^2$ on A_1 and A_2 using the measure μ.

Exercise 9.16 Let $a \in \mathbb{R}$. Compute the Riemann integral of the function $f(x, y) = x^2 + axy + y^2$ on the set $A = \{(x, y) \mid 0 \leq x + y \leq 1, 0 \leq x - y \leq 1\}$.

Exercise 9.17 Consider the function $f(x_1, x_2, x_3) = \exp \sum_{i=1}^3 \beta_i x_i$ with real β's and the set $A = \left\{ (x_1, x_2, x_3) \mid x_1, x_2, x_3 \geq 0, \sum_{i=1}^3 x_i \leq 1 \right\}$. Compute the integral $\int_A dl \, f$.

Exercise 9.18 Compute the area of the bounded part of the plane defined by the parabola $y = ax^2$ and the line $y = b$, with $a, b > 0$.

Exercise 9.19 Using the theory of integration, find the volume of the pyramid with height h and square base of length l.

Exercise 9.20 Using the theory of integration, find the volume of the cone with height h and base radius ρ.

Exercise 9.21 Let $B_n \subset \mathbb{R}^n$ be the closed n-ball or radius 1. Prove that $\int_{\partial B_m} dl_n = 0$.

Exercise 9.22 Let $f : \mathbb{R}_{\geq 0} \to \mathbb{R}$ be an integrable function, prove that in $\mathbb{R}^n$ for any $a > 0$,

$$\int_{\|\mathbf{x}\| \leq a} dl_n(\mathbf{x}) f(\|\mathbf{x}\|) = \int_0^a dB_n(\rho) f(\rho)$$

where $B_n(\rho)$ is the volume of the n-dimensional ball of radius ρ. *Hint: In $\mathbb{R}^n$ for any partition $\{a_1 = 0, a_2, \ldots, a_l = a\}$ of $[0, a]$ the function $s(\mathbf{x}) = \sum_{i=1}^l I\{a_{i-1} \leq \|\mathbf{x}\| < a_i\} f(a_i)$ is a simple function $s(x) \leq f(\|\mathbf{x}\|)$.*

Exercise 9.23 With reference to Example 9.40, how is the definition of the measure space modified if the probability to obtain head with each coin is 0.6?

Exercise 9.24 If $(\Omega, \mathcal{F}, p)$ is the probability space defined in Example 9.44, prove that $\int_X dP = \sum_{x \in \Omega} P(x) = 1$.

Exercise 9.25 Let X_1 and X_2 be two random variables with distribution function $F_i(x) = 1 - e^{-a_i x}$ if $x > 0$ and zero otherwise, with $i = 1, 2$ and $a_i \geq 0$. Compute the probability that the realisation of X_1 is greater than the realisation of X_2.

Exercise 9.26 The skewness s and kurtosis k of a random variable X are, respectively, defined as

$$s = \frac{\int dP \, (X - E[x])^3}{V[x]^{3/2}} \quad \text{and} \quad k = \frac{\int dP \, (X - E[x])^4}{V[x]^2}.$$

With reference to Example 9.45, prove that $s = C_3(X)/C_2(X)^{3/2}$ and $k = C_4(X)/C_2(X)^2 + 3$.

Exercise 9.27 Consider the measure on $\mathbb{R}$ defined by $\mu(dx) = e^{-x^2/2}/\sqrt{2\pi}dx$. Using the result in Example 9.38, prove that it is a probability measure. Its name is *Gaussian probability distribution*. With reference to Example 9.45, find its cumulant generating function *(Hint: Apply a change of variable)*.

Exercise 9.28 Let X be a random variable and c a constant. With reference to Example 9.45, prove that $K_{cX}(t) = K_X(ct)$. Use this result to show that if $X_1 \ldots X_n$ are independent random variables with bounded cumulants, then for $Y_n = \sum_{i=1}^n X_i/n$, $\lim_{n \to \infty} C_l(Y_n) = 0$ for any $l > 1$.

Exercise 9.29 Let X be a random variable and c be a constant. With reference to Example 9.45, prove that $K_{X+c}(t) = K_X(t) + ct$. Use this, together with the result in Exercise 9.28, to show that if $X_1 \ldots X_n$ are independent random variables with bounded cumulants, then for $Y_n = (S_n - C_1(S_N))/\sqrt{n}$, with $S_n = \sum_{i=1}^n X_i$, $\lim_{n \to \infty} C_l(Y_n) = 0$ for $l = 1$ and $l > 2$.

Exercise 9.30 Let X_1, X_2, and X_3 be vector random variables in $(\mathbb{R}^2, \mathcal{B}_2)$. X_1 takes with equal probability any value in $[0, 1] \times [0, 2]$. X_2 takes with equal probability any value in $[0, 2] \times [0, 1]$. The first component of $X_3 = (X, Y)$ takes with equal probability any value in $[0, 1]$ and the second component takes with equal probability any value in $[0, x]$, where x is the realisation of the first component. Consider the function $U : \mathbb{R}^2 \to \mathbb{R}$, $U(x, y) = x^\alpha y^\beta$ with $\alpha, \beta > 1$ and compute $E_{X_i}[U]$, with $i = 1, 2, 3$.

Cauchy Initial Value Problem

In the study of differential equations and dynamical systems, a *initial value problem* describes a situation in which a dynamic variable, represented by a function depending on a scalar parameter, often identified with time, has a specific initial value and then evolves according to some given rule. The rule is expressed as a relation between the derivative of the variable, its value, and the value of the scalar parameter. The specific rule can come from some optimisation argument, or it can simply describe some sort of "law of motion". The existence of a solution to the problem is not guaranteed and depends on the variable initial condition and the rule that describes its evolution. The search for a solution can be carried out on a global or local scale. In the first case, a possible solution is typically sought among a predefined set of functions, like those defined over a desirable domain. In the second case, one is interested in finding if there exists a suitable domain in which the problem admits a possible unique solution. The following analysis pertains to the second case. Specifically, we consider a vector-valued function $\mathbf{x} : \mathbb{R} \to \mathbb{R}^n$, depending on a real parameter t. The evolution of the variable and its initial value are described by the system

$$
\begin{cases}
\dot{\mathbf{x}}(t) = \mathbf{f}(\mathbf{x}(t), t), \\
\mathbf{x}(t_0) = \mathbf{x}_0,
\end{cases}
$$

where we have adopted the convention of indicating with a point, or "dot", the derivative of the variable with respect to the scalar parameter, $\dot{\mathbf{x}} = d\mathbf{x}/dt$. The solution $\mathbf{x}(t)$ must be a differentiable, and hence continuous, function. If $\mathbf{f} : \mathbb{R}^{n+1} \to \mathbb{R}^n$ is continuous in a neighbourhood of $(\mathbf{x}_0, t_0)$, then, in that neighbourhood, the right-hand side is a continuous function of t. In this case, we are, in fact, looking for a continuously differentiable solution. Since both sides are continuous, we can integrate them with respect to the scalar variable t and, for the fundamental theorem of calculus, we find an equivalent integral representation of the problem,

G. Bottazzi, *Advanced Calculus for Economics and Finance*, Classroom Companion: Economics, https://doi.org/10.1007/978-3-031-30316-6

$$\mathbf{x}(t) = \mathbf{x}_0 + \int_{t_0}^{t} d\tau \, \mathbf{f}(\mathbf{x}(\tau), \tau).$$

Note that the previous equation is implicit and somehow formal, as the function $\mathbf{x}$ appears on the left and right sides. Although the continuity of the function $\mathbf{f}$ is sufficient to establish the equivalence between the differential and integral versions of the problem, it is generally not sufficient to establish the existence and uniqueness of the solution.

▶ **Example A.1** (*A problem with no solution*) Consider the problem

$$\begin{cases} \dot{x}(t) = x(t), \\ x(0) = -1. \end{cases}$$

By dividing both sides by x, the first equation can be rewritten as $d \log x(t)/dt = 1$, which implies that $\log x(t) = t + c$, for some constant $c \in \mathbb{R}$. That is, $x(t) = \exp(t + c)$. The initial value condition requires that $e^c = -1$, which has no (real) solution.

▶ **Example A.2** (*A problem with multiple solutions*) Consider the problem

$$\begin{cases} \dot{x}(t) = 3|x(t)|^{2/3}, \\ x(0) = 0. \end{cases}$$

It is immediate to verify that both $x(t) = 0$ and $x(t) = t^3$ solve this problem.

In what follows, we want to derive sufficient conditions that guarantee the existence of a unique local solution to the initial value problem. Our strategy requires imposing some additional regularity conditions on the function $\mathbf{f}$. Then we build a suitable function based on the integral representation of the problem, and, finally, by applying the Banach fixed point theorem, we use this function to prove the existence and uniqueness of the solution. The additional regularity conditions are based on the notion introduced in Sect. 3.4.

Definition A.1 (*Uniform Lipschitz continuity*) The function $\mathbf{f}(\mathbf{x}, t)$ from $\mathbb{R}^{n+1}$ to $\mathbb{R}^n$, with $\mathbf{x} \in \mathbb{R}^n$ and $t \in \mathbb{R}$, is *uniformly Lipschitz continuous* in $\mathbf{x}$ with respect to t if $\exists K > 0$ such that for any two points $\mathbf{x}, \mathbf{y} \in \mathbb{R}^n$ and any $t \in \mathbb{R}$, $\|\mathbf{f}(\mathbf{x}, t) - \mathbf{f}(\mathbf{y}, t)\| \leq K \|\mathbf{x} - \mathbf{y}\|$.

The function is *locally uniformly Lipschitz continuous* in $(\mathbf{x}_0, t_0)$ if there exist a neighbourhood $N(\mathbf{x}_0) \subseteq \mathbb{R}^N$ of $\mathbf{x}_0$ and a neighbourhood $N(t_0) \subseteq \mathbb{R}$ of t_0 such that the previous statement is true $\forall \mathbf{x}, \mathbf{y} \in N(\mathbf{x}_0)$ and $\forall t \in N(t_0)$.

Next, we define a function, specifically built for the problem at hand, that maps a continuous function into a continuous function.

Definition A.2 *(Picard map)* Consider a continuous function $\mathbf{f}(\mathbf{x}, t)$ from $\mathbb{R}^{n+1}$ to $\mathbb{R}^n$, with $\mathbf{x} \in \mathbb{R}^n$ and $t \in \mathbb{R}$, and a point $(\mathbf{x}_0, t_0) \in \mathbb{R}^{n+1}$. The associated *Picard map* sends the continuous function $\mathbf{z} : \mathbb{R} \to \mathbb{R}^n$ into the continuous function $L\mathbf{z} : \mathbb{R} \to \mathbb{R}^n$ defined as

$$L\mathbf{z}(t) = \mathbf{x}_0 + \int_{t_0}^{t} d\tau \, \mathbf{f}(\mathbf{z}(\tau), \tau).$$

Note that $L\mathbf{z}(t)$, being the integral of a continuous function, is, in fact, continuously differentiable.

Now assume that the function $\mathbf{f}(\mathbf{x}, t)$ of the initial value problem is continuous in a neighbourhood of t_0 and locally Lipschitz continuous in $\mathbf{x}$ with respect to t in $(\mathbf{x}_0, t_0)$. Then there exist an interval $I = (t_0 - \delta_t, t_0 + \delta_t)$, a ball $B(\mathbf{x}_0, d_x) = \{\mathbf{z} \in \mathbb{R}^n \mid \|\mathbf{z} - \mathbf{x}_0\| < d_x\}$, and a constant $K > 0$ such that $\forall \mathbf{x}, \mathbf{y} \in B(\mathbf{x}_0, d_x)$ and $\forall t \in I$, $\|\mathbf{f}(\mathbf{x}, t) - \mathbf{f}(\mathbf{y}, t)\| \leq K \|\mathbf{x} - \mathbf{y}\|$. We take δ_t such that $\delta_t K < 1$. The Lipschitz continuity guarantees that the function is continuous and bounded in these neighbourhoods. Specifically, the diameter of the image of $\mathbf{f}$ is bounded above by $K d_x$ (see Corollary 3.2). Let $C(I, B)$ be the set of continuous bounded functions defined over I with images in $B(\mathbf{x}_0, d_x)$. As discussed in Example 5.19, $(C(I, B), \|.\|_\infty)$ is a Banach space.

We now have all the pieces in place and we are able to proceed with the proof of the existence and uniqueness of the local solution of the initial value problem. We start with a preliminary result.

Lemma A.1 (Picard) *The Picard map sends an element of $C(I, B)$ into an element of $C(I, B)$.*

Proof We have to prove that if, for a continuous function $\mathbf{z}$, $\forall t \in I$, $\|\mathbf{z}(t) - \mathbf{x}_0\| < \delta_x$, then $\forall t \in I$, $\|L\mathbf{z}(t) - \mathbf{x}_0\| < \delta_x$. In fact,

$$\|L\mathbf{z}(t) - \mathbf{x}_0\|^2 = \sum_{i=1}^{n} \left(\int_{t_0}^{t} d\tau f_i(\mathbf{z}(\tau), \tau) \right)^2 \leq$$

$$\leq (t - t_0) \int_{t_0}^{t} d\tau \sum_{i=1}^{n} f_i(\mathbf{z}(\tau), \tau)^2 \leq (t - t_0)^2 K^2 d_x^2 < \delta_x^2.$$

The first inequality is the Jensen inequality of Corollary 8.5 applied to the function $g(x) = x^2$ and the last inequalities follow from $\|\mathbf{f}\|^2 \leq K^2 \delta_x^2$, $|t - t_0| \leq \delta_t$ and our assumption $\delta_t K < 1$. $\qquad\square$

Once we have proved that the Picard map is a well-defined function in $C(I, B)$, we can state our main result.

Theorem A.1 (Cauchy–Lipschitz) *The Picard map is a contraction in $C(I, B)$.*

Proof Consider $\mathbf{x}, \mathbf{y} \in C(I, B)$. Note that

$$\|L\mathbf{x}(t) - L\mathbf{y}(t)\|^2 = \sum_{i=1}^{n} \left(\int_{t_0}^{t} d\tau f_i(\mathbf{x}(\tau), \tau) - f_i(\mathbf{y}(\tau), \tau) \right)^2 \leq$$

$$\leq (t - t_0) \int_{t_0}^{t} d\tau \sum_{i=1}^{n} (f_i(\mathbf{x}(\tau), \tau) - f_i(\mathbf{y}(\tau), \tau))^2 \leq$$

$$K(t - t_0) \int_{t_0}^{t} d\tau \sum_{i=1}^{n} (x_i(\tau) - y_i(\tau))^2 = K(t - t_0) \int_{t_0}^{t} d\tau \|\mathbf{x}(\tau) - \mathbf{y}(\tau)\|^2,$$

where we have used the Jensen inequality of Corollary 8.5 applied to the function $g(x) = x^2$ and the Lipschitz continuity of $\mathbf{f}$. From the property of the supremum,

$$(t - t_0) \int_{t_0}^{t} d\tau \|\mathbf{x}(\tau) - \mathbf{y}(\tau)\|^2 \leq (t - t_0)^2 \sup_{\tau \in I} \left\{ \|\mathbf{x}(\tau) - \mathbf{y}(\tau)\|^2 \right\},$$

so that

$$\|L\mathbf{x} - L\mathbf{y}\|_{\infty}^2 = \sup_{\tau \in I} \left\{ \|L\mathbf{x}(\tau) - L\mathbf{y}(\tau)\|^2 \right\} \leq$$

$$\delta_t^2 K^2 \sup_{\tau \in I} \left\{ \|\mathbf{x}(\tau) - \mathbf{y}(\tau)\|^2 \right\} = \delta_t^2 K^2 \|\mathbf{x} - \mathbf{y}\|_{\infty}^2,$$

and, remembering that $\delta_t K < 1$, the statement follows. $\square$

Finally, using the Banach fixed point theorem, we can state the result we were looking for.

Corollary A.1 *Consider the Cauchy initial value problem*

$$\begin{cases} \dot{\mathbf{x}}(t) = \mathbf{f}(\mathbf{x}(t), t), \\ \mathbf{x}(t_0) = \mathbf{x}_0, \end{cases}$$

with $\mathbf{f} : \mathbb{R}^{n+1} \to \mathbb{R}^n$, $\mathbf{x} : \mathbb{R} \to \mathbb{R}^n$, *and* $\mathbf{x}_0 \in \mathbb{R}^n$. *If* $\mathbf{f}(\mathbf{x}, t)$ *is continuous in a neighbourhood of* t_0 *and locally Lipschitz continuous in* $\mathbf{x}$ *with respect to* t *in* $(\mathbf{x}_0, t_0)$, *then there exists an interval* $I = (t_0 - \delta_t, t_0 + \delta_t)$ *in which the problem admits an unique continuously differentiable solution.*

Proof Given the hypothesis, one can build the Banach space $(C(I, B), \|.\|_{\infty})$. In this space, the Picard map is a contraction. Thus, according to Theorem 5.9, it has a unique fixed point $\mathbf{x}$ that satisfies the equation $\mathbf{x} = L\mathbf{x}$, that is $\mathbf{x}(t) = \mathbf{x}_0 + \int_{t_0}^{t} d\tau \mathbf{f}(\mathbf{x}(\tau), \tau)$. Since the function $\mathbf{f}$ is continuous by hypothesis, this is the unique solution of the initial value problem. $\square$

Brouwer Fixed Point Theorem

B

Every continuous function from a convex compact subset $K \subseteq \mathbb{R}^n$ to itself has at least one fixed point. This is the Brouwer fixed point theorem that is referred to in basically all courses of Microeconomic Theory, but a proof of which is almost never provided. Indeed, this theorem has a very natural proof in algebraic topology. However, that proof requires notions that are likely unavailable to students of economics. On the other hand, the proof I provide below is based on quite elementary notions introduced in any course of advanced calculus.[1]

In the discussion below, I will denote by $\bar{B}_n$ the closed ball of unit radius, that is, $\bar{B}_n = \{\mathbf{x} \in \mathbb{R}^n \mid \|\mathbf{x}\| \leq 1\} \subset \mathbb{R}^n$. I will denote by B_n its interior and with ∂B_n its boundary. The proof of Brouwer's theorem relies on a couple of lemmas concerning the properties of the unit ball. The first lemma characterises a *retraction*, that is, a function $\mathbf{r} : \bar{B}_n \to \partial B_n$ such that its restriction to the boundary is the identity: if $\mathbf{x} \in \partial B_n$, $r(\mathbf{x}) = \mathbf{x}$.

Lemma B.1 *There are no continuously differentiable retractions.*

Proof Assume that there exists a retraction $\mathbf{r}$ such that $r \in C^1(\bar{B}_n)$. We will see that this assumption will lead to a contradiction. For any $\lambda \in [0, 1]$, consider the function

$$\mathbf{g}_\lambda(\mathbf{x}) = (1 - \lambda)\mathbf{x} + \lambda\mathbf{r}(\mathbf{x}) .$$

The function $\mathbf{g}_\lambda$ is a convex combination of continuously differentiable functions, specifically $\mathbf{r}$ and the identity map. Thus, it is continuously differentiable. Since the ball is a convex set, it is $\mathbf{g}_\lambda(\bar{B}_n) \subseteq \bar{B}_n$. Also notice that the restriction of $\mathbf{g}$ to the boundary of the ball is the identity map. Now consider the differential of $\mathbf{g}_\lambda$,

---

[1] The proof presented here is loosely based on *A less strange version of Milnor's proof of Brouwer fixed-point theorem* by C.A. Rogers that appeared in *The American Mathematical Monthly* in 1980. I replaced the Weistrass polynomial approximation with a mollification kernel approach and I exploited the Lebesgue integration theory and the operator norm to simplify the argument.

© The Editor(s) (if applicable) and The Author(s), under exclusive license to Springer Nature Switzerland AG 2023
G. Bottazzi, *Advanced Calculus for Economics and Finance*, Classroom Companion: Economics, https://doi.org/10.1007/978-3-031-30316-6

$dg_\lambda = I + \lambda(dr - I)$, where I is that $n \times n$ identity matrix and dr is the differential of r. Its inverse dg_λ^{-1} can be formally written as

$$dg_\lambda^{-1} = \frac{1}{I + \lambda(dr - I)} = \sum_{i=0}^{\infty} \lambda^i (dr - I)^i .$$

Let $M = \max_{x \in \bar{B}_n} \|dr - I\|_{op}$ where $\|.\|_{op}$ stands for the operator norm in Definition 4.4. Notice that $M > 0$, because the matrix dr cannot be the identity on the whole ball. Then if $\lambda < 1/M$ it is

$$\|dg_\lambda^{-1}\|_{op} \leq \sum_{h=0}^{\infty} \lambda^h \|dr - I\|_{op}^h \leq \sum_{h=0}^{\infty} \lambda^h M^h = 1/(1 - \lambda M) .$$

Thus, for $\lambda \in [0, 1/M)$, the differential dg_λ has an inverse on the whole ball $\bar{B}_n$, which for the inverse function theorem (see Sect. 7.5) implies that $\mathbf{g}_\lambda$ is invertible and that its image is the whole $\bar{B}_n$. The differential of $\mathbf{g}_\lambda$ is a continuous function so that one can define the integral

$$\mu(\lambda) = \int_{\bar{B}_n} dl_n |dg_\lambda| = \int_{g_\lambda(\bar{B}_n)} dl_n,$$

where the second equality comes directly from the theory of the change of variable (Theorem 9.15). We have seen that for $\lambda \in [0, 1/M)$ it is $g_\lambda(\bar{B}_n) = \bar{B}_n$, thus $\mu(\lambda) = V_n$, where V_n is the volume of the nth dimensional ball of radius 1. At the same time, $|dg_\lambda|$ is just the sum of products of n linear expressions in λ, thus $\mu(\lambda)$ is a polynomial of degree n in λ. Being a polynomial, if it is constant on an interval, it is constant everywhere. This is absurd, because clearly $\mu(1) = \int_{\partial B_n} dl_n |dg_\lambda| = 0$ as ∂B_n is a zero measure set. $\qquad\square$

The second lemma somehow connects continuous maps with continuously differentiable maps.

Lemma B.2 *If there exists a continuous map from the compact ball to itself without fixed points, then there exists a continuously differentiable map from the compact ball to itself without fixed points.*

Proof Let $f : \bar{B}_n \to \bar{B}_n$ be a continuous map from the compact ball to itself. Consider the ball with radius $1 + \delta_1$, $\bar{B}_n(1 + \delta_1) \supset \bar{B}_n$ and the function $f_1 : \bar{B}_n(1 + \delta_1) \to \bar{B}_n$ defined as $f_1(\mathbf{x}) = f(\mathbf{x})$ if $\|\mathbf{x}\| \leq 1$ and $f_1(\mathbf{x}) = f(\mathbf{x}/\|\mathbf{x}\|)$ otherwise.

The function f_1 has no fixed points. Because it is continuous, it is uniformly continuous on $\bar{B}_n(1 + \delta_1)$. Let $\epsilon = \min_{\mathbf{x} \in \bar{B}_n(1+\delta_1)} \|f_1(\mathbf{x}) - \mathbf{x}\| > 0$ and $\delta_2 > 0$ be such that $\|f_1(\mathbf{x}) - f_1(\mathbf{y})\| < \epsilon/2$ if $\|\mathbf{x} - \mathbf{y}\| \leq \delta_2$. Define $\eta = \min\{\delta_1, \delta_2\}$ and consider a continuously differentiable function $\phi : \mathbb{R}^n \to \mathbb{R}_{\geq 0}$ such that $\phi(\mathbf{x}) = 0$ if $\|\mathbf{x}\| \geq \eta$ and $\int dl_n(\mathbf{x})\phi(\mathbf{x}) = 1$.

For any $\mathbf{x} \in \bar{B}_n$, define $f_2(\mathbf{x}) = \int dl_n(\mathbf{y}) f_1(\mathbf{y}) \phi(\mathbf{x} - \mathbf{y})$ (see Example 9.26). The function f_2 is continuously differentiable, $df_2(\mathbf{x}) = \int dl_n(\mathbf{y}) f_1(\mathbf{y}) d\phi(\mathbf{x} - \mathbf{y})$ and $f_2(\bar{B}_n) \subseteq \bar{B}_n$ as

$$\| f_2(\mathbf{x}) \| \leq \int dl_n(\mathbf{y}) \| f_1(\mathbf{y}) \| \phi(\mathbf{x} - \mathbf{y}) \leq 1 .$$

Note also that

$$\| f_2(\mathbf{x}) - f_1(\mathbf{x}) \| \leq \int dl_n(\mathbf{y}) \| f_1(\mathbf{y}) - f_1(\mathbf{x}) \| \phi(\mathbf{x} - \mathbf{y}) \leq \epsilon/2,$$

where the last inequality comes from the fact that the function ϕ is different from zero only when $\| \mathbf{x} - \mathbf{y} \| \leq \eta \leq \delta_2$. Moreover for any $\mathbf{x} \in \bar{B}_n$ it is $\| f(\mathbf{x}) - \mathbf{x} \| \leq \| f(\mathbf{x}) - f_2(\mathbf{x}) \| + \| f_2(\mathbf{x}) - \mathbf{x} \|$ that is $\| f_2(\mathbf{x}) - \mathbf{x} \| \geq \| f(\mathbf{x}) - \mathbf{x} \| - \| f(\mathbf{x}) - f_2(\mathbf{x}) \|$. But in $\bar{B}_n$, $f = f_1$ and using the previous inequalities one gets $\| f_2(\mathbf{x}) - \mathbf{x} \| \geq \epsilon/2$. This implies that f_2 is a continuously differentiable map from the compact ball to itself without fixed points. □

Now we are ready for the final theorem.

Theorem B.1 (Brouwer's fixed point theorem) *Every continuous function from a convex compact subset $K \subseteq \mathbb{R}^n$ to itself has at least one fixed point.*

Proof Recall that any convex compact subset $K \subseteq \mathbb{R}^n$ is homeomorphic to the compact ball $\bar{B}$ of radius 1 and centre $\mathbf{0}$. That is, there exists a continuous invertible one-to-one function $\phi : K \to \bar{B}$ (see Example 4.4).

Then if there exists a continuous function $f : K \to K$ with no fixed points, the function $f_1 = \phi \circ f \circ \phi^{-1}$ is a continuous function from $\bar{B}_n$ to $\bar{B}_n$ with no fixed points. Thus, for Lemma B.2, there exists a continuously differentiable map $g : \bar{B}_n \to \bar{B}_n$ with no fixed points.

On $\bar{B}_n$ consider the function $r : \bar{B}_n \to \partial B_n$ defined as $r(\mathbf{x}) = \mathbf{x} + \lambda(\mathbf{x})(g(\mathbf{x}) - \mathbf{x})$ where

$$\lambda(\mathbf{x}) = \frac{(g(\mathbf{x}) - \mathbf{x}) \cdot \mathbf{x}}{\| g(\mathbf{x}) - \mathbf{x} \|^2} - \frac{\sqrt{((g(\mathbf{x}) - \mathbf{x}) \cdot \mathbf{x})^2 - (\| \mathbf{x} \|^2 - 1) \| g(\mathbf{x}) - \mathbf{x} \|^2}}{\| g(\mathbf{x}) - \mathbf{x} \|^2}.$$

Because for any $\mathbf{x}$ it is $g(\mathbf{x}) \neq \mathbf{x}$, the function r is well defined. The value of $\lambda(\mathbf{x})$ is chosen such that $g(\mathbf{x}) \in \partial B_n$. Thus r represents a continuously differentiable retraction. But from Lemma B.1 we know that such a retraction does not exist. Thus, a function like g cannot exist and, consequently, a function like f cannot exist. □

Index

A

Absolute convergence, 109
Absolute value, 11
Algebra, 245
Almost everywhere, 249
Almost surely, 288
AM–GM inequality, 16, 95
Analytic function, 150
Angular coefficient, 135
Archimedean property, 12
Asymptotic approximation, 147
Asymptotic equivalence, 102

B

Banach fixed point, 90
Banach space, 107
Base, 29
Bernoulli inequality, 22
Bijective function, 3
Binary operation, 8
Binomial coefficient, 22, 174
Binomial distribution, 291
Bolzano–Weierstrass theorem, 37, 74
Borel measurable function, 258, 259
Borel set, 250
Borel σ-algebra, 250
Boundary candidate solution, 190
Boundary point, 26
Bounded
   function, 55
   norm, 66
   set, 55
Bounded above

function, 5
set, 4
Bounded below
   function, 5
   set, 4

C

Candidate solution, 190
   boundary, 190
   internal, 190
Cartesian product, 2
Cauchy condensation test, 111
Cauchy–Schwarz inequality, 67
Cauchy sequence, 85
Cauchy theorem, 137
Central moment, 292
Cesàro mean, 101
Cesàro summable, 115
Chain rule, 133, 160
Characteristic function, 261
Chebyshev distance, 57
Chebyshev inequality, 267
Closed, 8, 26
Closed map, 45
Closure, 26
Cofinite topology, 50
Combination
   conical, 21
   convex, 18
Comparison test
   series, 110
Comparison theorem
   sequences, 92

Complement, 26
Complete measure space, 249
Complete metric space, 88
Completion
  measure, 249
Concave function, 19, 49, 132
Condensation test
  Cauchy, 111
  Schlömilch, 110
Conical combination, 21
Connected set, 40
Constraint, 190
  active, 190
  binding, 190
  qualification, 199
Continuous
  function, 44
  Lipschitz, 61, 228, 302
  uniform, 59
Continuously differentiable, 166
Contraction, 61
Converging
  sequence, 81
  series, 108
Convex combination, 18
Convex cone, 20
Convex function, 19, 49, 132
Convex set, 18
Convolution, 272
Countable additivity, 246
Countable set, 6
Countable topology, 32
Counting measure, 247
Cover, 34
Critical point, 169
Cumulant, 292
Cumulant generating function, 292
Curve
  differential, 225
  length, 226
  non-singular, 200
  regular, 200

D
Decreasing function, 11
Dedekind cut, 10
Definite integral, 217
Dense set, 26
Dependent function, 188
Derivative, 130
  directional, 162

left, 131
partial, 162
product rule, 132
quotient rule, 133
reciprocal rule, 133
right, 131
Derivative set, 26
Diameter, 55
Differentiable, 130, 131, 158
Differential, 158
  multivariate, 158
Differential curve, 225
Dini theorem, 118
Directional derivative, 162
Dirichlet function, 218, 262
Discontinuity
  first kind or trivial, 128
  second kind or jump, 128
  third kind or essential, 128
Discrete σ-algebra, 245
Discrete distribution, 291
Discrete topology, 25
Disjoint set, 1
Distance, 55
  Chebyshev, 57
  Hamming, 56
Distribution function, 239, 290
Distribution law, 290
Divergent
  sequence, 98
  series, 108
Domain, 2
Dominated convergence theorem, 268
Dual
  Lagrangian, 201
  problem, 201
Dual cone, 21
Duality
  strong, 201
  weak, 201
Dual norm, 80, 214

E
Empty set, 1
Envelope theorem, 209
Equivalence class, 2
Equivalence relation, 2
Equivalent norm, 76
Euclidean metric, 56
Euclidean norm, 67
Euclidean topology, 32, 51

Expected value, 291
Exponential distribution, 241
Exponential function, 15, 98, 114
Exponential inequality, 98
Exterior point, 26
Extrema, 28
Extremal point, 28
Extremal value, 28
Extreme value theorem, 49

F
Factorial, 12
Farkas lemma, 195
Fatou's lemma, 266
Feasible region, 190
Feasible set, 190
Fibonacci numbers, 87
Field, 8
  opposite, 8
  reciprocal, 8
  scalar, 230
Finite measure, 248
Finite measure space, 248
First-order condition, 169, 190
Fixed point, 84, 90, 128
Flex, 140
Floor function, 45
Fritz John theorem, 198
Fubini's theorem, 284
Function, 3
  analytic, 150
  bijective, 3
  bounded, 5, 55
  bounded above, 5
  bounded below, 5
  closed map, 45
  concave, 19
  convex, 19
  convolution, 272
  decreasing, 11
  dependent, 188
  distance, 55
  exponential, 15, 98, 114
  floor, 45
  implicit, 182
  increasing, 11
  independent, 188
  injective, 3
  integrable, 266
  into, 3
  Jacobian matrix, 163

  linearly dependent, 188
  logarithm, 94, 95
  lower semicontinous, 49
  mollification, 272
  monotonic, 11
  one-to-one, 3
  onto, 3
  open map, 45
  power, 15
  smooth, 177
  step, 45
  strictly concave, 19
  strictly convex, 19
  surjective, 3
  upper semicontinuous, 49
  value, 210
Functional space, 57, 71

G
Gaussian distribution, 298
Geometric progression, 12
Geometric series, 109
GM-HM inequality, 22
Golden ratio, 87, 91
Gradient, 164
Greatest lower bound property, 5

H
Hamming distance, 56
Harmonic series, 111
Hausdorff topology, 35
Heaviside theta function, 142, 228
Heine–Borel, 39, 75, 78
Heine-Cantor
  theorem, 60
Hessian matrix, 175
Holder's inequality, 69
Homeomorphism, 47
Hyperbolic cosine, 133
Hyperbolic sine, 133
Hyperharmonic series, 111

I
Image, 2
Image measure, 258
Implicit function, 182
Improper integral
  bounded, 225
  unbounded, 223
Increasing function, 11
Indefinite integral, 227

Independent events, 288
Independent function, 188
Independent random variable, 293
Indicator function, 261
Inequality
  AM–GM, 16, 95
  Bernoulli, 22
  Cauchy–Schwarz, 67
  Chebyshev, 267
  difference of powers, 12
  exponential, 98
  GM-HM, 22
  Holder, 69
  Jensen, 20, 241, 269
  logarithm, 94
  max-min, 5
  Minkowski, 70
  Peter–Paul, 22
  power mean, 17
  triangle, 11, 68
  Young, 17
Infimum, 4
Infinite limit, 126
Infinitesimal, 103
Inflection point, 140
Initial value problem, 301
Injective function, 3
Integer part, 10
Integers modulo p, 9
Integrable function, 266
Integral, 217, 233
  definite, 217
  improper, 223, 225
  indefinite, 227
  line, 230
Integral function, 226
Integrand, 217
Integration by part, 230
Integration by substitution, 229
Intercept, 135
Interior point, 26
Internal candidate solution, 190
Inverse image, 2
Inverse relation, 2
Irrational number, 10

J
Jacobian matrix, 163
Jensen's inequality, 20, 241, 269

K
Karush-Kuhn-Tucker theorem, 199
Kronecker delta, 56
Kronecker's lemma, 115

L
Lagrange error bound, 144
Lagrange theorem, 199
Lagrangian function, 192, 194, 200
Lagrangian multiplier, 192
Landau symbols, 103
Lattice, 218
Least upper bound property, 5
Lebesgue measurable function, 259
Left continuous, 129
Left derivative, 131
Left differentiable, 131
Left limit, 126
Left-order topology, 35
Leibnitz criterion, 112
Leibnitz integral rule, 231
Length of a curve, 226
L'Hopital's rule, 140
Limit
  at infinity, 126
  inferior, 99
  infinite, 126
  left, 126
  right, 126
  sequence, 81
  superior, 99
Limit of a function, 42
Limit point, 26
Linearly dependent function, 188
Line integral, 230
Lipschitz continuity, 61, 228, 302
Local maximum, 11, 28
  strict, 28
Local minimum, 11, 28
  strict, 28
Logarithm
  function, 94, 95
  natural, 94
Logarithm inequality, 94
Lower bound, 4
Lower integral, 217, 234
Lower limit, 99
Lower limit topology, 51
Lower semicontinuity, 49
Lower sum, 215, 232

M
Maclaurin polynomial, 144
Map, 3
Maximal element, 5
Maximum, 5
Max-min inequality, 5
Mean, 241, 291
  weighted arithmetic, 16
  weighted geometric, 16
  weighted harmonic, 22
  weighted power, 17
Mean value theorem, 137, 170, 171
Measurable function, 256, 259
  Borel, 258, 259
  Lebesgue, 259
Measurable rectangle, 278
Measurable set, 246
Measurable space, 246
Measure, 246
  finite, 248
  image, 258
  product, 281
  pushforward, 258
Measure completion, 249
Measure space, 246
Metric function, 55
Metric space, 55
Minimal element, 5
Minimum, 5
Minkowski inequality, 70
Minus infnity limit, 126
Mollification, 272
Moment, 240
Monotone class, 280
Monotone convergence theorem, 93, 264
Monotonically decreasing, 11
Monotonically increasing, 11
Monotonic function, 11
Monotonic sequence, 93
Motzkin theorem, 196
M-test, 119
Multinomial coefficient, 174
Multiplier
  Lagrangian, 192
Multivariate, 292

N
N-cell, 74
Nearest point, 78
Negative definite, 176
Negative semi-definite, 176

Neighbourhood, 26
  left, 126
  right, 126
Nonlinear programming, 189
Non-singular curve, 200
Norm, 65
  continuity, 66
  dual, 80, 214
  equivalent, 76
  Euclidean, 67
  p-, 69
Normal convergence, 119
Normed space, 65
Null set, 249

O
Objective function, 189
Open ball, 57
Open map, 45
Open set, 25
Operation
  absorbing element, 8
  associative, 8
  close, 8
  commutative, 8
  distributive, 8
  neutral element, 8
Operator, 71
  bounded, 71
Operator norm, 72
Optimisation problem, 189
Ordered set, 4
Order relation, 4
Outer measure, 252

P
Partial derivative, 162
Partial sum, 101
Partition, 215
  equispaced, 218
  refinement, 216
Peter–Paul inequality, 22
Picard map, 303
Plus infnity limit, 126
p-norm, 69
Point
  critical, 169
  extremal, 28
  fixed, 84, 90, 128
  inflection, 140
  stationary, 137, 169

Point mass measure, 247
Pointwise convergence, 116
Polar cone, 21
Polar coordinates, 156, 287
Positive definite, 176
Positive homogeneity, 65
Positive semi-definite, 176
Power function, 15
Power mean inequality, 17
Power series expansion, 148
Power set, 1
Preimage, 2
Preorder relation, 4
Primal problem, 201
Primitive function, 226
Probability density, 291
Probability distribution, 290
Probability measure, 288
Probability space, 288
Product measurable space, 278
Product measure, 281, 282
Product σ-algebra, 278
Progression
    arithmetic, 121
    geometric, 121
Property
    Archimedean, 12
p-series, 111
Punctured neighbourhood, 43
Pushforward measure, 258

Q
Quotient set, 2

R
Radius of convergence, 149
Random variable, 239, 289
    independent, 293
Range, 2
Ratio test
    series, 113
Real numbers, 10
Rectangle, 278
    measurable, 278
Recursion, 84
Reduction formula, 230
Regular curve, 200
Regular grid, 218
Regularity condition, 199
Relation, 2
    equivalence, 2

    inverse, 2
    order, 4
    preorder, 4
    reflexive, 2
    strict order, 4
    symmetric, 2
    transitive, 2
Relative interior, 21
Remainder
    Taylor polynomial, 144
Retraction, 305
Right continuous, 129
Right derivative, 131
Right differentiable, 131
Right limit, 126
Right-order topology, 35
Rolle theorem, 137
Root test
    series, 112
Rule
    chain, 133, 160

S
Saddle point, 176
    Lagrangian, 193, 201
Scalar field, 155, 230
Scalar function, 155
Schlömilch condensation test, 110
Second-order condition, 177, 203, 205
Sequence, 81
    asymptotically equivalent, 102
    Cauchy, 85
    Cesàro summable, 115
    constant, 82
    decreasing, 93
    increasing, 93
    monotonic, 93
    summable, 108
Series, 108
    absolute convergence, 109
    power, 148
Set
    Borel, 250
    bounded above, 4
    bounded below, 4
    connected, 40
    countable, 6
    dense, 26
    derivative, 26
    disjoint, 1
    empty, 1

measurable, 246
unbounded, 55
σ-algebra, 245
  discrete, 245
  generated, 246
  trivial, 245
σ-finite measure space, 248
Simple function, 262
Singlet, 26
Slackness conditions, 192
Smooth function, 177
Space
  Banach, 107
  measurable, 246
  measure, 246
  metric, 55
  normed, 65
  topological, 25
Square-integrable, 295
Stationary point, 137, 169
Step function, 227, 235
Sticltjes measure, 232
Stiemke's lemma, 197
Strictly concave function, 19
Strictly convex function, 19
Strict order relation, 4
Strong duality, 201
Subcover, 34
Subsequence, 83
Subspace
  topological, 41
Substitution theorem, 270
Summation by parts, 12
Sup-norm, 71
Support, 240
Supremum, 4
Surjective function, 3
Sylvester's criterion, 176

T
Taylor polynomial, 143, 175
Theorem
  AM–GM inequality, 16, 95
  Bolzano–Weierstrass, 74
  Borel–Cantelli, 288
  Brouwer, 307
  Cauchy, 137
  Cauchy–Schwarz inequality, 67
  Cauchy-Hadamard, 149
  Cauchy–Lipschitz, 303
  Farkas' lemma, 195
  Fritz John, 198
  Heine–Borel, 39
  Heine-Cantor, 60

  Holder's inequality, 69
  Jensen's inequality, 20
  Karush-Kuhn-Tucker, 199
  Lagrange, 137, 199
  mean value, 137, 170, 171
  Minkowski inequality, 70
  Motzkin, 196
  Rolle, 137
  Stiemke's lemma, 197
  Stolz-Cesàro, 100
  Young inequality, 17
Theorems of the alternatives, 195
Thomae's function, 219
Topological space, 25
Topology, 25
  cofinite, 50
  countable, 32
  Euclidean, 32, 51
  left-order, 35
  lower limit, 51
  right-order, 35
  upper limit, 51
Triangle inequality, 11, 55, 65, 68
Trivial σ-algebra, 245
Trivial topology, 25

U
Unbounded set, 55
Uniform continuity, 59
Uniform convergence, 117
Uniform norm, 71
Upper bound, 4
Upper integral, 217, 234
Upper limit, 99
Upper limit topology, 51
Upper semicontinuity, 49
Upper sum, 215, 232
Utility function, 210

V
Value function, 210
Variance, 241, 291

W
Weak duality, 201
Weierstrass theorem, 49
Weighted arithmetic mean, 16
Weighted geometric mean, 16
Weighted harmonic mean, 22
Weighted power mean, 17

Y
Young inequality, 17